Drinking Water Treatment 2

Drinking Water Treatment 2

Chemical and Physical Elimination of Organic Substances and Particles

Kader Gaid

iSTE

WILEY

First published 2023 in Great Britain and the United States by ISTE Ltd and John Wiley & Sons, Inc.

ISTE Ltd
27-37 St George's Road
London SW19 4EU
UK

www.iste.co.uk

John Wiley & Sons, Inc.
111 River Street
Hoboken, NJ 07030
USA

www.wiley.com

Library of Congress Control Number: 2022942618

British Library Cataloguing-in-Publication Data
A CIP record for this book is available from the British Library
ISBN 978-1-78630-784-2

Contents

7

Removal of Natural Organic Matter

7.1. Natural organic matter: humic substances

Organic matter can be divided into three categories: particulate matter, colloidal matter and dissolved matter. In general, zooplankton, phytoplankton and most bacteria are included in particulate matter. Dissolved matter is defined as matter whose compounds pass through a filter with a porosity of 0.45 μm.

Humic substances (HS) are the main organic constituents of soils and sediments. They are described as polymers with a molecular weight (MW) of 1,000–350,000. HS are formed by the decomposition of plant and animal tissues, and by chemical and biological processes that tend to produce complex chemical structures that are more stable than the parent material from which they are derived. Dissolved organic substances encompass a wide range of compounds, including HS, amino acids, fatty acids, phenols, sterols, carbohydrates and porphyrins.

There are two sources of natural organic matter (NOM): allochthonous NOM and autochthonous NOM. Allochthonous NOM (known as HS) is derived from decaying plant material. Soil humus, plant litter, microbial biomass and root exudates contribute to allochthonous NOM. They typically have a high molecular weight (>1,000 Daltons; 1 Dalton = 1 $g{\cdot}mol^{-1}$) and are hydrophobic.

Autochthonous NOM is derived from algae, phytoplankton and materials present in wastewater discharges. They are hydrophilic.

Figure 7.1. *Vegetation–decomposition and water pollution. For a color version of this figure, see www.iste.co.uk/gaid/watertreatment2.zip*

If allochthonous inlets are high, such as in colored springs or during precipitation/snowmelt events, the proportion of autochthonous NOM is generally low. Conversely, if allochthonous inlets are low, such as in clear springs or during drought periods when there is very little runoff, the proportion of autochthonous NOM is generally high.

In surface waters, dissolved HS originate from soils that come from plant weathering, chemical polymerization or microbial synthesis.

The identification of organic matter requires a fractioning step of these constituents.

This separation is based on the solubility of the molecules in water according to the pH level. HS are generally divided into three fractions (Figure 7.2) based on their solubility in acidic and alkaline solutions:

– humic acid, brown or black, soluble in a basic medium and insoluble in the acidic pH range (pH < 2);

– fulvic acid, yellow in color, soluble in water regardless of the pH value. It is the humic fraction that remains in acidified aqueous solutions and is soluble in bases and acids. It has a lower molecular weight than humic acid but has more hydrophilic functions than humic acid or humin. Their average length is 60 nm, and their average diameter is 2 nm;

– humins, which are black in color, are composed of bitumen, fatty acids and humic acids. They represent the insoluble humic fraction of soils that cannot be extracted by either an acid or a base.

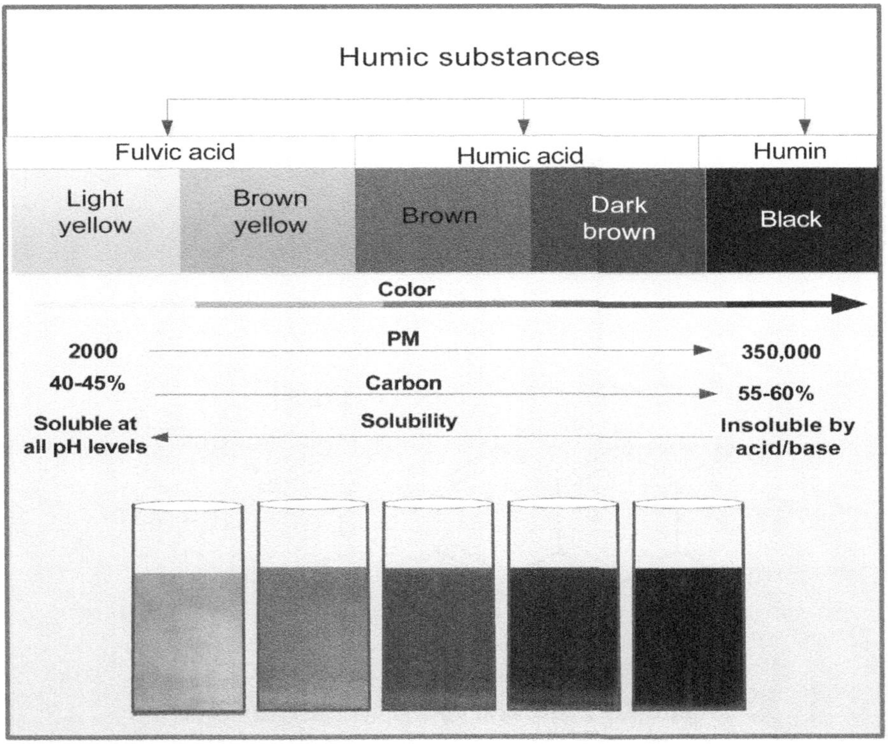

Figure 7.2. *Characteristics of humic substances. For a color version of this figure, see www.iste.co.uk/gaid/watertreatment2.zip*

The elemental analysis of HS shows that they are composed of carbon (C), hydrogen (H), nitrogen (N), sulfur (S) and oxygen (O). The major constituents of humic and fulvic acids are carbon and oxygen. The presence of carbon and nitrogen is higher in humic acids than in fulvic acids. Conversely, the fulvic fraction contains more oxygen. The O/C ratio makes it possible to differentiate humic acids (O/C = 0.5) from fulvic acids (O/C = 0.7). The H/C ratio is inversely proportional to the aromaticity or the degree of condensation.

Proteins and polysaccharides are grouped together under the term *biopolymers*. They can have both negatively and positively charged sites. Their net charge depends on the pH of the water. They are large molecules with high molecular weights in the range of a few hundred to several tens of thousand Daltons.

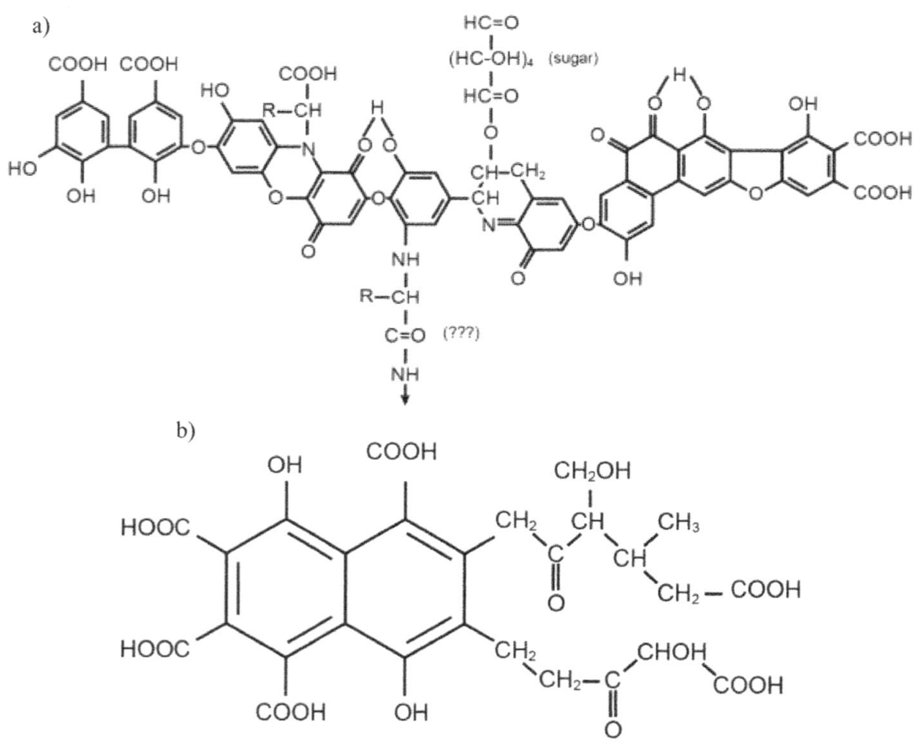

Figure 7.3. *Humic acid (a) and fulvic acid (b) model*

More globally, NOM is mainly composed of humic acids, fulvic acids, hydrophilic acids, carboxylic acids, amino acids, carbohydrates and polysaccharides.

HS are dark in color, mostly made up of aromatic acids (containing one or more benzene rings [C_6H_6]) and hydrophilic acids (polar molecules with an affinity for water), with molecular weights ranging from a few hundred to several thousand

Daltons. They contain phenolic OH and carboxylic groups, with a lower number of aliphatic OH groups.

Carboxyl, phenol, alcohol, carbonyl, quinone and methoxyl groups are the main functional groups of HS. Ether, ester and ketone groups may also be present. The total acidity of fulvic acids is higher than that of humic acids.

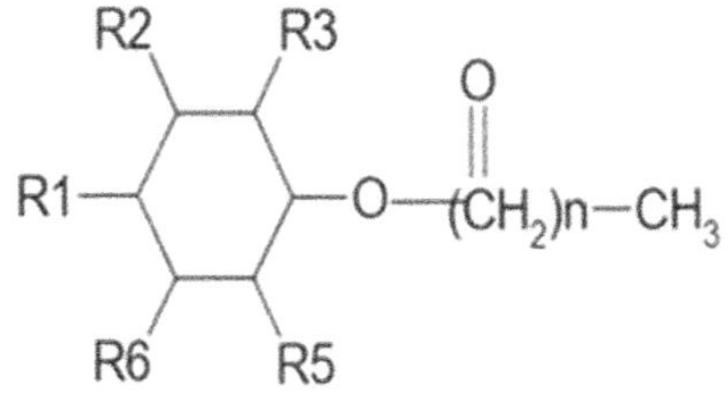

Figure 7.4. *Monomer of humic substances*

R1 = -COOH or $-COCH_3$ or -OH

R2 = -H or -OH or -COOH

R3 = -H or -OH or $-OCH_3$ or -COOH

R5 = -H or -OH or $-OCH_3$

R6 = -H or $-COOCH_3$

n = 14 and n = 15 for the AH and n = 14, 15, 16 and 18 for the AF

Humic and fulvic acid particles vary in shape and size depending on the pH. They tend to aggregate into long fibers or fiber bundles at low pH. However, at high pH, they disperse, and the molecular arrangement becomes smaller but better oriented. Humic acids are more aromatic and have a higher molecular weight than fulvic acids. Ozone tends to react more readily with humic acids. Fulvic acids are slightly more biodegradable than humic acids. Biodegradable organic matter is more hydrophilic than non-biodegradable matter.

A common approach to its characterization is to divide the mixture into hydrophilic and hydrophobic fractions. The hydrophilic fraction includes, for example, carboxylic acids, carbohydrates and proteins, while the hydrophobic fraction includes HS.

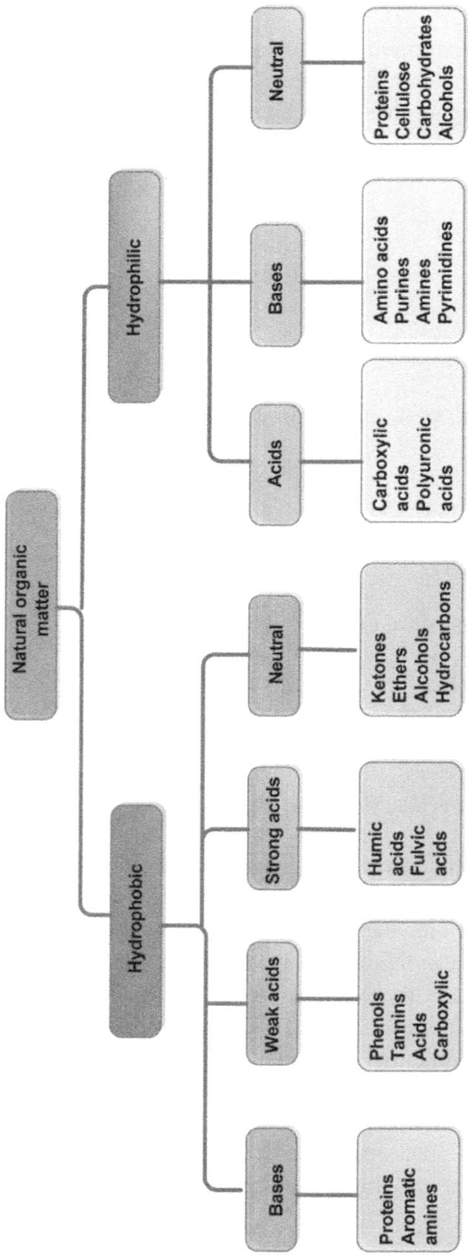

Figure 7.5. *Classification of natural organic matter. For a color version of this figure, see www.iste.co.uk/gaid/watertreatment2.zip*

The analysis methods of the functional groups show a great diversity of functions in HS, such as carboxylic, phenolic, alcoholic, quinone and hydroquinone groups. The relative distribution of these groups varies according to the types of substances. The total acidity of fulvic acids is higher than that of humic acids. Fulvic acids have more oxygen atoms per unit weight than humic acids, which coincides with a greater number of carboxyl groups COOH, hydroxyl groups OH, aldehydes and ketones C=O. The reactivity of HS is partly due to the presence of oxygen in the carboxylic, phenolic or alcoholic functional groups. Moreover, among the groups contributing to the acidic character of HS, carboxylic acids (-COOH) are more important than alcoholic functions (-OH). The charge of the HS is always negative or null, of variable intensity according to the pH of the medium, and comes from the dissociation of the functional groups. In general, the color of the surface water is attributable to the organic matter.

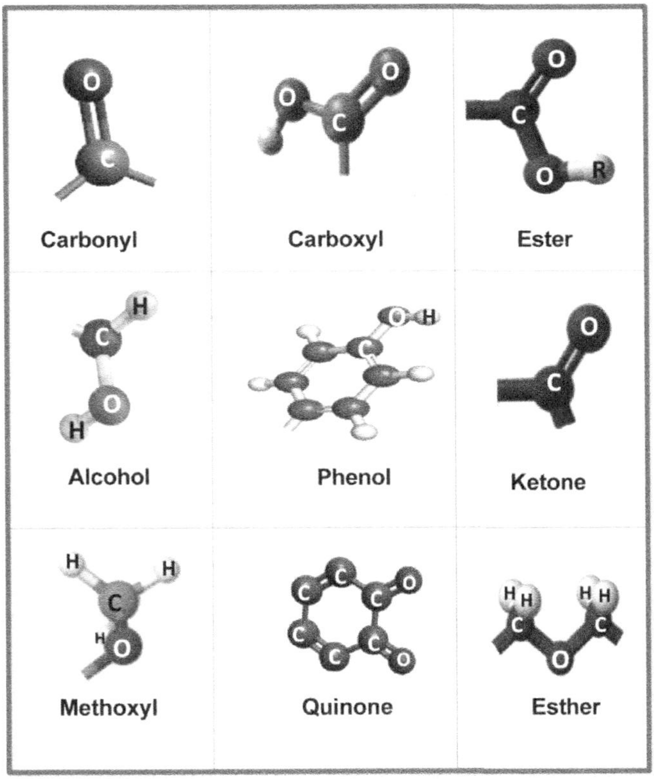

Figure 7.6. *Some functional groups of humic substances. For a color version of this figure, see www.iste.co.uk/gaid/watertreatment2.zip*

One of the most important characteristics of HS is their ability to form water-soluble and insoluble complexes with metal ions and hydrated oxides, as well as interact with clay minerals and various organic compounds such as alkanes, fatty acids and other organic substances such as pesticides. The formation of water-soluble complexes of HS with metal ions is of particular interest because complexation can increase the concentrations of these ions in natural waters.

The types of interactions that occur between metal ions and organic acids found in natural waters range from complexation reactions to colloid formation. Many complexes between HS and metals are chelates.

Thus, the quantity, character and properties of NOM differ from one water to another in the same region and depend on the biogeochemical cycles of the elements in its environment.

In addition, the various molecules that make up NOM vary seasonally within a site due to, for example, precipitation, runoff from melting snow, flooding, or drought. Floods and droughts are the primary impacts of climate change on water availability and quality. It is suggested that these climate changes may be causing an increase in the total amount of NOM in surface waters.

The consequences of the presence of organic matter in treated water include the following:

– effects on the organoleptic qualities of the water, such as color, odor and taste;

– reactions with chemical disinfectants, therefore controlling the demand for disinfectants;

– reactions with metals, inducing their precipitation;

– the formation of disinfection by-products, such as trihalomethanes and halogenated acetic acids;

– a contribution to corrosion in the distribution networks;

– a contribution to the development of biofilm in the distribution networks;

– a reduction in the life span of activated carbons, requiring frequent renewal.

7.2. Methods of quantification and assessment of organic substances in water

Methods for quantifying and assessing NOM generally include:

– total organic carbon (TOC) and dissolved organic carbon (DOC);

– UV absorbance at 254 nm;

– specific UV absorbance (SUVA);

– liquid chromatography to qualitatively assess the distribution of different organic substances in a water sample;

– fluorescence.

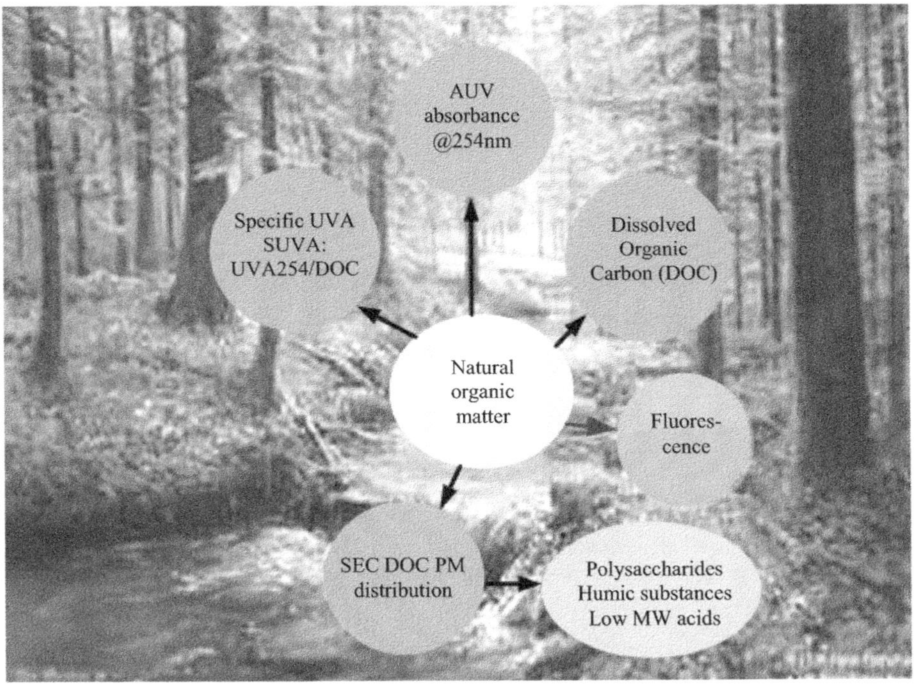

Figure 7.7. *Methods of NOM quantification. For a color version of this figure, see www.iste.co.uk/gaid/watertreatment2.zip*

7.2.1. *Total organic carbon*

HS represent approximately 60–80% of the organic carbon in surface waters. They are found in highly variable concentrations in natural surface waters but are much less concentrated in groundwater.

The main elements of NOM are carbon, oxygen, nitrogen, sulfur and hydrogen. Carbon is the dominant element in terms of weight, accounting for 40–50% of the weight of NOM. For this reason, NOM is quantified by TOC analysis. TOC can be fractionated into particulate organic carbon (POC) and DOC. The separation (Figure 7.8) between these two forms is performed by vacuum or pressure filtration (0.45 μm porosity filter). DOC is the portion of organic carbon that is completely dissolved in water. To measure DOC, the particulate portion of the organic carbon must be removed before analysis.

DOC is characterized by a biodegradable dissolved organic carbon (BDOC) and a fraction that is resistant to biodegradation (RDOC).

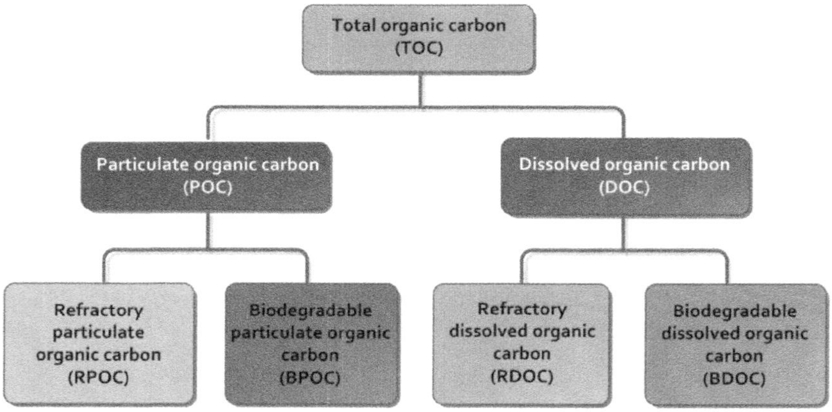

Figure 7.8. *Fractionation of total organic carbon. For a color version of this figure, see www.iste.co.uk/gaid/watertreatment2.zip*

BDOC can also be fractionated into a rapidly biodegradable fraction and a slowly biodegradable fraction. The rapidly biodegradable BDOC is available as a source of carbon and energy for microorganisms. It represents approximately 5–30% of the DOC. Slowly biodegradable BDOC is more likely to be found in the distribution system when it has not been degraded or eliminated in the treatment process. Assimilable organic carbon (AOC) is similar to BDOC in terms of availability to microorganisms, but the analytical methods are different. BDOC represents the fraction of DOC that can be mineralized by heterotrophic microorganisms, whereas AOC is a portion of DOC that can be converted to biomass and expressed as carbon concentration.

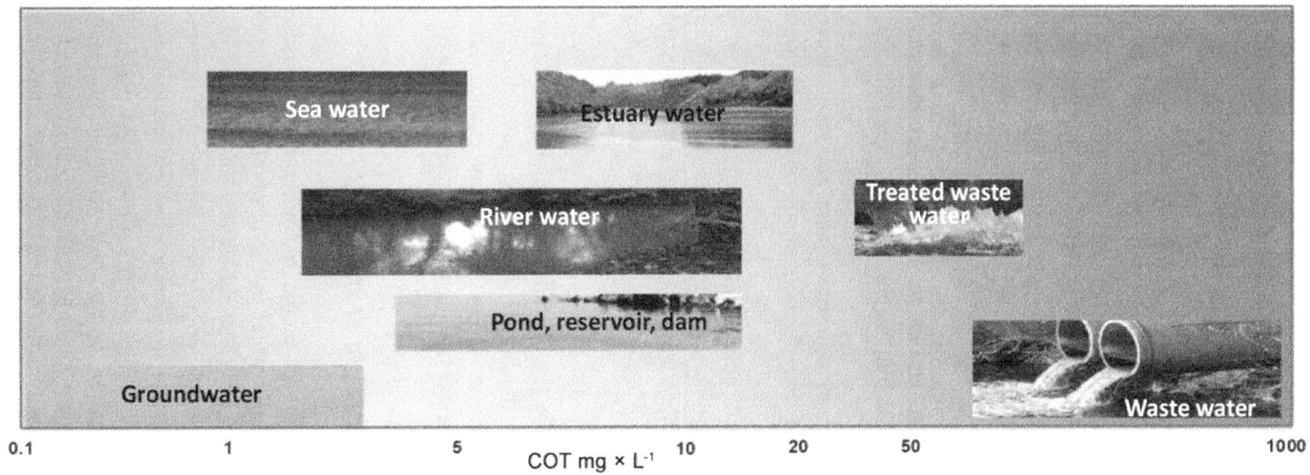

Figure 7.9. *Total organic carbon concentrations in different types of water. For a color version of this figure, see www.iste.co.uk/gaid/watertreatment2.zip*

BDOC also presents problems for the quality of distributed water. Specifically, when the temperature is favorable and the residual chlorine concentration is low, it becomes a nutrient-rich source for many bacteria and promotes biofilm formation in the system.

Biofilms can contribute to the survival of pathogens that have successfully penetrated drinking water treatment barriers or directly entered the distribution system due to a breakdown in pipe integrity. The most important elements in controlling bacterial growth in distribution systems are maintaining a residual concentration of disinfectant, limiting BDOC and controlling corrosion. A BDOC concentration of less than 0.2 $mgC \cdot L^{-1}$ is recommended for treated water to be considered biologically stable and not conducive to bacterial growth.

Most natural waters contain TOC and DOC. TOC concentrations in waters vary from resource to resource, as illustrated in Figure 7.9.

Seawater has a lower TOC content than most lakes and rivers, except in a few special cases where treated wastewater is discharged not far from the water inlet.

Groundwater most often has TOC concentrations <1 $mg \cdot L^{-1}$. HS in groundwater are mainly derived from the leaching of humic matter from soils or from the leaching of fine organic particles that were deposited together with the sediments that make up the aquifer rock. They can be contaminated by surface water infiltration.

River waters are contaminated by the degradation of plants, while estuary waters are mostly polluted by wastewater discharges. Generally, more than 70% of DOC constitutes humic and fulvic substances. However, this percentage varies according to the geographical area and the manner in which the plants decompose.

7.2.2. Absorbance of ultraviolet light at 254 nm (UV 254)

Ultraviolet spectroscopy is a spectroscopic technique that measures the light absorbed by compounds in solution in water in the ultraviolet range. This is in the wavelength range of 200–400 nm. When molecules in solution in water are subjected to radiation in this wavelength range, they undergo one or more electronic transitions. They absorb light at certain frequencies that represent the exact energy required to excite an electron to a higher energy state. Since light absorption is essentially related to electron density, compounds with electron-rich functional groups will absorb more light than compounds without such groups. The electronic spectrum depends on the light intensity absorbed by the sample, analyzed according to a given wavelength, known as UV absorbance. This provides correlations with the concentration of the compounds present in the water.

UV 254 is the amount of light, with a wavelength of 254 nanometers, that a sample will absorb. The principle of this method is that UV absorbing constituents, such as humic or fulvic acids, absorb UV light in proportion to their concentration. UV 254 samples must be measured in water that does not contain any oxidants or disinfectants. This is necessary because oxidants react with organic compounds and cleave the double bonds that absorb UV 254. Samples for UV 254 must be filtered to remove particulates. UV 254 is representative of the existence of unsaturated carbon bonds, including aromatic compounds, which are generally resistant to biodegradation. For the same amount of DOC, a decrease in UV absorbance at 254 nm usually results in an increase in the biodegradability of NOM.

Figure 7.10. *Concentration of humic substances on some surface waters (France and Africa). For a color version of this figure, see www.iste.co.uk/gaid/watertreatment2.zip*

7.2.3. *Specific UV absorbance*

SUVA is an indicator of the aromatic fraction of NOM that contains chromophores (groups of atoms with one or more carbon double bonds). The SUVA parameter is an indicator of the humic content of water.

It is a calculated parameter that is equal to the UV absorption at 254 nm (measured in cm^{-1}) divided by the DOC (measured in $mg{\cdot}L^{-1}$).

Waters with low SUVA values mostly contain non-humic matter that is difficult to remove with advanced coagulation. On the other hand, the TOC present in waters with high SUVA values is generally easier to remove by advanced coagulation.

The equation to calculate the SUVA value is:

$$\text{SUVA}\ (\text{L}{\cdot}\text{mg}^{-1}{\cdot}\text{m}^{-1}) = 100(\text{cm}{\cdot}\text{m}^{-1})\frac{\text{UV 254 (cm}^{-1})}{\text{DOC (mg}{\cdot}\text{L}^{-1})}$$

Generally speaking, organic compounds with high SUVA values have low biodegradability due to the increased presence of aromatic groups and other unsaturated configurations.

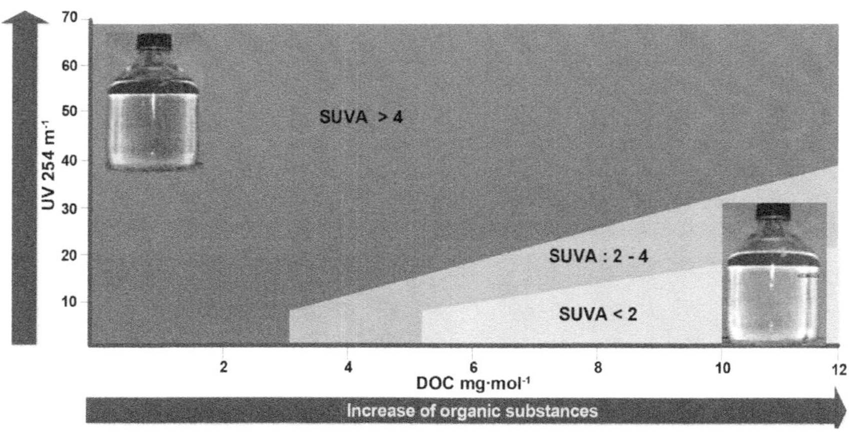

Figure 7.11. *UV 254, color, DOC and SUVA. For a color version of this figure, see www.iste.co.uk/gaid/watertreatment2.zip*

SUVA is a judging parameter for deciding between conventional and advanced coagulation. The general rule is that the DOC of water with an SUVA of less than 2 $L{\cdot}mg^{-1}{\cdot}m^{-1}$ is mainly composed of non-humic (and lower hydrophobicity) substances with low molar weights. This indicates that a large fraction of the matter is hydrophilic and non-humic, with a low UV absorbance, low chlorine demand and low potential for THM formation.

A SUVA greater than 4 $L{\cdot}mg^{-1}{\cdot}m^{-1}$ indicates that the DOC is predominantly made up of HS with high molar weights. In this case, we are in the presence of hydrophobic aromatic HS associated with a high UV absorbance, a high chlorine demand and a high potential for THM formation.

A SUVA that is between 2 and 4 $L \cdot mg^{-1} \cdot m^{-1}$ indicates a DOC consisting of a mixture of non-humic hydrophilic matter and humic hydrophobic matter with a medium UV absorbance, higher chlorine demand and higher potential for THM formation.

The different NOM fractions have different properties in terms of treatability by coagulation, adsorption with activated carbon, chemical oxidation, reactivity to chlorine and ozone, and potential for the formation of disinfection by-products.

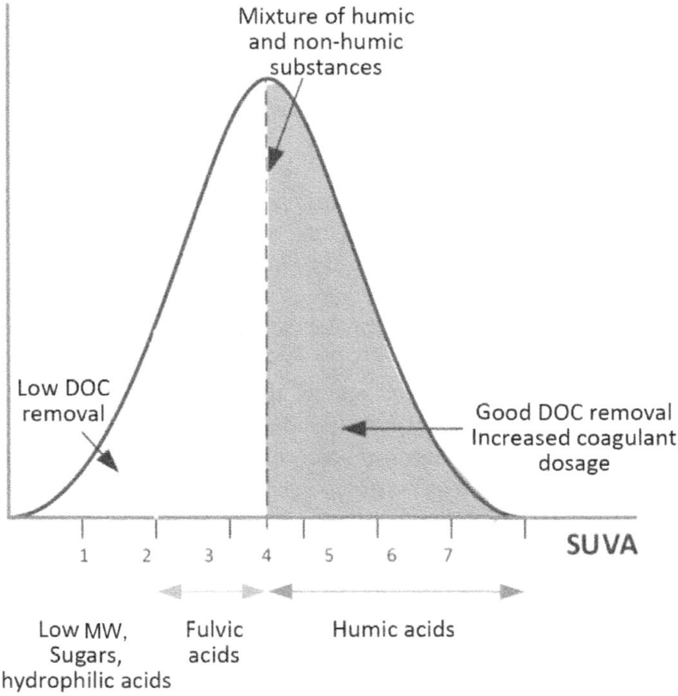

Figure 7.12. *SUVA and coagulation efficiency. For a color version of this figure, see www.iste.co.uk/gaid/watertreatment2.zip*

SUVA is a quantitative measure of the unsaturated bonds and/or aromaticity of organic matter in water. Overall, an increase in SUVA reflects a greater humification, aromaticity and hydrophobicity of the NOM, which leads to a lower biodegradability.

7.2.4. *Liquid chromatography*

DOC, UV 254 absorbance, pH, turbidity and color are common water quality parameters assessed in drinking water plants.

However, they do not give accurate information on the character of the NOM, such as molar weight (MW) or hydrophobicity.

The liquid chromatography-organic carbon detection (LC-OCD) technique has been developed for the characterization of NOM contained in surface waters. The organic fractions described below are usually found in surface waters.

The qualitative determination is based on the separation of hydrophilic molecules according to their hydrodynamic molecular size, performed by the size exclusion chromatography (SEC). High molecular weight molecules, which do not penetrate the pores of the column material, appear first on the chromatogram, while smaller molecules are eluted after.

The quantification is based on the determination of TOC in the Gräntzel reactor. The device is equipped with three detectors: an organic carbon detector (organic carbon detection [OCD]), a UV detector at 254 nm and an organic nitrogen detector (organic nitrogen detection [OND]). Figure 7.13 shows the specific chromatogram of a natural surface water (water from the Seine river). The separation principle (SEC) and triple detection (OCD, UVD, OND) allow the LC-OCD system to determine the nature of the different organic fractions present in the sample to be studied, as well as their quantification. It should be noted that in UV light, some peaks are not visible in OCD because some molecules do not have chromophore groups and are therefore not visible in UV light (such is the case of polysaccharides). The peak elution between 60 and 75 min for OND corresponds to the oxidation of low molecular weight nitrogenous materials into nitrates and to mineral species. The quantification limit of LC-OCD is 0.1 $mgC \cdot L^{-1}$ per compound. All samples were analyzed with an SEC HW-50S column for an analysis time of 150 min. The main organic matter fractions for a natural water sample are as follows.

7.2.4.1. *Polysaccharides and biopolymers (response time Tr = 25–35 min)*

This fraction represents the group of polysaccharides (such as starch, cellulose, chitosan, which are often associated with amino acids and amino sugars), proteins (such as gelatin, casein) and colloids (both organic and inorganic), with approximate molecular weights >> 20,000 $g \cdot mol^{-1}$. In the composition of natural waters, this fraction is only present in surface waters. Polysaccharides are hydrophilic substances that do not absorb UV light. They can be formed by fermentation of sugars (such as –polyhydroxybutyrate [PHB]) or be the result of bacterial metabolism or even debris of cell walls of algae and/or bacteria. Polysaccharides synthesized by bacteria are also called extracellular polymeric substances (EPS). The differences in OCD/OND/UVD responses allow for a more accurate characterization of the components of this fraction.

Polysaccharides and organic colloids only respond to OCD, inorganic colloids only respond to UV light, amino sugars respond to OND and finally, proteins respond to all three detections. The P fraction represents approximately 5–10% of the DOC in surface waters.

7.2.4.2. *HS and building blocks (Tr = 35–50 min)*

This corresponds to the majority fraction that is present in natural waters (generally representing ~60% of TOC). The HS group generally corresponds to organic molecules with a high molecular weight (~1,000 $g \cdot mol^{-1}$). They are mainly formed of aromatic macromolecules such as humic acids (not soluble in acids but soluble in bases) and fulvic acids (soluble in acids and bases). The aromaticity index of the HS fraction is generally given according to the presence of HS in the sample. This index estimates the content of aromatic molecules (aromatic carbon–carbon double bonds) in the HS fraction of the sample. The BB group (representing 15–25% of TOC in a surface water) corresponds to the intermediate products of the degradation of the HS (equivalent structure, lower molecular weight ~300–500 $g \cdot mol^{-1}$). BBs are therefore highly aromatic molecules (hydrolysis products of HS) and contain many acid functions.

7.2.4.3. *LMW (low molecular weight) organic acids (Tr = 50–60 min)*

This fraction represents the group of molecules with the lowest molecular weight (<350 $g \cdot mol^{-1}$). In OCD, this fraction factors the 52-min peak corresponding to low molecular weight organic (and aliphatic) acids, such as volatile fatty acids (in this last case, no response to UV light). They are assimilated to the final products of NOM degradation. When there is a significant response to UV light (aromatic character), this peak can sometimes be associated with humic-type compounds of lower molecular weight. Depending on the nature of the sample, the acidic LMW fraction could also represent bacterial or algal products.

7.2.4.4. *Neutrals (after Tr = 60 min)*

This fraction represents 10–30% of the TOC. It is only composed of very low molecular weight, very weakly charged molecules, which can be considered non-biodegradable and are associated with substances of low hydrophobicity (amphiphiles), such as mono-oligosaccharides, alcohols, aldehydes, ketones and certain amino acids. Their hydrophobic character increases proportionally with their response time Tr (pentanol Tr = 120 min, octanol Tr = 240 min); products that are not eluted after 200 min are considered hydrophobic substances.

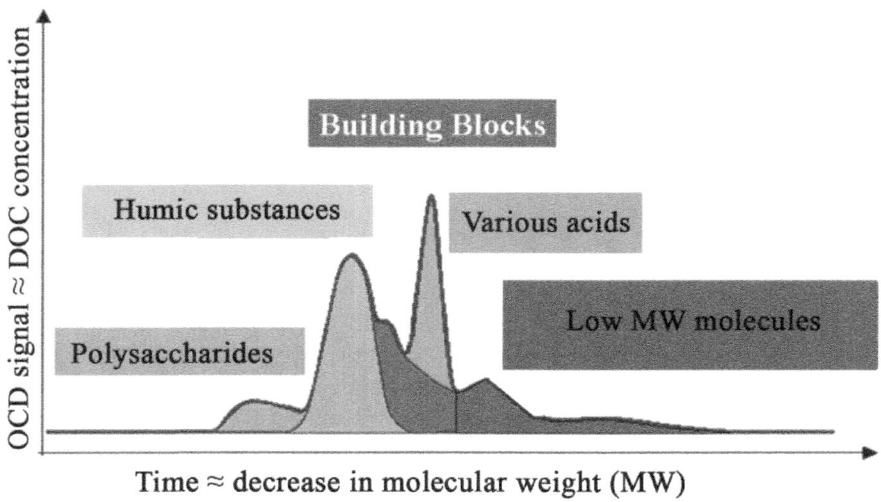

Figure 7.13. *Typical chromatogram of a natural surface water sample. For a color version of this figure, see www.iste.co.uk/gaid/watertreatment2.zip*

7.2.5. *Hydrophobic organic carbon*

This corresponds to the difference between the TOC value and the sum of the fractions constituting the chromatogram in OCD detection (CDOC: Chromatographable Dissolved Organic Carbon). The hydrophobic organic carbon (HOC) fraction represents the organic compounds of hydrophobic and non-chromatographable nature, which are retained on the chromatographic column (probably natural lipids, lipoids with their esters and hopanoids). They are generally found between 0 and 25% in TOC in natural surface waters.

The results in organic carbon, expressed in $\mu gC \cdot L^{-1}$, represent the values obtained during the analysis through OCD and particularly:

– TOC: this value is obtained after filtration on a 10 μm filter through a direct injection into the Gräntzel reactor (where the oxidation of organic carbon takes place) without passing through the steric exclusion column;

– DOC: the sample is filtered through a 0.45 μm filter before being injected directly into the Gräntzel reactor without passing through the steric exclusion column;

– CDOC: the sample is filtered through a 0.45 μm filter and then injected into the Gräntzel reactor to determine the DOC distribution of each fraction. The total CDOC is obtained by adding the CDOC of the different fractions.

The OND detector makes it possible to determine the dissolved organic nitrogen (DON) content expressed in $\mu g \cdot L^{-1}$ for the P and HS fractions.

The UVD detector makes it possible to measure the UV absorbance at 254 nm and thus obtain the following information:

– the global UV absorbance of the sample is expressed in m^{-1} (UVD@254 SAC) before injection in the column (ByPass). By determining the UV 254/DOC ratio, we obtain the SUVA (specific UV absorbance) of the sample;

– for the biopolymer fraction: an assessment of the inorganic colloid content expressed in m^{-1} (UVD inorganic colloids);

– for the HS fraction: the aromaticity of HS (UVD aromaticity) corresponding to the UV/DOC ratio (SUVA is a widely used indicator to assess the aromaticity of carbon contained in NOM);

– chromatographic separation makes it possible to assess the molecular weight of the HS fraction of the sample because of a prior calibration with natural HS.

Thus, using LC-OCD, it is also possible to assess the content of HMW polysaccharides and biopolymers in the sample, which are not usually visible or traceable by UV detectors. Chromatographic profiles provide very interesting information for comparing changes in the quality of NOM during the different stages of a treatment process and their degree of removal.

7.2.5.1. *Example of the characterization of DOC composition on surface water (southwest France) carried out by VERI-Veolia: river water (southwest France)*

The results obtained in OCD, expressed in terms of $\mu g\text{-}C \cdot L^{-1}$, are presented in Table 7.1.

River water	DOC Dissolved	HOC Hydrophob.	CDOC Hydrophil.	BIO polymers	DON (Norg)	Humic Subst. (HS)	DON (Norg)	Aromati. (SUVA-HS)	Mol-Weig. (Nn)	Building Blocks	Neutrals	Acids	Inorg. Colloid SAC	SAC	SUVA (SAC/DOC)
				Approx. Molecular Weights in g/mol:									(UV@254 nm)		
				>>20.000		*~1000 (see separate HS-Diagram)*				*300-500*	*<350*	*<350*			
	ppb-C	*ppb-C*	*ppb-C*	*ppb-C*	*ppb-N*	*ppb-C*	*ppb-N*	*L/(mg m)*	*g/mol*	*ppb-C*	*ppb-C*	*ppb-C*	*(m^{-1})*	*(m^{-1})*	*L/(mg m)*
	% DOC	% DOC	% DOC	% DOC	–	% DOC	–	–	–	% DOC	% DOC	% DOC	–	–	–
River water	3209	46	3163	145	30	2080	69	4.94	766	463	451	24	0.50	14.13	4.40
15664	100	1.4	98.6	4.5	–	64.8	–	–	–	14.4	14.1	0.7	–	–	–

Table 7.1. *DOC characterization results*

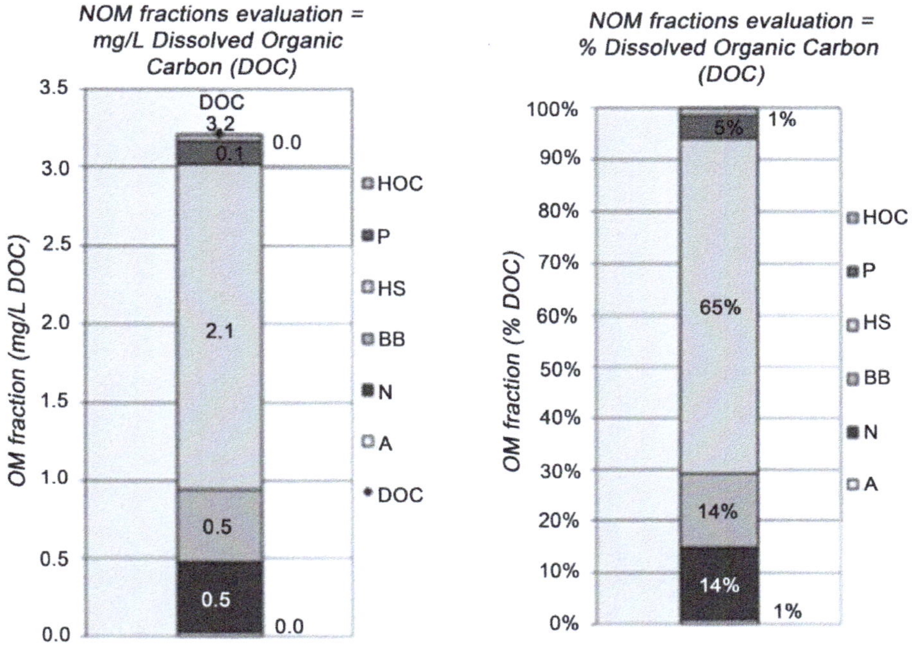

Figure 7.14. *Distribution of organic matter in terms of DOC (mg·L^{-1}) and DOC %. For a color version of this figure, see www.iste.co.uk/gaid/watertreatment2.zip*

COMMENT ON TABLE 7.1 AND FIGURE 7.14.– *DOC: dissolved organic carbon; CDOC: chromatographable, hydrophilic organic carbon; DON: organic nitrogen detection; HOC: non-chromatographable, hydrophobic organic carbon; SAC: spectral absorption coefficient (λ = 254 nm); SUVA: specific ultraviolet absorbance; P or biopolymers: polysaccharides, proteins, amino sugars, colloids; BB: building blocks – degradation products of HS; HS: humic substances; Acids: sum of monoprotic acids (<350 Da), small acid molecules or HS; Neutrals: includes mono-oligosaccharides, alcohols, aldehydes, ketones and amino sugars.*

The chromatograms obtained in the OCD/UVD/OND signal are presented in Figures 7.15–7.17.

The chromatographic profile of the sample is similar to that generally seen for most river waters, exhibiting all of the biopolymer, HS, BB, A and N organic fractions (Figure 7.15). The sample is characterized by 3.2 ppm-C in terms of DOC (Table 7.1).

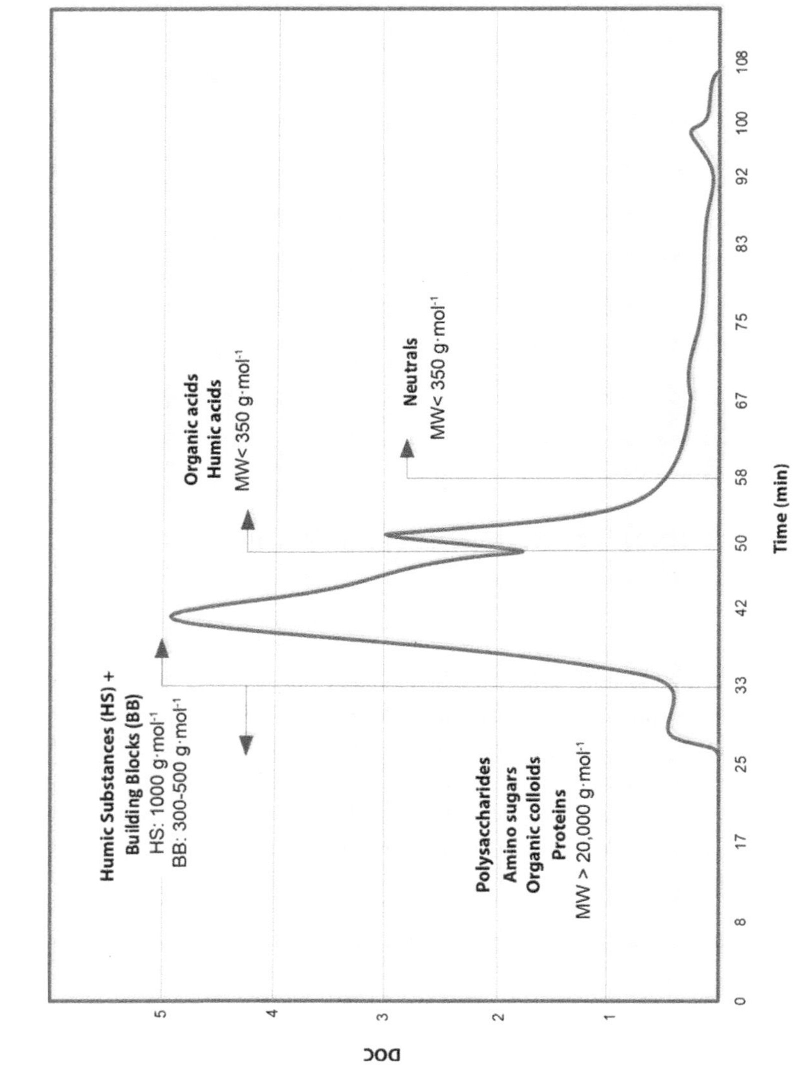

Figure 7.15. *Chromatogram for OCD (DOC). For a color version of this figure, see www.iste.co.uk/gaid/watertreatment2.zip*

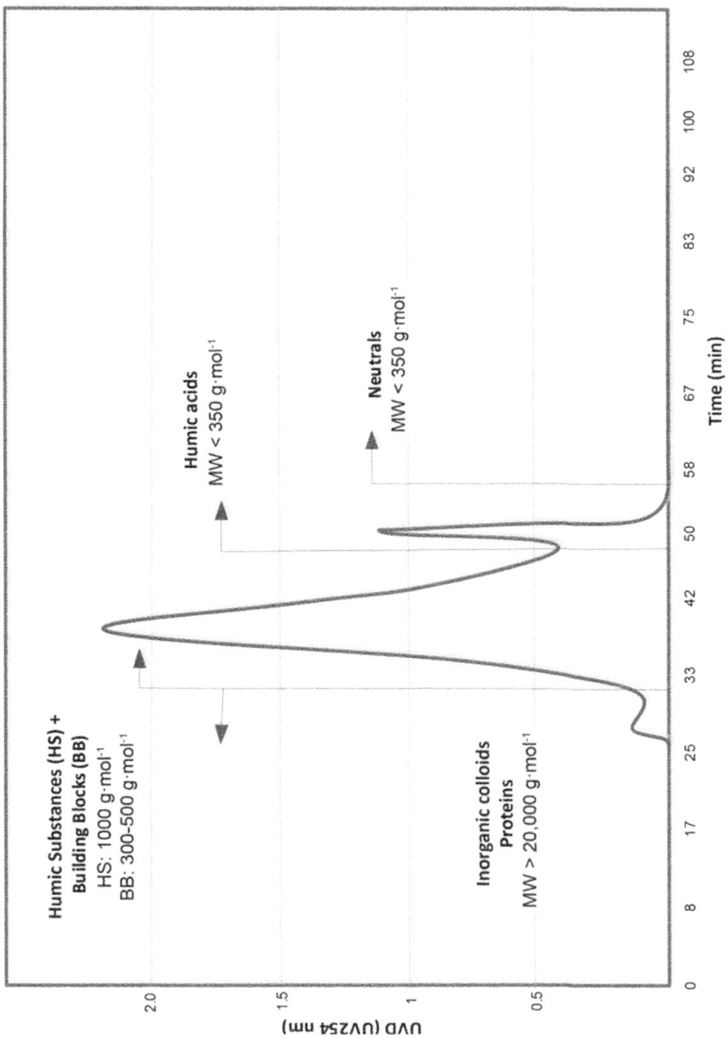

Figure 7.16. *Chromatogram for UVD (UV 254 nm). For a color version of this figure, see www.iste.co.uk/gaid/watertreatment2.zip*

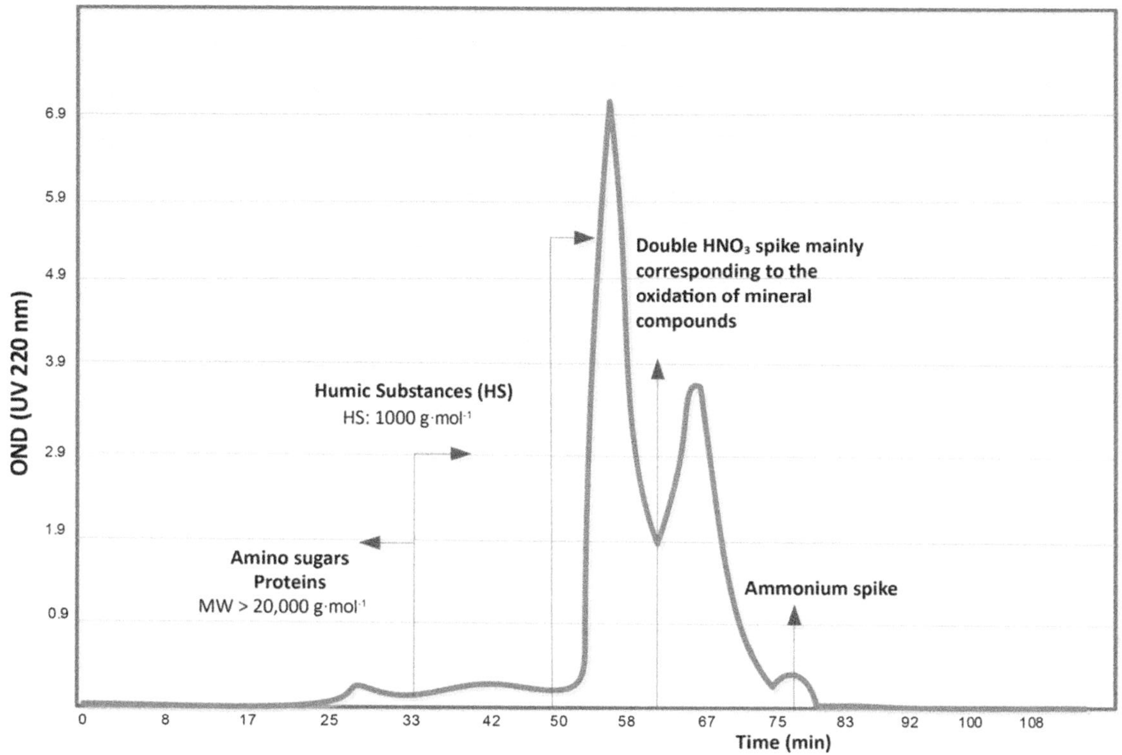

Figure 7.17. *Chromatogram for OND (UV 220 nm). For a color version of this figure, see www.iste.co.uk/gaid/watertreatment2.zip*

A slightly larger than usual OCD/UVD signal ratio was noted. This suggests that the overall raw water matrix is a slightly more aromatic matrix than that encountered in most river waters.

The majority fraction of the sample consists of HS and BB (high molecular weight molecules), representing approximately 2.5 ppm CDOC, or ~80% of the total sample (Table 7.1). This percentage remains quite typical for river water. The characteristic of the biopolymer fraction of the water is its composition: the N/C ratio indicates that this fraction is composed of polysaccharides and proteins at equivalent levels, which differentiates this sample from most river waters. However, the rate of CDOC for the biopolymer fraction remains quite common for river water. Another characteristic of this sample concerns fraction A (small organic acid molecules), which is present at a low level. This fraction contains the coexistence of organic acids and HS debris (Figure 7.16). The overall aromaticity of the sample (Table 7.1) is mainly due to the HS fraction. In OND (Figure 7.17), there is a presence of HS and biopolymer fractions associated with nitrogenous compounds, as well as the presence of nitrates and ammonium.

All the results obtained show that this river water is a fairly aromatic matrix, with a high potential for coagulation–flocculation treatment.

7.2.6. *Fluorescence*

Fluorescence is a spectrochemical method in which molecules are excited by irradiation at a certain wavelength, and the emitted radiation is measured at a different wavelength. The specific wavelengths of excitation and emission are the characteristics of a particular molecular conformation, known as fluorophore. These fluorophores are useful for describing the structural compositions of humic materials. Fluorescence is recommended as a tool to estimate the presence of biodegradable NOM, as good correlations are found between fluorescence and biological oxygen demand values.

The excitation-emission matrix (EEM) spectrophotometric technique allows visualization of a range of different fluorophores by scanning excitation and emission wavelength bands from ~200 to ~500 nm. Raw water can show main fluorescence peaks for HS, fulvic substances, tryptophan and protein molecules. Conjugated double bonds and aromatic rings, -OH and $-NH_2$, increase fluorescence, while -COOH decreases it.

Excitation wavelengths (nm)	Emission wavelengths (nm)	Types of molecules
270–280	310–320	Tyrosine-like, protein-like
270–285 (220–235)	340–360	Similar to tryptophan, proteins
320–350	400–450	Fulvic
310–320	380–420	Humic (marine humic)
330–390	420–500	Humic

Table 7.2. *Emission and excitation wavelengths used in fluorescence for the characterization of molecules present in natural organic matter*

7.3. Conditions for the removal of NOM

TOC removal is required to reduce the potential for disinfection by-product formation (THMs, AHAs, etc., see Chapter 22).

Chlorine disinfection is an important step in protecting the consumer from pathogens. Unfortunately, while chlorination inactivates pathogens, it also reacts with NOM to form disinfection by-products (DBPs). Some DBPs, such as trihalomethanes (THMs), are a health concern. TOC is used as a measure of NOM concentration and serves as an indicator to estimate the potential for disinfection by-product formation. A TOC requirement is then set in the produced water, limiting the formation of THMs and other by-products.

Thus, the removal of NOM (and therefore TOC) reduces the potential for THM formation during the final disinfection stage.

All surface water treatment plants must comply with TOC requirements.

The DOC reductions shown in Table 7.3 are dependent on the dosage of coagulant and the pH. No acid or base is expected to be added. TOC removal is generally more difficult as alkalinity increases because the optimum pH ranges are not reached.

TOC raw water ($mg \cdot L^{-1}$)	Alkalinity of the raw water ($mg \cdot L^{-1}$ $CaCO_3$)		
	0–60	>60–120	>120
$2 \leq TOC \leq 4$	42% average	35% average	30% average
$4 \leq TOC \leq 8$	54% average	47% average	44% average
$TOC > 8$	66% average	54% average	52% average

Table 7.3. *Estimation of DOC removal according to water alkalinity*

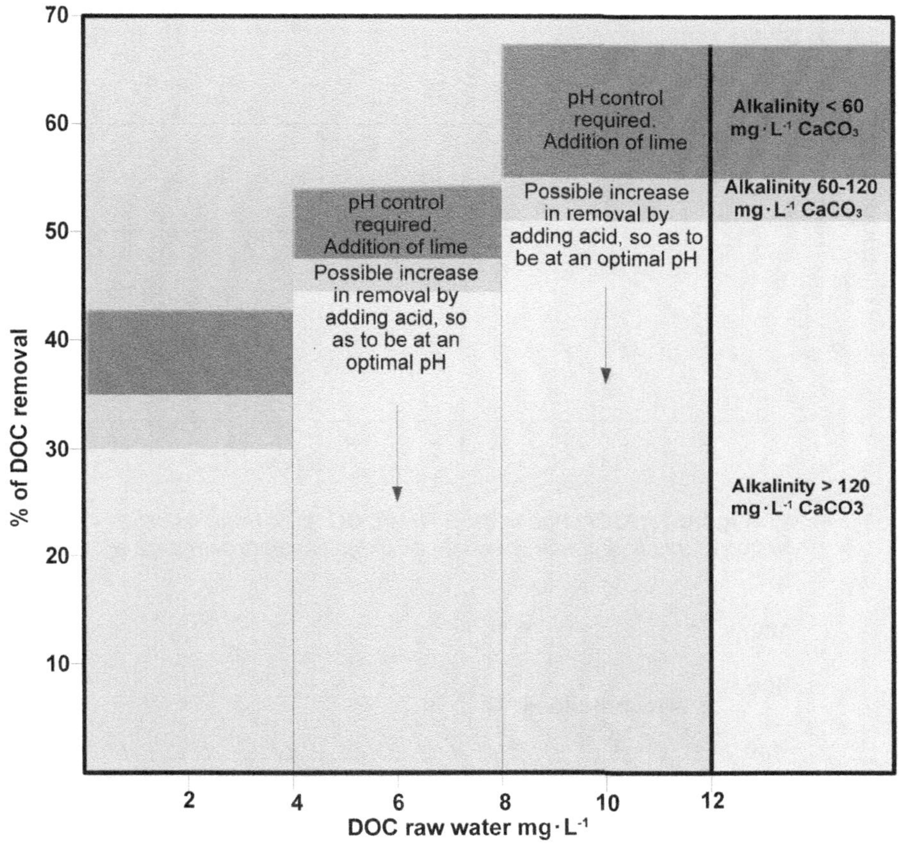

Figure 7.18. *DOC removal according to alkalinity. For a color version of this figure, see www.iste.co.uk/gaid/watertreatment2.zip*

The optimal pH for coagulation with alum is 6.0–6.5, and it is 5.5–6.5 with ferric chloride. When alkalinity is high, a higher dosage of acid must be injected to lower the pH because an increase in coagulant alone is not sufficient (Figure 7.18). Periodic variations in TOC and/or raw water alkalinity require operators to modify their operating conditions with respect to coagulant dosage, pH control and addition of reagents (acid or lime).

As SUVA increases, the TTHM/DOC ratio increases (Figure 7.19). Thus, for an equivalent concentration of DOC, the addition of chlorine to water containing (autochthonous) NOM forms less chlorinated DBPs. Its application to water containing allochthonous NOM leads to the formation of THMs at higher concentrations due to the presence of aromatic molecules (Figure 7.20).

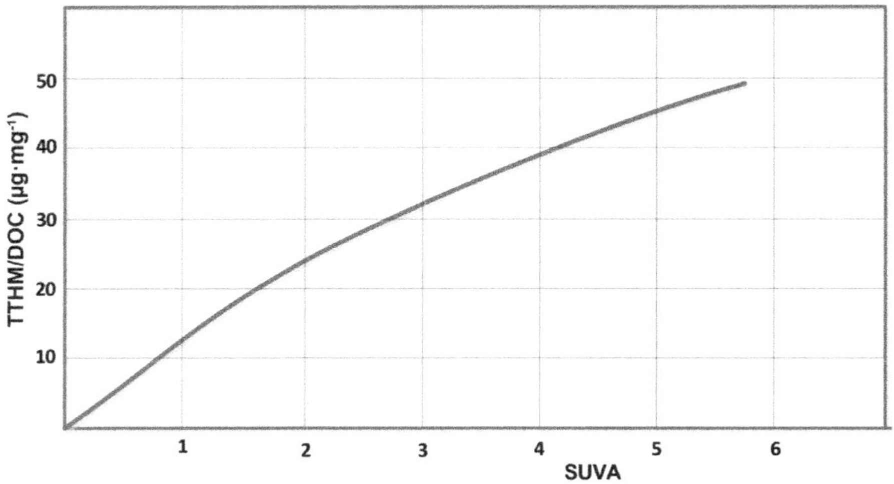

Figure 7.19. *Change in the TTHM/DOC ratio. For a color version of this figure, see www.iste.co.uk/gaid/watertreatment2.zip*

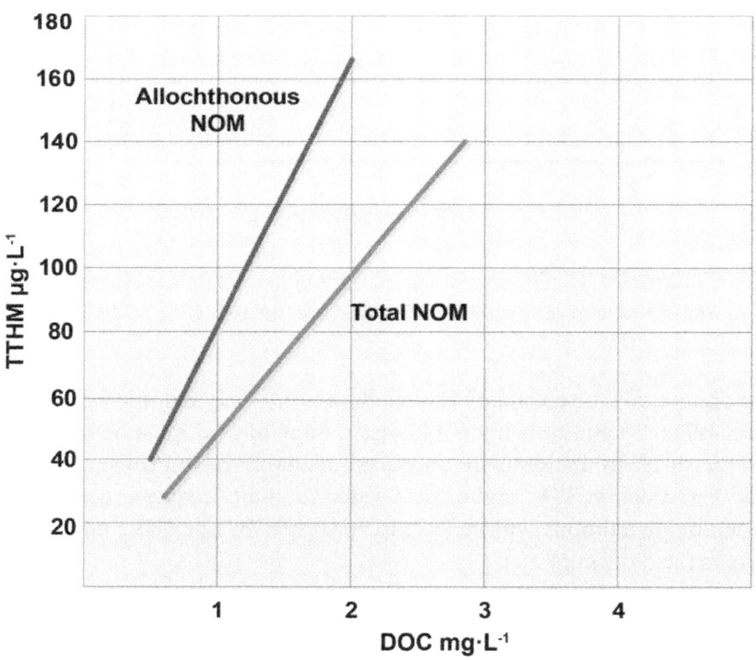

Figure 7.20. *TTHM formed according to the residual SUVA DOC. For a color version of this figure, see www.iste.co.uk/gaid/watertreatment2.zip*

7.4. NOM removal techniques

NOM removal techniques are based on physical, chemical and biological processes:

– *advanced coagulation*: coagulants are used to remove high and medium molecular weight NOM. However, removal of NOM by advanced coagulation is not very effective in waters with high alkalinity and low SUVA values;

– *activated carbon adsorption*: the use of activated carbon is recommended, especially to eliminate small neutral hydrophobic molecules. It is complementary to coagulation–flocculation;

– *nanofiltration or low-pressure reverse osmosis*: these high-pressure membranes eliminate most of the compounds that form NOM by molecular sieving;

– *chemical oxidation and biofiltration*: NOM is characterized by C=C double carbon bonds. A strong oxidant such as ozone is used to break these bonds and form smaller molecules. They are partly eliminated when this oxidant is combined with other treatment techniques, such as biofiltration, coagulation or activated carbon adsorption;

– *biofiltration*: biofiltration is associated with chemical oxidation that forms small biodegradable molecules. These are then degraded by bacteria that have been developed on a medium installed in a filter;

– *ion exchange*: this solution consists of using anionic resins to eliminate small anionic molecules and hydrophilic charged molecules;

– *softening*: eliminates part of the NOM by adsorption on the carbonates.

Organic matter removal varies between 10 and 95%, depending on the process applied. The degree of organic carbon removal depends on several parameters, such as the quality and quantity of organic matter in the source water, the design of the treatment process and the operating conditions.

Selection of the most appropriate treatment requires knowledge of the variations in the concentration and characteristics of the NOM over the course of the year. Parameters to consider in a treatability study include chemical reagent dosages, turbidity, organic content (such as DOC, UV 254, color), organic properties (hydrophobicity/hydrophilicity ratio, ratio of humic to non-humic substances, SUVA), pH and alkalinity, and DBP formation potential.

7.4.1. *Coagulation–flocculation*

7.4.1.1. *Principles of coagulation*

The purpose of coagulation is to destabilize colloidal particles so that they agglomerate during flocculation and can subsequently be removed by clarification (decantation or flotation) and filtration. Coagulation is also involved in the removal of NOM through a phase change that converts dissolved organic matter into particles directly through precipitation or adsorption onto particles created by the coagulant. If metal coagulants are added to the water, chemical reactions occur with both the particles and the NOM. Therefore, when a coagulant is added, NOM acts as a ligand that associates with positively charged metal ions. It exerts a coagulant demand that must be met before flocculation occurs. The coagulation process is affected by the characteristics of NOM. For example, the coagulant dosage must be doubled to promote coagulation of fulvic acids compared to humic acids. Therefore, coagulation is seen as an "integrated" process that considers both the NOM and the particles (in other words, turbidity) while taking their particularities toward coagulation into account. Both NOM and particles have a negative charge that becomes even more negative with increasing pH. The negative charge of NOM is generally between 5 and 15 $\mu eq \cdot mg^{-1}$ of carbon, while that of particles ranges from 0.05 to 0.5 $\mu eq \cdot mg^{-1}$ of carbon. In most cases, the coagulant dosage is controlled by NOM, not by turbidity.

Two main mechanisms govern coagulation: one consists of charge neutralization and the formation of insoluble precipitates; the other involves the adsorption of NOM onto a floc of aluminum hydroxide or iron hydroxide. The effectiveness of each of these mechanisms depends on the operating conditions related to pH, coagulant dosage and alkalinity.

The charge neutralization mechanism involves the neutralization of negatively charged NOM molecules by the positively charged species of the coagulant before metal precipitation occurs. Initially, there is little or no complex formation between the coagulant and NOM. This concept not only applies to iron or aluminum salt coagulants and polyaluminum chloride (PACl) but can also be applied to organic cationic polymers used as coagulants.

The second mechanism involves the precipitation of metals, with the formation of complexes between the metal coagulant and organic ligands such as HS.

Its application must be well analyzed because it can only remove certain fractions of NOM, leaving some substances behind and forming DBPs during chlorine disinfection.

The "dissolved" HS can be considered true solutions of negatively charged hydrophilic macroions or colloids. Due to their negative charge, one of their properties is coagulation by electrolytes. Trivalent ions are more effective in coagulating HS than divalents. Since NOM consists of a mixture of compounds that react in different ways to the coagulant, it is important to characterize it beforehand and to perform jar tests in the laboratory.

In the presence of water with low DOC, coagulation is limited to destabilization and removal of particles (turbidity), with little DOC removal. DOC yields in the range of 10–25% are obtained. This is the so-called conventional coagulation.

Therefore, for waters containing DOC concentrations >2 $mg \cdot L^{-1}$, so-called advanced coagulation has been developed for NOM removal and has become the first treatment step for better control of NOM and the formation of disinfection by-products.

7.4.1.2. *Advanced coagulation*

Advanced coagulation is based on the application of higher dosages of coagulant to achieve the objective of good removal of NOM with yields between 40 and 70%. *Advanced coagulation* is the term used to define the process of obtaining a high NOM removal compared to a conventional treatment. The aim is to reduce the concentration of NOM in water by means of chemical reactions. This can be followed by an additional treatment using another technique.

With conventional coagulation, only processes involving high-pressure membranes can achieve very low residual NOM concentrations.

Advanced coagulation requires high coagulant dosages that can be up to 10–15 times the DOC concentration of the raw water.

Table 7.4 summarizes the performance of conventional and enhanced coagulation.

Coagulation efficiency is strongly related to coagulant type, coagulant dosage, pH and coagulation temperature. The optimal coagulation conditions for turbidity reduction are generally not the same as those for NOM removal. This complicates the choice of coagulation conditions when both particles and NOM are to be removed. Coagulation at a low pH and with a relatively high coagulant dosage favors NOM removal. For turbid and colored water, NOM usually determines the coagulation conditions because of the stoichiometry between NOM and the coagulant dosage.

Conventional coagulation		Advanced coagulation
Metal salt coagulant	**Polymerized coagulant**	**Metal salt coagulant**
10–30% DOC removal 20–45% UV 254 removal	Less than 15% DOC removal	40–70% DOC removal 60–90% UV 254 removal
10–20% BDOC removal	No effect on BDOC	80–90% color removal 50–70% BDOC removal
THMs formed above standards without effective additional treatment	THMs formed well above standards and additional treatment may not be sufficient	THMs formed well under standards
Good turbidity removal	Little control over pH	Additional operating cost for treatment
Low to medium coagulant dosage	No addition of acid	Good turbidity removal
Essential removal of turbidity and much less for NOM	Medium turbidity removal	High to very high coagulant dosage
No need to add an acid		Good removal of NOM
		An acid is often needed depending on the alkalinity of the raw water
Less rigorous pH control		Mandatory pH control

Table 7.4. *Performance of conventional coagulation and advanced coagulation*

7.4.1.3. *Which is the most efficient: Al-based or Fe-based coagulant?*

7.4.1.3.1. Aluminum-based coagulants

The most commonly used aluminum-based coagulant is aluminum sulfate, although other polymerized coagulants are becoming more widely used.

Aluminum sulfate

The significant increase in the aluminum sulfate dosage (120–150 $mg \cdot L^{-1}$) makes it possible to increase NOM removal but to a limit that does not generally exceed 40%. Some specific cases show 50–60% of DOC removal. However, depending on the composition of the water and the characterization of the NOM fractions, higher yields can be obtained.

The coagulation pH has an important effect on coagulation efficiency, and the optimum pH is between 6.0 and 6.5. At lower pH levels, a dissolved aluminum residual is obtained in the water.

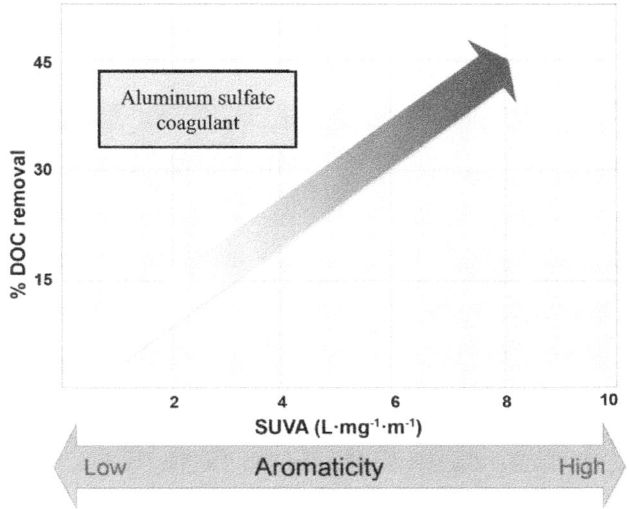

Figure 7.21. *DOC removal based on SUVA for aluminum sulfate. For a color version of this figure, see www.iste.co.uk/gaid/watertreatment2.zip*

The mechanism of charge neutralization with aluminum sulfate (Figure 7.22) involves the neutralization of negatively charged NOM molecules by the positively charged species of the coagulant. According to pH, coagulant dosages are higher. The ratios applied are:

– for a pH of 6.0–6.5: 0.5 mg Al/mg DOC;

– for a pH of 7: this value increases to 1 mg Al/mg DOC.

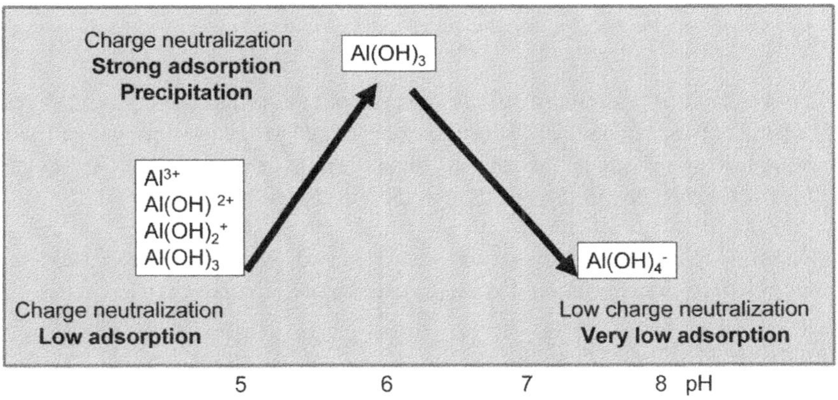

Figure 7.22. *Overall mechanism of NOM removal with aluminum salt coagulant. For a color version of this figure, see www.iste.co.uk/gaid/watertreatment2.zip*

Figure 7.23 shows that both the coagulation pH and aluminum sulfate dosage dictate the target area of the coagulation process for NOM removal.

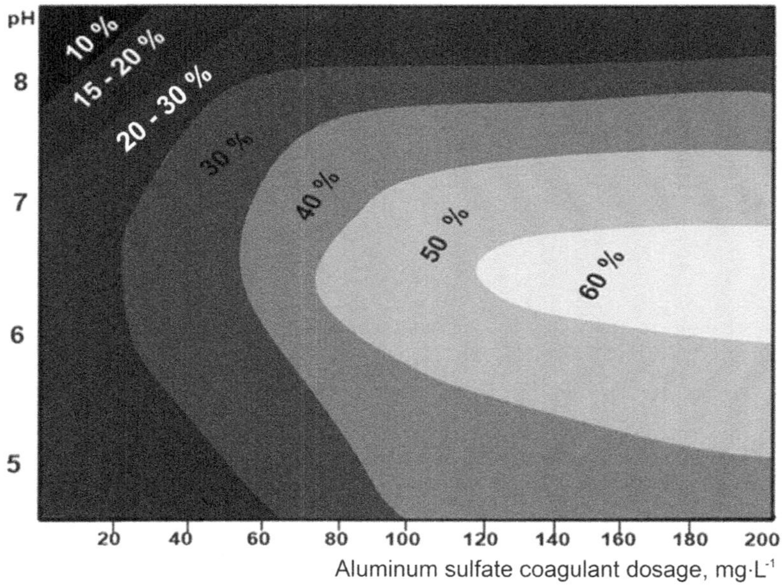

Figure 7.23. *NOM removal efficiency according to pH and aluminum sulfate dosage. For a color version of this figure, see www.iste.co.uk/gaid/watertreatment2.zip*

The maximum percentage of DOC removal obtained with aluminum sulfate (10% Al_2O_3) is 60% for a dosage equal to 180 $g \cdot m^{-3}$ and a pH of 6.0. This gives a DOC concentration in the treated water of 2 $mg \cdot L^{-1}$ for an inlet concentration of 5.0 $mg \cdot L^{-1}$.

This variation in aluminum sulfate dosages suggests that there is an interesting correlation between coagulant dosage, operating pH and removal efficiency. However, for high dosages, the total residual aluminum (soluble and particulate) is high for pH levels below 6.0, even though there is a high NOM removal.

Overall, Figure 7.23 shows that the removal efficiency decreases as the aluminum sulfate concentration decreases, and the pH increases.

Two distinct areas are visible.

At pH values above 6.5, the gradient becomes more vertically oriented, showing greater variability between removal percentage and aluminum sulfate dosages.

Increasing the aluminum sulfate concentration has little effect on the DOC removal percentages. This gives aluminum sulfate the lowest efficiency compared to other coagulants. A clear decrease in the removal percentage appears as the pH increases from 6.0 to 7.0, and the aluminum sulfate concentration increases from 140 to 80 $mg \cdot L^{-1}$.

Residual aluminum in NOM removal

The residual aluminum concentration is estimated from the aluminum hydroxide solubility diagram (see section 7.4.1). When the operating point is in the amorphous aluminum hydroxide precipitate zone (pH 6–6.5), the solubility of Al is lowest due to the main $Al(OH)_3$ species. Residual aluminum is minimal.

Outside this zone, the residual aluminum concentration in the treated water increases, since all hydrolysis products are in soluble form. The control of the coagulation pH must be adjusted, so that the hydrolysis products of aluminum sulfate are mostly particulate, and therefore easy to remove during the decantation and filtration steps.

For $Al_2(SO_4)_3$, the pH influence on the coagulation mechanisms correlates with the hydrolyzed coagulant species. When the pH is below 6.0, the hydrolyzed species are mainly positively charged. The mechanism is therefore charge neutralization (Figure 7.22). The positive charge density is sufficient to neutralize low molecular weight molecules because these molecules have a relatively small negative surface charge. However, the charge density is insufficient for medium- and high molecular weight molecules. This leads the lowest residual concentration of Al to move to the region of lowest Al solubility (pH 6–6.5).

As a result, the characterization of NOM could be a parameter for the choice of coagulant, as low molecular weight substances can favor a high residual Al.

During the removal of NOM, the residual aluminum concentration is lower in the presence of high molecular weight organic substances. This is because the charge neutralization mechanism is the factor that governs the concentration of residual aluminum. Since this mechanism neutralizes most of the charges of the Al species present in solution, the flake structures formed by the charge neutralization mechanism are more compact for medium and high molecular weight NOM.

NOM fractions with high molecular weights are well removed by Al sulfate, but since the chemical composition of the water contains organic substances with different molecular weights, this may impact the residual Al concentration in the treated water.

Low molecular weight substances have a lower negative charge density, so increasing the dosage of coagulant (positive charge of Al^{3+}) results in a residual of excess positive charges and thus residual aluminum.

SUVA	Characterization of NOM	Coagulation	% DOC removal
4	Humic substances with high hydrophobicity and high molecular weight	NOM controls the mechanisms Good DOC removal, but high coagulant dosage	>50%
2–4	Mixture of humic and fulvic acids Mixture of hydrophobic and hydrophilic substances Mixture of molecular weights	NOM influence Good to medium DOC removal	25–40%
<2	Mainly non-humic substances, low hydrophobicity Low molecular weight	NOM have little influence Low DOC removal	<25%

Table 7.5. *Relation of SUVA to the efficacy of aluminum-based coagulants*

Polyaluminum chloride

PACl increases the amount of Al_{13} (AlO_4Al_{12} $(OH)^{7+}$), which is the most effective species for organic matter removal. These coagulant species (Al_{13}) are considered the most effective Al species due to their larger size and higher positive charges. PACl leads to a lower concentration of residual Al, especially of dissolved monomeric Al. This is particularly the case in coagulants with a basicity of 2.0.

Operating in a pH range of 6–7, DOC removal generally varies between 20 and 50%, depending on the coagulant dosage. A 50% yield is obtained with dosages between 80 and 100 mg·L^{-1}. Color removal is higher with yields of 80%.

7.4.1.3.2. Iron-based coagulants

The first mechanism related to electrostatic adsorption results in a variation in the electrokinetic potential. The adsorption of negatively charged organic molecules Y^{n-}, by the preformed floc $[XOH]^+$, is written as:

$$[XOH]^+ + Y^{n-} \rightarrow [XOHY]^{(n-1)-}$$

The second mechanism corresponds to a slow kinetic reaction on the flocs and consists of an exchange of hydroxide ions:

$$[XOH]^{+} + Y^{n-} \rightarrow [XY]^{(n-2)-} + OH^{-}$$

The adsorption-based mechanism of dissolved organic matter removal suggests that an approach including floc recirculation is beneficial. Indeed, the recirculation of flocs increases the number of particles and thus the efficiency of coagulation/flocculation. Moreover, contact with the already formed flocs, having a significant adsorption surface, favors the fixation mechanisms of the organic compounds.

Organic substances with many conjugated bonds ionize in an aqueous medium and typically present a negative charge at acidic pH. Therefore, during the hydrolysis of ferric ions, hydroxo-metallic complexes ($Fe(H_2O)_{6-n}(OH)^{+3-n}$) are formed, which react rapidly with HS to form an insoluble compound:

$$Fe(H_2O)^{3+} + n\ OH^{-} \rightarrow Fe(H_2O)_{6-n}\ (OH)^{+3-n}$$

hydrolysis of Fe^{3+}

Reaction with HS:

$$Y^{n-} + Fe(H_2O)_{6-n}\ (OH)^{+3-n} \rightarrow (Y)\ Fe(H_2O)_{6-n}\ (OH)^{+3-n}$$

insoluble complex

This explains why Fe salts (in an acid medium) are more efficient than Al salts.

Ferric salts commonly used in coagulation processes include ferric chloride ($FeCl_3$) and ferric sulfate ($Fe_2(SO_4)_3$), with yields ranging from 40 to 70%, depending on the coagulant dosage. Figure 7.24 shows the consequences of overdosing or underdosing the Fe coagulant. An overdosage (200–300 $mg \cdot L^{-1}$) leads to good yields at a pH of 4.5-6, but a high residual Fe gives a red color to the water that must then be precipitated in the filtration process. The molecules affected at a pH < 5 are those with a molecular weight below 750 Daltons (Figure 7.24). Underdosing implies poor NOM removal at any pH because there are not enough adsorption sites provided by the coagulant.

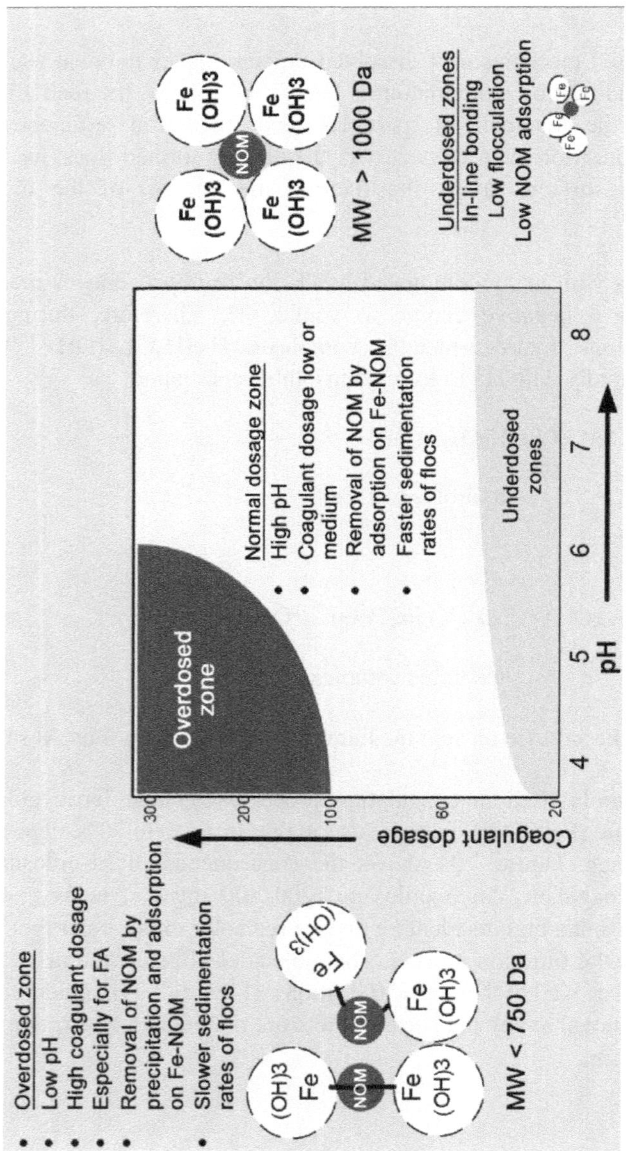

Figure 7.24. *Control of NOM removal with ferric chloride. For a color version of this figure, see www.iste.co.uk/gaid/watertreatment2.zip*

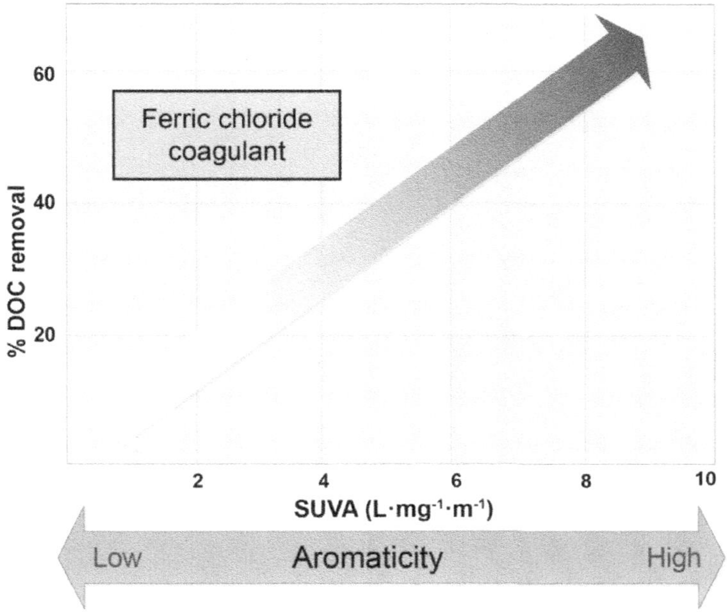

Figure 7.25. *Percentage of DOC removal based on SUVA for ferric chloride. For a color version of this figure, see www.iste.co.uk/gaid/watertreatment2.zip*

At low dosages of coagulant, the anionic humic network reorganizes by association with cationic coagulant species, giving compact structures. However, at high dosages, aggregation of humic acids with hydrolyzed Fe species is attributed to competition between the rate of humic network re-conformation and the rate of collision of destabilized colloids. Figure 7.25 shows that increasing SUVA results in better DOC removal. The aromatic configuration of the organic substances facilitates their adsorption and the formation of insoluble complexes.

The percentage of NOM removal depends on the pH and the coagulant dosage (Figure 7.26). The maximum percentage of DOC removal obtained with ferric chloride (commercial product at 42%) is often 70% for a dosage equal to 160 $g{\cdot}m^{-3}$ and a pH of 5.5. In the treated water, this gives a DOC concentration of 2 $mg{\cdot}L^{-1}$ for an inlet concentration of 6.7 $mg{\cdot}L^{-1}$. Higher dosages may be needed, depending on the characterization of the NOM.

This wide variation in ferric chloride dosages suggests that there is a strong correlation between coagulant dosage, operating pH and removal efficiency. Removals are highest for pH values below 6.0. However, a weak correlation is observed between coagulant dosage and NOM removal if the operating pH is not controlled.

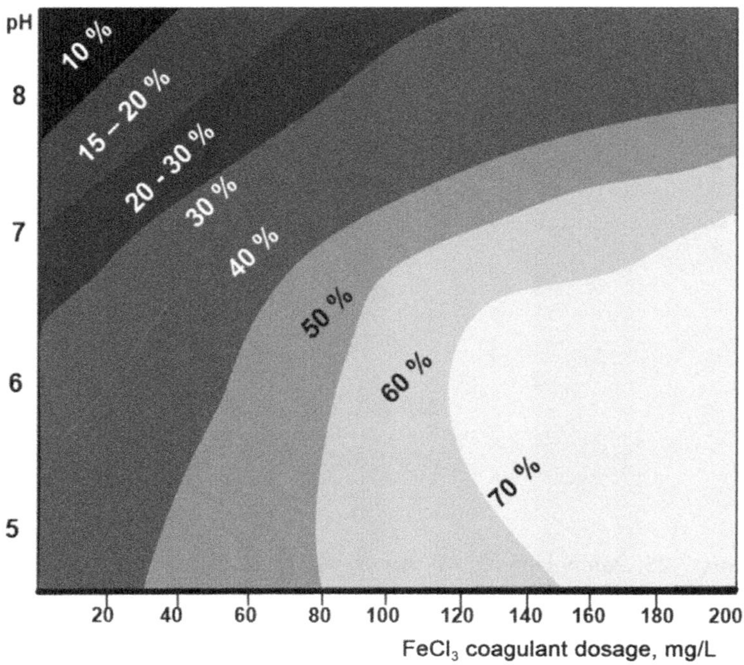

Figure 7.26. *NOM removal efficiency according to coagulant dosage ($FeCl_3$) and pH. For a color version of this figure, see www.iste.co.uk/gaid/watertreatment2.zip*

Overall, Figure 7.26 shows that the removal efficiency decreases as the ferric chloride concentration decreases, and the pH increases.

In fact, three distinct areas are visible:

– at pH values above 5, the gradient becomes more vertically oriented, showing greater variability between removal percentage and ferric chloride dosages;

– a decreasing trend in the removal percentage appears as the pH increases from 5.0 to 7.0, and the ferric chloride concentration increases from 160 to 80 $mg \cdot L^{-1}$;

– a third region is observed at a pH above 6, where ferric chloride dosages are higher than 160 $mg \cdot L^{-1}$. DOC removal is less affected by pH and is more sensitive to ferric chloride concentration.

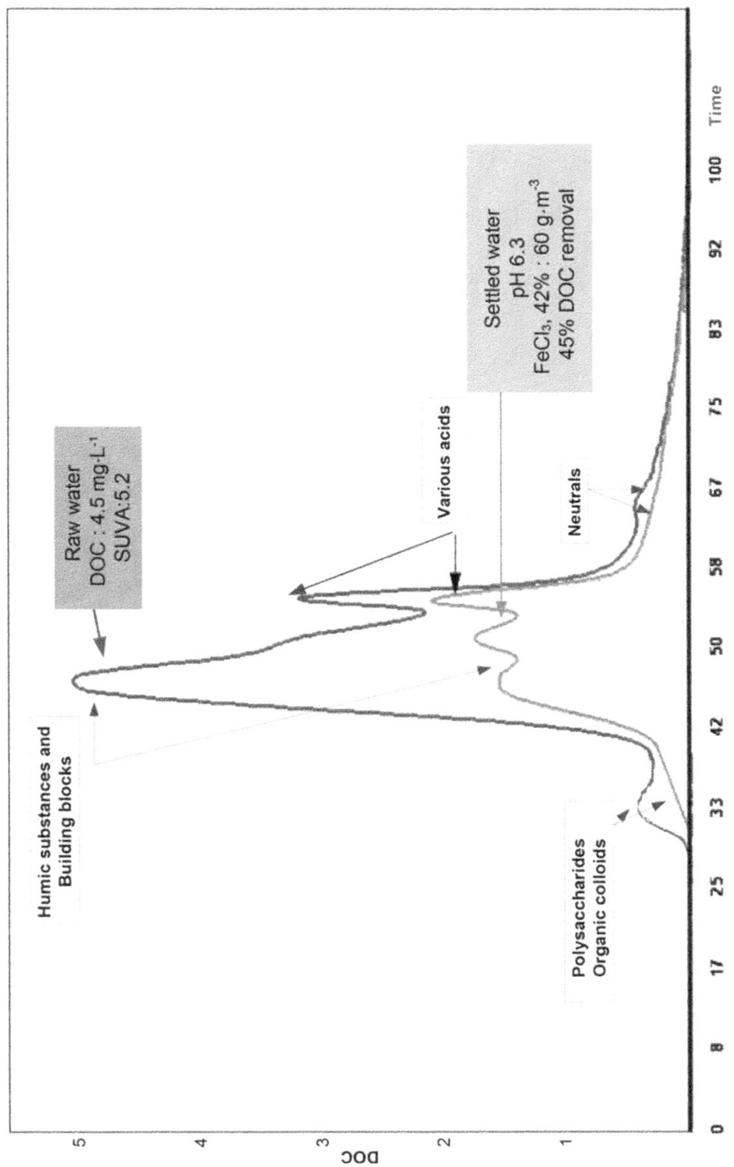

Figure 7.27. *LC-DOC on raw water and decanted water with $FeCl_3$. For a color version of this figure, see www.iste.co.uk/gaid/watertreatment2.zip*

The LC-DOC chromatogram (Figure 7.27) clearly illustrates the molecules targeted by the addition of Fe coagulant, with a strong decrease in high molecular weight substances (polysaccharides, HS and building blocks). Organic acids (medium MW) are less affected, while neutral molecules are not. This chromatogram clearly shows that the coagulation efficiency of high molecular weight molecules is greater than that of other molecules present in the medium. This clearly shows the influence of a high SUVA (Figure 7.25) and that it is important to provide additional treatment to remove residual substances after the coagulation stage.

Thus, the adsorption and desorption of H^+ at the interface between the iron oxide and the solution depends on the pH of the medium (Figure 7.28). DOC removal at pH 6.5 is 46% for this type of water, while the removal percentage is 64% at pH 5.

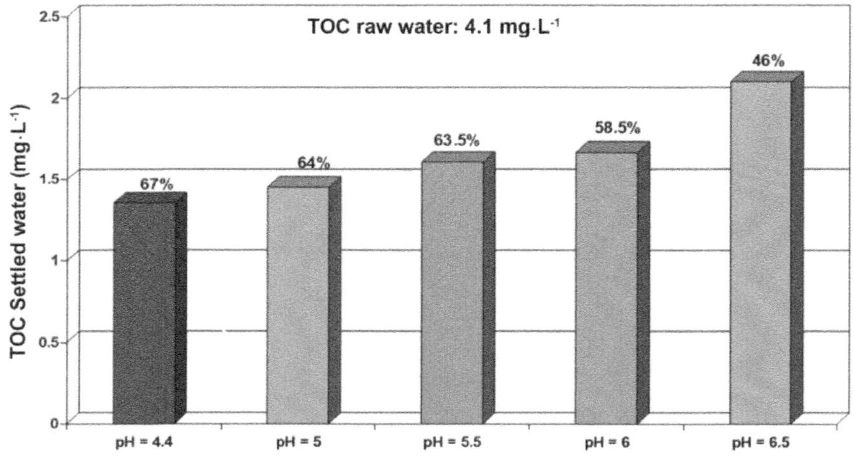

Figure 7.28. *$FeCl_3$ efficiency (42%) at different pH values (Veolia France site). For a color version of this figure, see www.iste.co.uk/gaid/watertreatment2.zip*

At high pH, FeOH loses its H^+ proton and forms FeO^-:

$$FeOH + OH^- \rightarrow FeO^- + H_2O$$

At low pH, the oxygen present in FeOH can be considered a Lewis base that provides an electron pair and combines with H^+ to form positively charged $FeOH^{2+}$ on the surface of the flocs:

$$FeOH + H^+ \rightarrow FeOH^{2+}$$

As a result, the removal of HS decreases with increasing pH due to the repulsion between HS and the surface of hydroxide flocs $(FeOH)_3$.

Figure 7.29 shows the difference in volume of $Fe(OH)_3$ flocs compared to that of $Al(OH)_3$ for nearly equivalent dosages. This property gives iron hydroxides higher adsorption capacities.

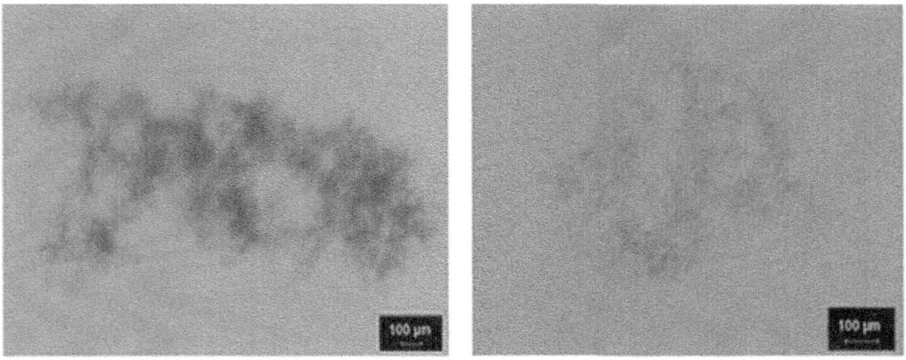

a) b)

Figure 7.29. *$Fe(OH)_3$ (a) and $Al(OH)_3$ (b) flocs. For a color version of this figure, see www.iste.co.uk/gaid/watertreatment2.zip*

The evolution of SUVA on water coagulated with ferric chloride is shown in Figure 7.30. The DOC of the raw water varies between 4 and 5 $mg \cdot L^{-1}$, with peaks at 10 $mg \cdot L^{-1}$. The UV 254 nm of the raw water varies between 18 and 30 m^{-1}, with peaks at 65 m^{-1}. The turbidity of the raw water is between 5 and 20 NTU. The SUVA is high and evolves between 3.8 and 5 $L \cdot mg^{-1} \cdot m^{-1}$, with a peak at 6.5 $L \cdot mg^{-1} \cdot m^{-1}$. These high values indicate that coagulation is favorable for good NOM removal. Coagulation with $FeCl_3$ (42%) at pH 6 is performed with a coagulant dosage varying between 60 and 100 $g \cdot m^{-3}$. TOC removal is between 55 and 75%.

The settled water has SUVA values that vary between 2 and 3 $L \cdot mg^{-1} \cdot m^{-1}$. The UV 254 nm value is strongly reduced. The coagulation was very efficient, with removal of most of the high and medium molecular weight substances. The UV 254 nm values are in the range of 1.8–3.5 m^{-1}.

Ferric sulfate

Ferric sulfate shows a similar correlation to ferric chloride at pH values above 5.5. The DOC removal percentages are between 40 and 60% for dosages varying between 140 and 250 mg·L^{-1}. Unlike ferric chloride, increasing ferric sulfate dosages for pH values above 6.5 only results in a 10% increase in DOC removal. The required coagulant concentration of ferric sulfate is 50% higher and removes nearly 15% less DOC compared to ferric chloride.

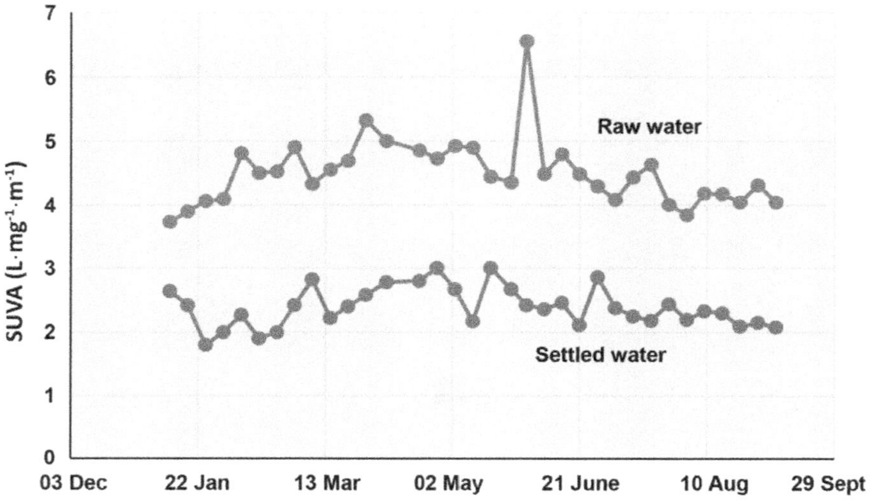

Figure 7.30. *Evolution of SUVA between raw water and decanted water on a plant, including $FeCl_3$ coagulation (Veolia site) . For a color version of this figure, see www.iste.co.uk/gaid/watertreatment2.zip*

7.4.1.4. *Ferric chloride–aluminum sulfate efficiency comparison*

The choice between the two reagents mostly depends on the mineralization of the water, and the pH conditions are related to the solubility of the two metals. DOC removal during coagulation is defined as an adsorption equilibrium between the surface of the hydroxides (adsorbent) [$Fe(OH)_3$ or $Al(OH)_3$] and the adsorbable DOC fraction (adsorbate).

In the presence of organic matter, a salt known as Al or Fe humate is formed, and the coagulation efficiency is associated with a stoichiometric relationship between the coagulant dosage and the amount of substances to be removed. However, the isotherm defining the adsorption equilibrium between DOC and the hydroxide surface only depends on the type of coagulant and the final pH.

This adsorption mechanism can be expressed according to the Freundlich or Langmuir relation.

$Al_2(SO_4)_3,14H_2O$ mg·L^{-1}	$Al_2(SO_4)_3,18H_2O$ mg·L^{-1}	$FeCl_3,6H_2O$ mg·L^{-1}	$FeCl_3$ mg·L^{-1}	$Fe_2(SO_4)_3,9H_2O$ mg·L^{-1}	$FeSO_4,7H_2O$ mg·L^{-1}
10	11.2	9.1	5.5	9.5	9.4
20	22.4	18.2	11	19	18.8
30	34.6	27.3	16.5	28.5	28.2
40	44.8	36.4	22	38	37.6
50	56	45.5	27.5	47.5	47
60	67.2	54.6	33	57	56.4
70	78.4	63.7	38.5	66.5	65.8
80	90.6	72.8	44	76	75.2
90	100.8	81.9	49.5	85.5	84.6
100	112	91	55	95	94

Table 7.6. *Equivalent dosages of coagulants*

For example, the application of Langmuir's relation gives:

$$\frac{[\text{Removed DOC}]}{[\text{Fe3 + or Al3+}]} = \frac{\text{qm b [ads DOC]}}{1 + \text{b [ads DOC]}}$$

with:

– ads DOC: adsorbed DOC;

– qm: the max adsorbed DOC, expressed in mg DOC·mg^{-1} Fe^{3+} or Al^{3+};

– b: the adsorption constant of adsorbable DOC in relation to the hydroxide surface, expressed in L·mg^{-1} DOC.

The DOC affinity for the hydroxide surface and the number of sites available for adsorption are strongly related to pH. As a result, the qmax value varies with the coagulation pH. Comparative tests between ferric chloride and Al sulfate show that the value of b is two times lower with $FeCl_3$ than with $Al_2(SO_4)_3$.

Three comparative examples of operating plants are presented.

7.4.1.4.1. Water from the Rophemel dam (Veolia site, western France)

These results (Table 7.7) show better NOM removal with ferric chloride (66.3%) than with aluminum sulfate (53.6%) at a pH between 5.9 and 6.0. The dosages of the two coagulants are 105 g·m^{-3}.

Parameters	Raw water	Settled water	
		$Al_2(SO_4)_3$ (8.5%) Dosage (g·m^{-3})	$FeCl_3$ (41 %) Dosage (g·m^{-3})
		105	105
TOC (mg·L^{-1})	11	5.1	3.7
Turbidity (NTU)	32–40	1.2	1.0
$KMnO_4$ Ox. (mg·L^{-1})	14.3	4.2	2.9
pH	6.95	5.9–6.0	5.9
UV 254 nm (cm^{-1})	0.610	0.095	0.085
SUVA	5.5	–	–

Table 7.7. *Comparison of DOC and turbidity removal efficiency with $Al_2(SO_4)_3$ and $FeCl_3$*

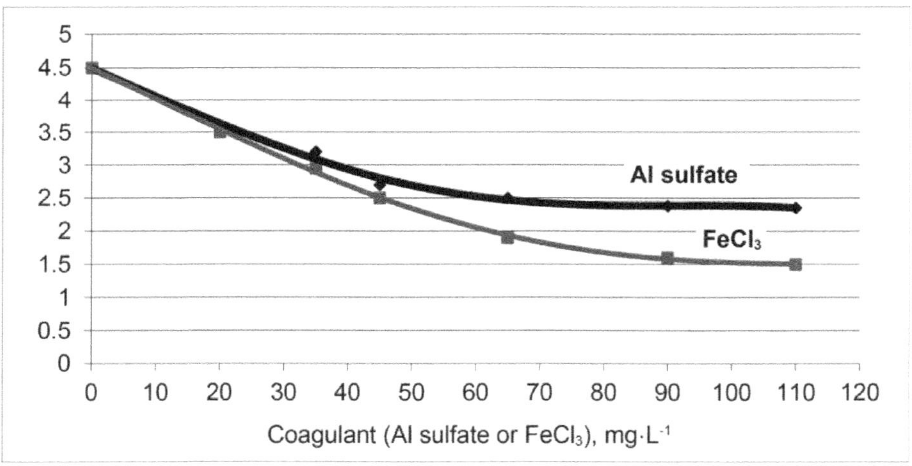

Figure 7.31. *Comparison of DOC removal with $Al_2(SO_4)_3$ and $FeCl_3$ at different dosages. For a color version of this figure, see www.iste.co.uk/gaid/watertreatment2.zip*

Other comparative results are presented in Figures 7.31 and 1.32, showing the better efficiency of ferric chloride.

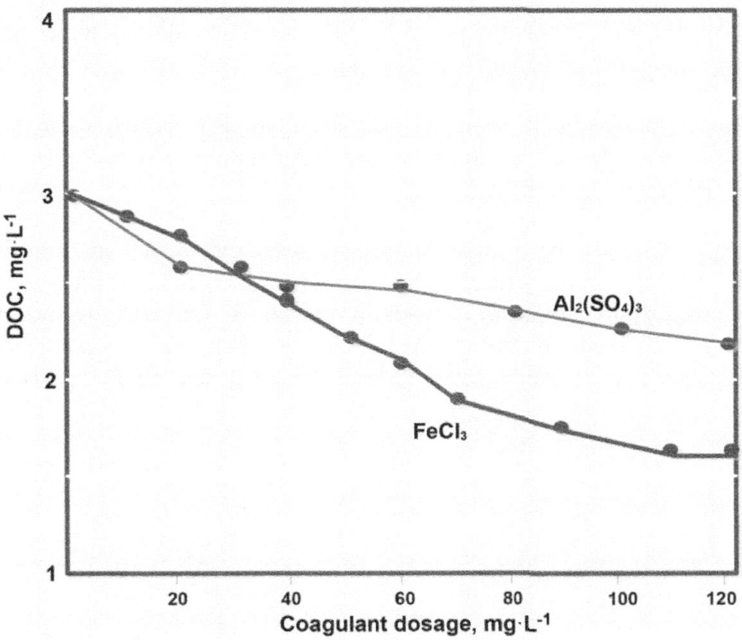

Figure 7.32. *Comparison of DOC removal with $Al_2(SO_4)_3$ and $FeCl_3$ at different dosages. For a color version of this figure, see www.iste.co.uk/gaid/watertreatment2.zip*

7.4.1.4.2. Water from the Marillet reservoir (Veolia site, western France)

The treatment process includes a clarification stage with $FeCl_3$ as a coagulant in an acidic medium. During coagulation, the ferric chloride that is introduced into the medium precipitates in the form of hydroxide, whose adsorptive properties allow the trapping of colloidal particles and organic matter. In practice, we try to determine the optimum pH of coagulation–flocculation, which, for a given $FeCl_3$ rate, makes it possible to obtain the best removal efficiency of colloidal particles and organic matter.

The coagulant dosage is between 70 and 90 $mg \cdot L^{-1}$, intended to treat raw water with a variable turbidity between 30 and 40 NTU, with peaks of more than 60 NTU. The concentration of organic matter expressed in $KMnO_4$ oxidability is between 8 and 10 $mg \cdot L^{-1}$, with peaks of 16 $mg \cdot L^{-1}$.

The results obtained on the organic matter show that the removal efficiencies are approximately 80% given that, on average, the residual organic matter concentration is between 2 and 2.6 mg·L^{-1}. The high rates of $FeCl_3$ used in the treatment are justified by this high concentration of NOM. The pH conditions for coagulation–flocculation are between 5.5 and 6.0.

Systematic monitoring of residual iron concentrations in treated water shows that the threshold of 0.2 mg·L^{-1} of iron in treated water is rarely reached, given that it is between 0.05 and 0.15 mg·L^{-1}.

Figure 7.33 shows the results obtained on reservoir water that is loaded with seasonally varied organic matter (expressed in $KMnO_4$ oxidizability) with ferric chloride coagulation in an acid medium, which is followed by flocculation-decantation and sand filtration. The yields are between 75 and 85%. The characterization of the organic matter showed that this water mainly consists of high molecular weight HS.

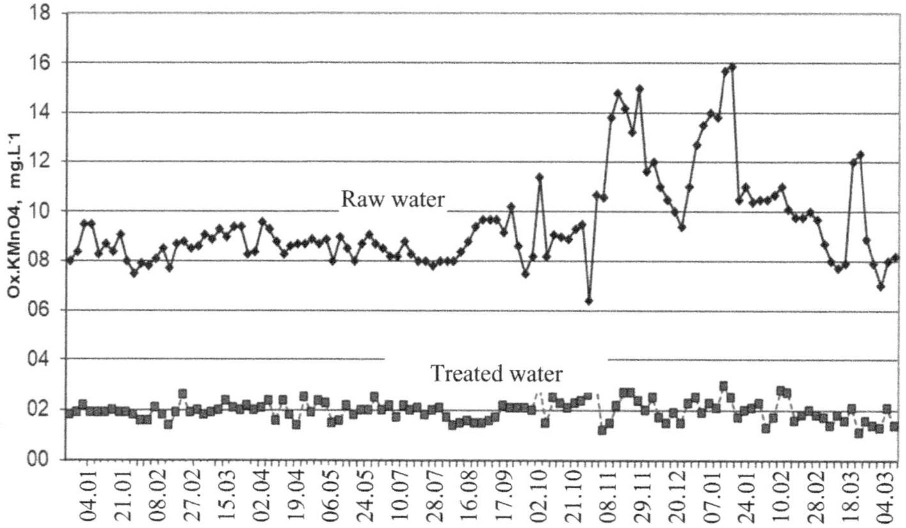

Figure 7.33. *Evolution of organic matter removal ($FeCl_3$ coagulant: 70–90 g·m^{-3}, pH 5.5–6). For a color version of this figure, see www.iste.co.uk/gaid/watertreatment2.zip*

7.4.1.4.3. Water from the Urne River (West France)

The Urne River (west) has average TOC concentrations equal to 3.3 mg·L^{-1}, with peaks at 8 mg·L^{-1} and a variable turbidity between 5 and 20 NTU. The comparative tests obtained with various coagulants are summarized in the table.

It includes an identical coagulant dosage of 50 g·m^{-3}. The coagulation pH is between 6.8–7.0. The various coagulants used are as follows:

– Aqualenc: Al polyhydroxychlorosulfate (d = 1.15);

– Aqualenc F1: Al polychloride (d = 1.25);

– $FeCl_3$: ferric chloride (42%) (d = 1.41);

– WAC HB: Al polyhydroxychlorosulfate (d = 1.17);

– WAC HBA: Al polyhydroxychlorosulfate (d = 1.20).

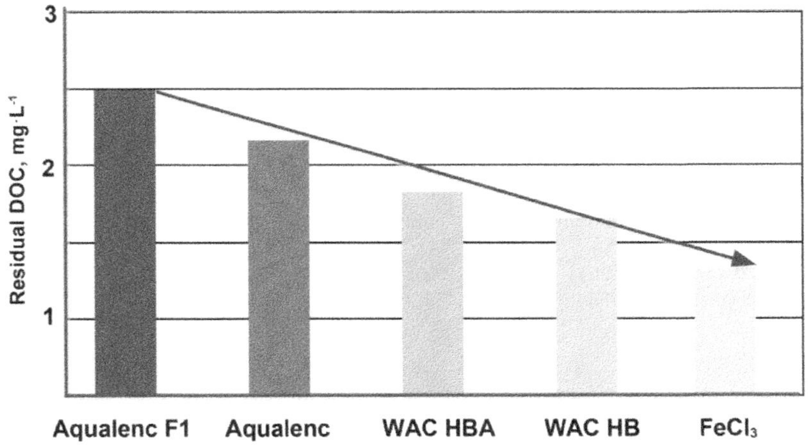

Figure 7.34. *The effectiveness of various coagulants against TOC (3.3 mg·L^{-1}) on the Urne River. For a color version of this figure, see www.iste.co.uk/gaid/watertreatment2.zip*

It appears that in this type of water, Aqualenc coagulants are not very efficient (24% and 36%), WAC HB and HBA coagulants are equivalent (45%) and $FeCl_3$ is the most efficient (60%). The ranking of coagulants based on their efficiency toward TOC removal is $FeCl_3$ > WAC HB and WAC HBA > Aqualenc > Aqualenc F1.

Figure 7.35 shows the 3-year monitoring of organic matter (TOC) removal results using advanced coagulation with ferric chloride on a river near Landivisiau (western France). These results show the importance of considering each of the objectives of the coagulation process (pH, alkalinity, TOC) when choosing the coagulant dosage. The primary objective is to obtain the lowest possible TOC concentration, depending on the physicochemical characteristics of the water and its variations.

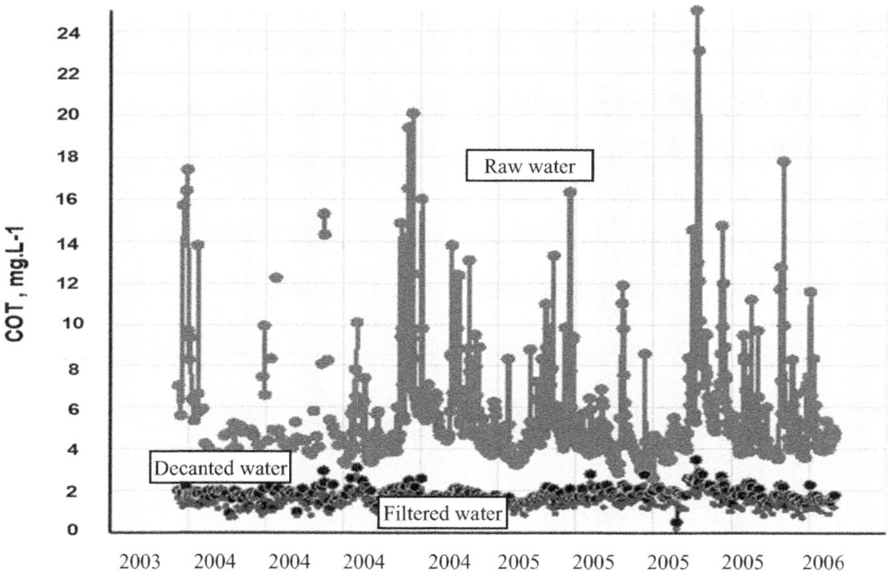

Figure 7.35. *Efficiency of ferric chloride for the removal of organic matter. For a color version of this figure, see www.iste.co.uk/gaid/watertreatment2.zip*

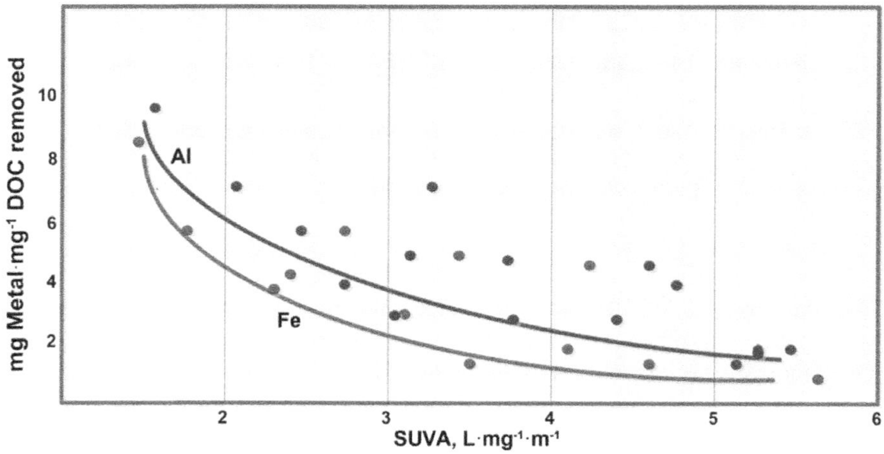

Figure 7.36. *SUVA relation according to the dosage of metal/DOC removed. For a color version of this figure, see www.iste.co.uk/gaid/watertreatment2.zip*

Figure 7.36 shows the ratio of mg metal to mg DOC according to SUVA. It is clear that coagulant dosages are higher when the SUVA value is lower than 3. This confirms that the coagulation of low molecular weight substances (neutral hydrophilic NOM and non-humic substances) is difficult and requires high dosages.

7.4.1.5. *Factors and parameters affecting coagulation–flocculation*

7.4.1.5.1. Influence of the nature (hydrophobic–hydrophilic) of organic matter

Allochthonous NOM tends to be hydrophobic in nature and is well removed by coagulation, whereas hydrophilic NOM is more difficult to treat. In the case of water rich in neutral hydrophilic NOM, coagulation is not very effective.

UV absorbance (254 nm) is relatively constant in winter and summer but increases in fall and spring due to the massive transport of allochthonous NOM in streams. In summer, the aquatic organism activity is higher and, therefore, more autochthonous NOM is produced. In winter, aquatic organism activity (and therefore the production of autochthonous NOM) slows down, and there is no transport of allochthonous NOM due to frozen ground and ice cover.

The nature of NOM has a significant effect on the coagulant dosage. Water containing HS with a high molecular weight requires a rather low coagulant dosage because the removal mechanism is mainly based on charge neutralization. If the water contains low molecular weight HS or non-humic substances, the removal mechanism is instead adsorption on the metal hydroxide surfaces, and the coagulant dosage is significantly higher.

Figures 7.37 and 7.38 illustrate the evolution of organic matter characteristics in surface water over the course of a year.

The hydrophobic fraction of NOM is, generally speaking, removed by coagulation more efficiently than the hydrophilic fraction. High molecular weight HS are also more easily removed than low molecular weight compounds, probably because they are quite hydrophobic, consisting of more aromatic compounds.

Hydrophobic NOM includes high levels of negative charge due to the presence of ionized groups, such as carboxylic and phenolic groups with higher MW, and is removed much more efficiently by coagulation than the hydrophilic fraction. Thus, the hydrophobic fraction dominates the specific colloidal charge nature of the water, and fractions with a higher charge are more likely to be properly removed.

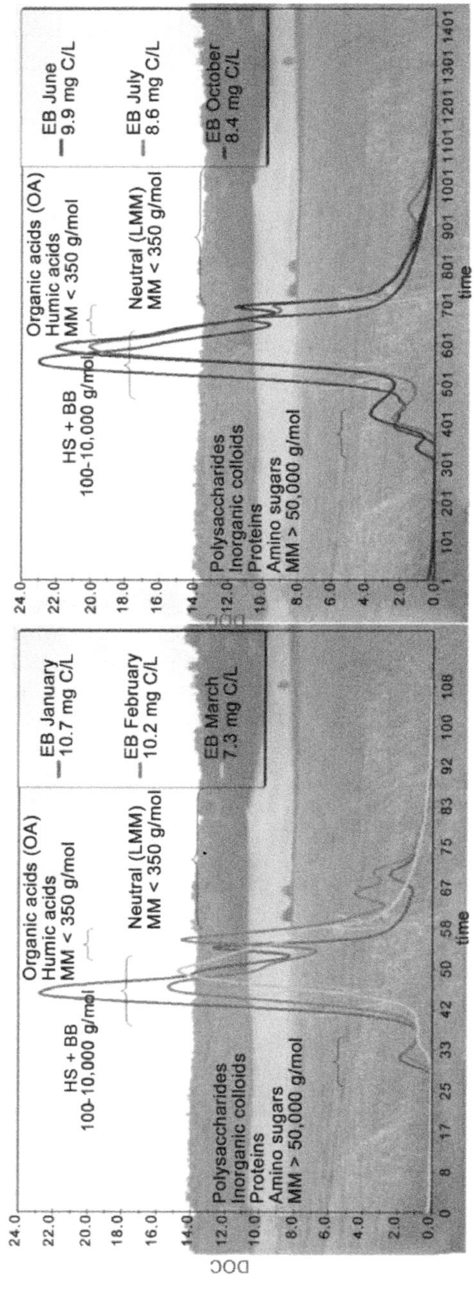

Figure 7.37. *Evolution (LC-DOC chromatograms) of organic matter over the course of a year (source: Veolia site, western France). For a color version of this figure, see www.iste.co.uk/gaid/watertreatment2.zip*

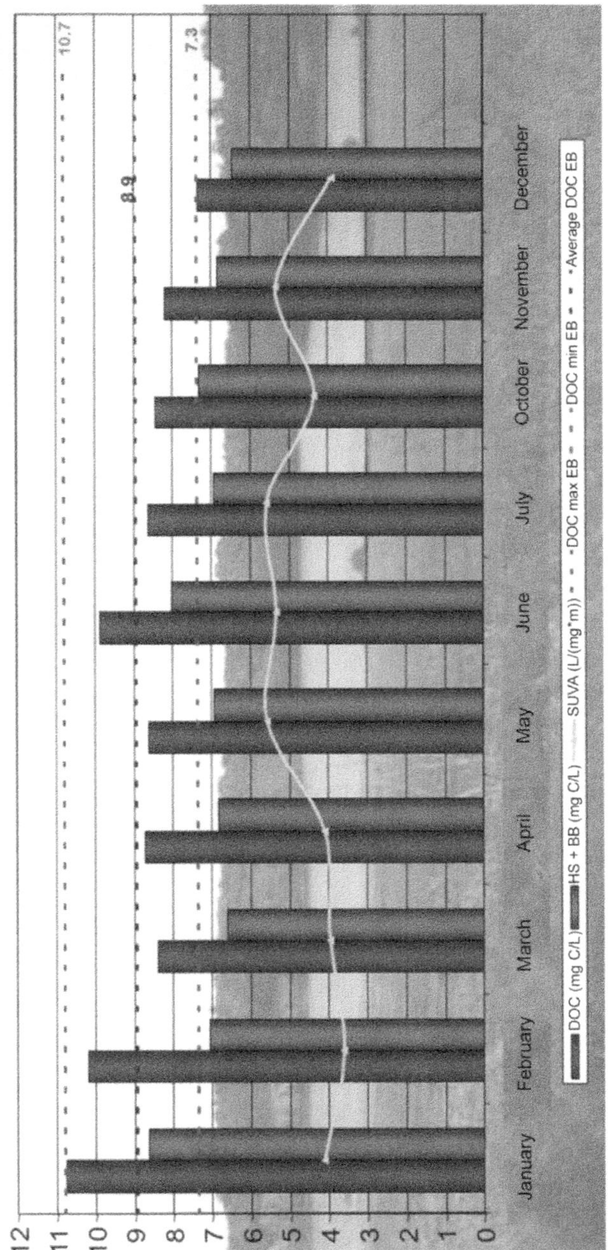

Figure 7.38. *Evolution of the composition of organic matter over a year (source: Veolia site, western France). For a color version of this figure, see www.iste.co.uk/gaid/watertreatment2.zip*

The hydrophilic fraction of NOM consists of compounds with aliphatic carbon branches and nitrogenous compounds such as proteins.

7.4.1.5.2. Influence of the type of coagulant

Chemical coagulation is achieved by the addition of inorganic coagulants, such as aluminum and iron salts. When added to water, the aqueous salts of aluminum sulfate and ferric chloride are dissociated into their respective trivalent ions, in other words, Al^{3+} and Fe^{3+}. After this, they are hydrolyzed and form several soluble complexes with high positive charges, thus attaching to the surface of negative colloids.

The mechanisms by which NOM is transformed into insoluble particulate aggregates include charge neutralization, entrapment, adsorption and complexation with the metal ion of the coagulants. NOM removal is primarily dictated by the coagulable (adsorbable) fraction of NOM, and SUVA is used to estimate this removal. The higher the specific UV absorbance (SUVA), the greater the NOM removal. The chemical nature of the NOM also influences its affinity to the coagulants used.

Ferric chloride and ferric sulfate, as well as alum, are acidic, thus lowering the pH after injection into the water. As a result, waters with low alkalinity may require the addition of a base, while high alkalinity is partially reduced by acidity. Waters with high alkalinity have a higher buffer capacity than low alkalinity waters with respect to pH decrease.

Iron-based coagulants are more effective, especially for removing NOM that has a molecular weight between 1,000 and 4,000 $g \cdot mol^{-1}$, because this fraction, which is mainly composed of fulvic acids, is subjected to a higher charge density provided by these coagulants. In addition, the flocs formed by iron-based coagulants are larger and more numerous than those formed by aluminum-based coagulants, thus increasing the probability of collisions, entrapment and adsorption. In most cases, coagulation chemistry governs the removal of NOM.

The most commonly used coagulants are aluminum or iron metallic salts, such as $Al_2(SO_4)_3$, xH_2O, $FeCl_3$, xH_2O (ferric chloride) and $Fe_2(SO_4)_3$, xH_2O (ferric sulfate). Prehydrolyzed metallic salts such as polyaluminum chloride are also used, as well as hydroxy aluminum polysulfate chloride. Field results show that iron salts are better than inorganic and prehydrolyzed aluminum salts or organic coagulants. The most commonly used coagulant in drinking water production is aluminum sulfate. However, the use of ferric salts is more common for NOM removal because the removal mechanism of ferric and alum coagulants is different.

Although organic coagulants have advantages such as not being pH dependent, lower sludge production, better sludge dewatering and lower residual aluminum concentration, they are not used in advanced coagulation because their DOC removal efficiencies are lower than metallic salts.

In addition, turbidity removal is also worse with residual turbidity and is often higher than that obtained with metallic and prehydrolyzed coagulants.

Figure 7.39 shows that the use of ferric chloride ($FeCl_3$, 41%) gives better DOC removal than PACl for equivalent dosages. An average of 47% DOC removal is obtained with $FeCl_3$ and 30% with PACl. The SUVA of raw water is 3.2 $L \cdot m^{-1} \cdot mg^{-1}$. It is therefore important, when possible, to test coagulants in the laboratory. Furthermore, considering that iron and aluminum-based coagulants work by the same mechanism, the difference between them lies in the optimal pH. A coagulant works most effectively at or near its isoelectric point. The isoelectric point of the Fe-based coagulant is approximately pH 4.5, while the Al-based isoelectric point is approximately pH 5.8.

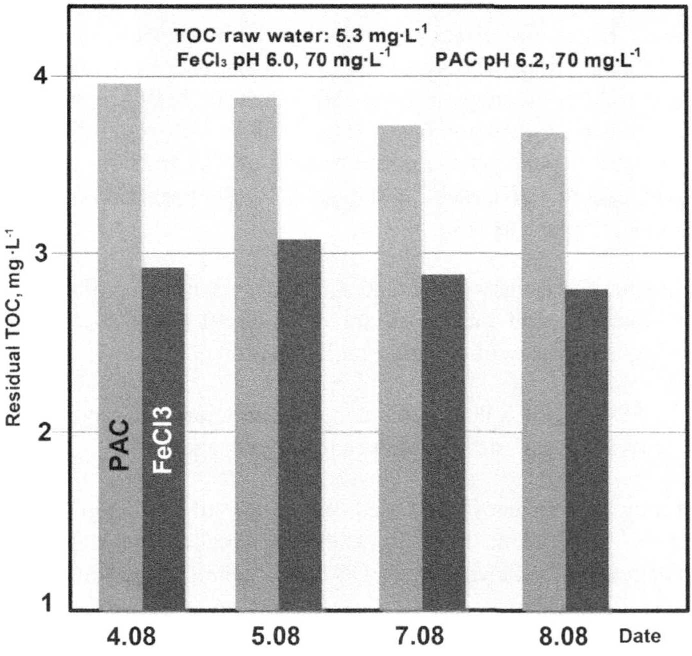

Figure 7.39. *Influence of the type of coagulant used for DOC removal. For a color version of this figure, see www.iste.co.uk/gaid/watertreatment2.zip*

Cationic polyelectrolytes such as polydiallyldimethylammonium chloride (PolyDADMAC) and cationic polyacrylamides (CPAM) are sometimes used for NOM removal. Indeed, being negatively charged, the mechanism of charge neutralization is involved in the cationic NOM-polyelectrolyte reaction over bridging.

Thus, polydadmac is interesting because of its very high charge density and its reaction with high molecular weight substances. Because of its cost, it is more often combined with a metal-based coagulant. However, it is advisable to carry out jar tests to ensure their effectiveness on the type of water to be treated.

7.4.1.5.3. Coagulant dosage

When the TOC is higher than 2 $mg \cdot L^{-1}$, coagulation is controlled by the NOM concentration.

As previously seen, NOM can be divided into hydrophobic and hydrophilic fractions. The hydrophobic fraction is less soluble, has a higher molecular weight and is more aromatic and humic. The hydrophilic fraction of low molecular weight is non-humic. Coagulation effectively removes the hydrophobic fraction, while the hydrophilic fraction is poorly removed by coagulation. As a result, the different fractions have a different coagulant demand. When the SUVA is greater than 4, the hydrophobic fraction is predominantly high and is well removed (60–80%), and coagulant dosages depend on the concentration of this fraction. A high coagulant dosage provides more active metal in the water for the formation of complexes with organic compounds, such as HS.

When the SUVA is between 2 and 4, there is a mixture of hydrophobic and hydrophilic fractions, and the yields are between 40 and 60%. In this case, the dosages are higher because of the hydrophilic fraction.

When the SUVA is less than 2, the dominant fraction is the hydrophilic fraction, and DOC yields are lower, despite high coagulant dosages.

DOC removal efficiency is based on ratios of 0.5 $mgDOC \cdot mg^{-1}Fe$ and 1 $mgDOC \cdot mg^{-1}Al$. With an MW of Fe that is equal to more than twice that of Al, the molar removal efficiency (mol DOC/mol metal) is the same for both ratios, approximately 2.3 $mol \cdot mol^{-1}$.

Under optimal operating conditions, a removal efficiency of 1 $mgDOC \cdot mg^{-1}Al$ results in 50% DOC removal, and a removal efficiency of 0.5 $mgDOC \cdot mg^{-1}Fe$ results in 60% DOC removal.

Sometimes, an increase in residual metal concentration (Fe^{3+} or Al^{3+}) in water is observed, even though the coagulant dosage is low. Indeed, when the dosage is lower than the optimum dosage, coagulation is not carried out efficiently, and residual metal concentrations remain in solution.

In general, coagulation at low pH requires special attention for potential issues of corrosion, as well as the need to correct the aggressiveness and corrosiveness of the treated water to ensure that it is not.

7.4.1.6. Mechanism of NOM-coagulant reactions

Coagulation–flocculation, combined with decantation, is a physical–chemical treatment that removes suspended particles, pathogenic microorganisms, organic particles and NOM. Coagulation corresponds to a destabilization of colloidal particles to make them suitable for agglomeration during flocculation. These particle agglomerates, known as flocs, are then separated from the water by sedimentation in a decanter.

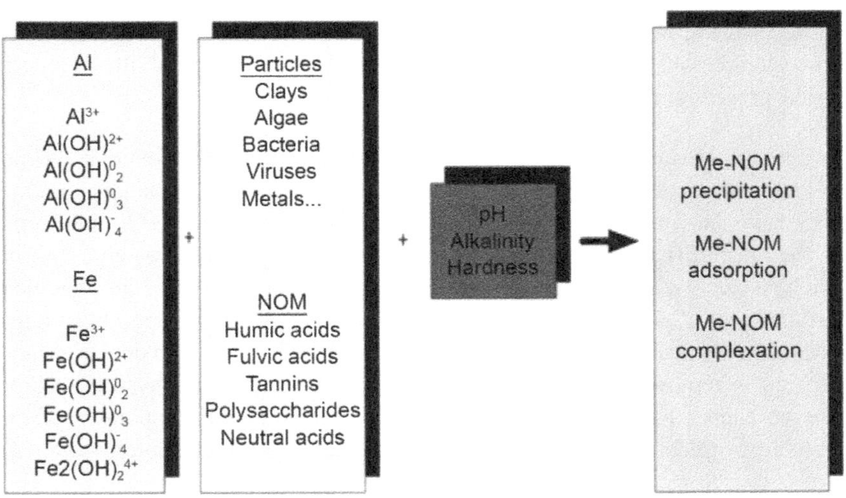

Figure 7.40. *Mechanism for removal of NOM and turbidity. For a color version of this figure, see www.iste.co.uk/gaid/watertreatment2.zip*

The coagulation reaction takes place almost instantaneously in the rapid mixing stage of the water treatment process. Coagulation mechanisms are different for particles and NOM. After the addition of a metal coagulant and its hydrolysis, different soluble species (hydrolyzed metal of higher or lower MW) or solid species (metal hydroxide) are formed that will interact with the particles and the NOM

(Figure 7.40). The different mechanisms of NOM coagulation are shown in a diagram in Figures 7.41 and 7.42.

Coagulation is probably the most common method used to remove NOM. The conventional process involves several physicochemical processes, including rapid mixing, slow mixing (flocculation), sedimentation, filtration and disinfection. Coagulation reactions occur almost instantaneously in the rapid mixing stage and continue until the water is filtered.

Effective coagulation is achieved by adding charged (or other destabilizing) species to the water source. This process can be accomplished using coagulants, the two most commonly used in practice being the hydrolysis of the Al^{3+} and Fe^{3+} metal ions, usually supplied as aluminum sulfate ($Al_2(SO_4)_3 \cdot 14H_2O$) and ferric chloride ($FeCl_3 \cdot 6H_2O$). Aluminum sulfate, commonly known as alum, is the most widely used coagulant in water treatment processes.

NOM removal, in this case, is due to several mechanisms that include double layer compression, charge neutralization, sweep coagulation and interparticle bridging. Several factors influence the efficiency of metallic salt coagulation. These factors include, but are not limited to, coagulant dosage, pH, alkalinity, temperature and ions present in the solution.

Ambient pH is critical to maximizing the efficiency of NOM removal. Although maximum adsorption of humic and fulvic acids occurs under acidic conditions, studies have shown that adsorption is the key mechanism involved in their removal over the entire pH range. NOM removal by inorganic adsorption occurs primarily due to Van der Waals forces or polarization resulting from the rearrangement of macromolecules. Through this mechanism, polar moments in two adjacent molecules will cause a net attractive force. Hydrophobic humic molecules, which are the easiest to remove by coagulation, are strongly influenced by physical absorption. However, there is a general consensus that natural organic compounds of higher molecular weight are more easily removed than their low MW counterparts.

Different mechanisms of NOM removal by coagulation are possible. These include:

– formation of interactions between the coagulant and the organic substance that makes the whole insoluble and makes it possible to treat it as a simple particle;

– complexation of NOM with dissolved metallic species (Al or Fe), leading to direct precipitation with the formation of a solid Me-NOM phase;

– adsorption of organic substances on the colloidal particles, which are then removed by flocculation-decantation;

– adsorption of organic substances on the surface of $Me(OH)_3$ hydroxides. This hypothesis is corroborated by the fact that neutral, and not very reactive, compounds are partially removed by coagulation;

– physical trapping of the molecules in flocs.

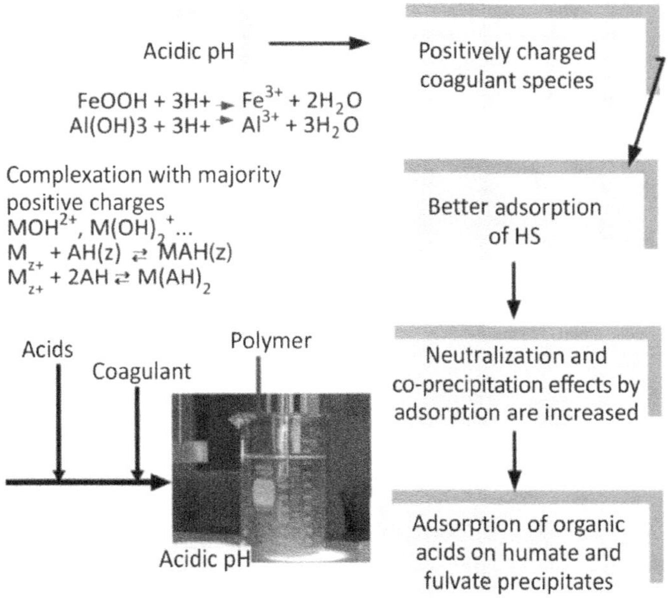

Figure 7.41. *Mechanism of NOM removal in an acidic medium. For a color version of this figure, see www.iste.co.uk/gaid/watertreatment2.zip*

Adsorption onto flocs and hydrolyzed species is the key mechanism involved in the removal of NOM over all pH ranges. It mainly occurs due to Van der Waals forces and polarization resulting from the rearrangement of macromolecules. Through this mechanism, the polar moments of two adjacent molecules cause an attractive force.

In an (acid solution) acidic medium (Figure 7.41), the mechanism of NOM removal follows a chain of reactions, including complexation of metal ions with HS, neutralization effects and coprecipitation by adsorption.

In a (basic solution) basic medium (Figure 7.42), the mechanism of NOM removal responds to competition between OH^- and NOM with respect to metal ions, followed by precipitation. The adsorption phenomenon exists but is less significant than in an acidic medium.

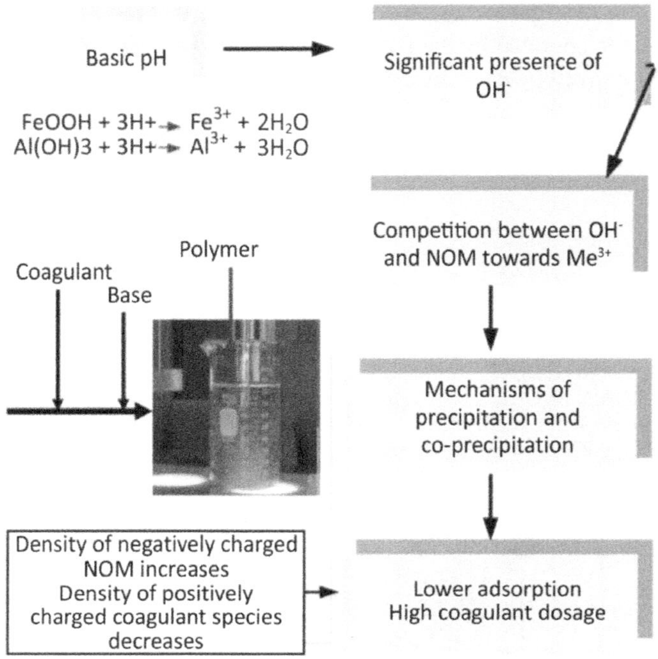

Figure 7.42. *Mechanism of NOM removal in a basic medium. For a color version of this figure, see www.iste.co.uk/gaid/watertreatment2.zip*

These mechanisms also include the coagulant dosage, alkalinity, temperature and nature of the metal ions, and their species present in the medium.

Figure 7.43 shows that some of the DOC is strongly complexed on the surface of the hydroxides formed during coagulation, while another portion remains in solution. This suggests the presence of compounds that are recalcitrant to removal by coagulation. High molecular weight substances (HMW) are better removed than low molecular weight substances (LMW). However, hydrophilic neutral NOM substances are difficult to remove, regardless of the coagulant dosage. This explains part of the DOC that is strongly complexed on the surface of hydroxides, formed during coagulation, and a second part that remains in solution.

A distinction is then made between an adsorbable DOC fraction and a non-adsorbable DOC fraction (Figure 7.44).

The contaminants that are most easily removed would be those with functional groups, such as alcohols and carboxylic acids, that are likely to react with specific sites on the surface of the particles formed by coagulation. However, low molecular weight molecules remain in solution and are not easily removed by coagulation–flocculation.

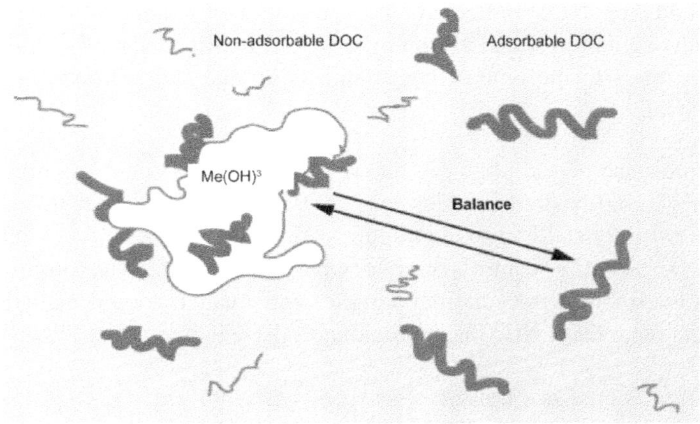

Figure 7.43. *Mechanisms of NOM coagulation. For a color version of this figure, see www.iste.co.uk/gaid/watertreatment2.zip*

Thus, there is an adsorbable DOC fraction and a non-adsorbable (or very poorly adsorbable) DOC fraction. Adsorbable DOC includes hydrophobic humic molecules that are strongly influenced by physical adsorption. The non-adsorbable (or minimally adsorbable) DOC includes hydrophilic neutral NOM substances of low molecular weight that are therefore difficult to remove, regardless of the coagulant dosage.

Additional coagulant must be added to neutralize the negative charge of the NOM before the coagulant reacts to form the metal hydroxides on which the adsorption takes place.

Operating parameters	Raw water quality parameters
Type of coagulant	NOM concentration
Coagulant dosage	NOM characteristics
pH control	Turbidity
Reagent injection order	Alkalinity
Aggressiveness and corrosiveness of the treated water	

Table 7.8. *Operating and control parameters for water containing NOM*

7.4.1.7. *Turbidity-NOM competition*

Coagulation is most often used to reduce turbidity and remove pathogens. However, the optimal conditions applied for turbidity are not always the same as those applied for NOM removal. In conventional coagulation, the operating conditions (coagulant dosage) are optimized for turbidity removal, while advanced coagulation refers to the type of coagulant, its dosage and pH control for organic matter removal.

In the presence of turbidity and NOM, two pathways occur in parallel (Figure 7.44). When turbidity predominates over NOM, the pathway that involves charge neutralization and metal hydroxide trapping prevails. For raw waters with low TOC (<3 mg·L^{-1}), turbidity controls conventional coagulation. The dosage must be sufficient to destabilize the colloidal particles and create a good decantation floc. In this case, the removal of NOM is approximately 10–25%.

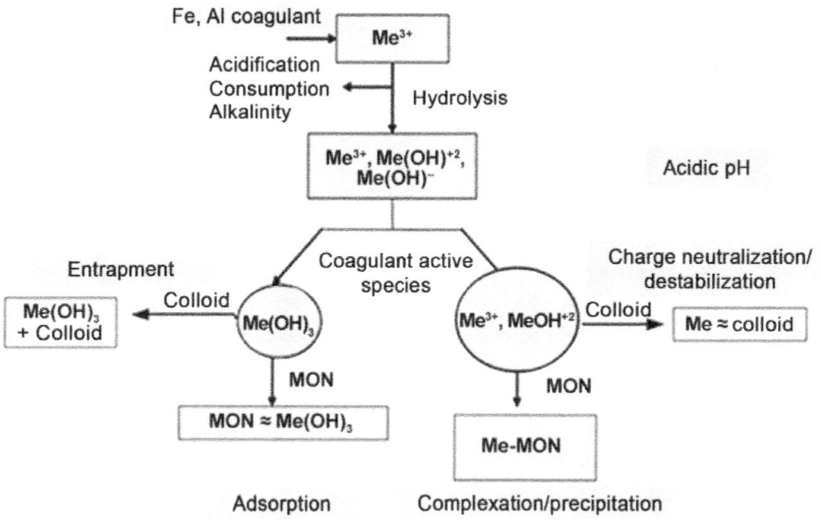

Figure 7.44. *Mechanism for removal of NOM and turbidity. For a color version of this figure, see www.iste.co.uk/gaid/watertreatment2.zip*

In the presence of medium turbidity and high NOM concentration, both pathways can exist with charge neutralization, complexation, adsorption and entrapment phenomena. In this case, coagulant dosages are significantly higher. NOM exerts a demand for coagulant that must be met before flocculation occurs and, as a result, interferes with turbidity removal. The effectiveness of coagulation to jointly remove NOM and turbidity depends on several factors, such as the type of coagulant and the dosage, the mixing conditions, pH, temperature, NOM properties (such as size, functionality, charge and hydrophobicity) and the presence of divalent cations and concentrations of destabilizing anions (such as bicarbonates).

For raw waters with TOC (>3–5 mg·L^{-1}), NOM controls the coagulant dosage, and advanced coagulation is implemented. The total dosage injected includes the coagulant demand by the NOM and the turbidity when it exceeds 10 NTU. NOM yields exceed 40%.

For raw waters with TOC greater than 5 mg·L^{-1}, the NOM controls the coagulant dosage, and advanced coagulation is implemented. However, it is not enough to obtain a final DOC that does not form THMs and other by-products during disinfection. In this case, an additional treatment should be proposed to remove the residual DOC. This residual is made up of low molecular weight molecules. Treatment by activated carbon, biological processes or high-pressure membranes is necessary. This is the case even if NOM removal by advanced coagulation exceeds 60% (Figure 7.45).

In most waters containing NOM concentrations, it is recommended to focus on the removal of NOM at the same time as turbidity, even if the latter is easier to remove.

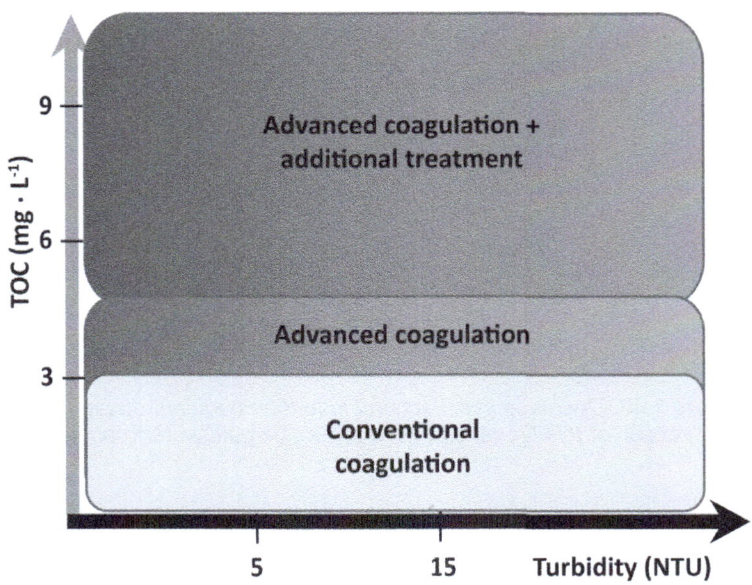

Figure 7.45. *Choice of the coagulation method to be used in the presence of natural organic matter and turbidity. For a color version of this figure, see www.iste.co.uk/gaid/watertreatment2.zip*

7.4.1.8. *Influence of the pH*

Figure 7.47 shows the reduction in pH according to coagulant dosage. With water having an alkalinity equal to 100 mg·L^{-1} $CaCO_3$, the pH decreases from 8.4 to 5.2 after the addition of 60 mg·L^{-1} $FeCl_3$ (41%). The addition of 60 mg·L^{-1} of aluminum sulfate to the same water leads to a pH decrease from 8.5 to 7.1.

The coagulation pH affects both the metal salt coagulant and the dissociation of fulvic and humic acids. As the pH decreases, the negative charge of the NOM, particularly the HS, increases, which promotes interactions with the positively charged metal hydroxides. Therefore, a decrease in coagulation pH results in a higher removal of organic matter and thus a lower DOC concentration in the decanted water.

A lower pH can be achieved, either by adding acid or by increasing the coagulant dosage. With a higher coagulant dosage, more metal hydroxides [$Fe(OH)_3$ or $Al(OH)_3$] are formed, releasing more H^+ ions.

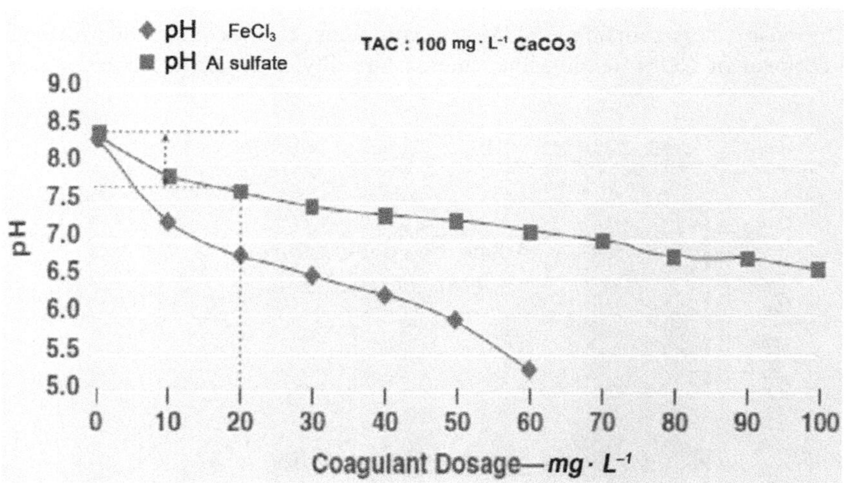

Figure 7.46. *Decrease in pH according to the coagulant dosage. For a color version of this figure, see www.iste.co.uk/gaid/watertreatment2.zip*

The chemistry of aluminum sulfate is described by an equilibrium between three species, Al^{3+}, $Al(OH)^{2+}$ and Al $(OH)_4$, and the aluminum hydroxide $Al(OH)_3$. The distribution of these species depends on the pH, and aluminum sulfate is less soluble at a pH of 6, which is very close to the minimum pH solubility. This means that at this pH, the maximum amount of coagulant is converted into particles in the form of flocs.

The best coagulation performance is usually at pH values that are as close as possible to this pH. A pH of 6.0 at 20°C and a pH of 6.5–6.7 at 4°C, is often recommended to maximize TOC removal with aluminum sulfate. This controls the residual dissolved Al concentration and maximizes the presence of flocs for NOM adsorption.

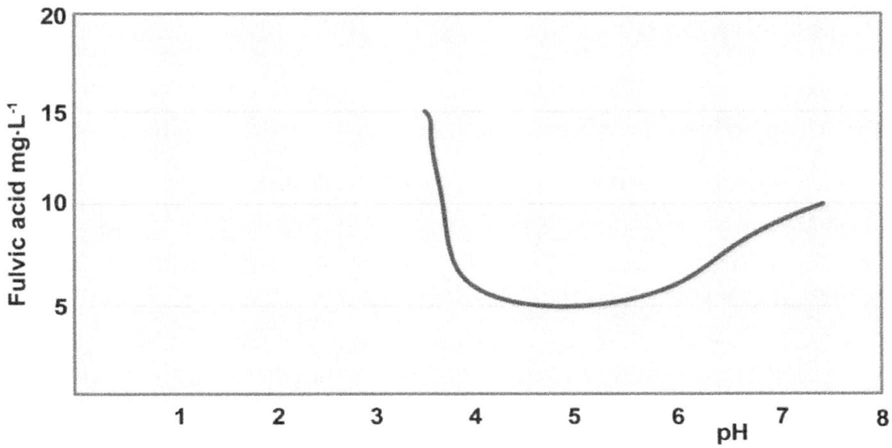

Figure 7.47. *Precipitation of fulvic acid according to pH with $FeCl_3$. For a color version of this figure, see www.iste.co.uk/gaid/watertreatment2.zip*

Given that the minimum solubility pH is higher at lower temperatures, a higher dosage of coagulant may be required to neutralize the higher negative charge of NOM. Continuous monitoring of the pH is necessary to achieve optimal coagulation.

As pH increases, coagulant hydrolysis products with a lower positive charge dominate. Thus, at a pH >7, the coagulant dosage must be quadrupled to overcome the negative charge of the NOM compared to a pH of 6.0.

PACl has a similar chemistry to aluminum sulfate, except that it contains highly polymerized aluminum species. PACl has a higher minimum solubility pH than aluminum sulfate. For different types of PACl, the minimum solubility pH increases with increasing basicity. For this reason, PACl is used at higher pH values without causing an increase in the residual dissolved Al concentration.

7.4.1.9. *Influence of the alkalinity*

The alkalinity of most natural raw waters is generally due to dissolved bicarbonate salts. Alkalinity is very variable depending on the region and can be between 10 and 250 mg·L^{-1} $CaCO_3$.

For low alkaline waters, the addition of coagulant may consume all the available alkalinity, lowering the pH to values that are too low for the effective treatment of NOM, while high alkaline waters may require acid additions or an increase in coagulant to lower the pH to values that are favorable to coagulation. For this reason, DOC removal is lower for strongly alkaline waters because of the buffering capacity it creates, and the lowering of pH to a favorable value for NOM removal is more difficult to achieve.

Aluminum sulfate and ferric chloride are more acidic than PACl and require more alkalinity after injection.

7.4.1.10. *Residual dissolved organic carbon modeling*

Online monitoring tools for NOM include TOC, DOC and UV absorbance. The simplified diagram (Figure 7.48) of the removal of the different organic carbon fractions present in the water shows the following:

– the fraction relating to particulate organic carbon is removed by coagulation–flocculation-decantation (or flotation) and filtration, according to the mechanism of trapping particles by chemical reagents;

– the fraction related to DOC includes a fraction of non-adsorbable DOC, as well as a fraction of adsorbable dissolved DOC (which integrates all the mechanisms of neutralization–entrapment–adsorption–precipitation in the adsorbable term).

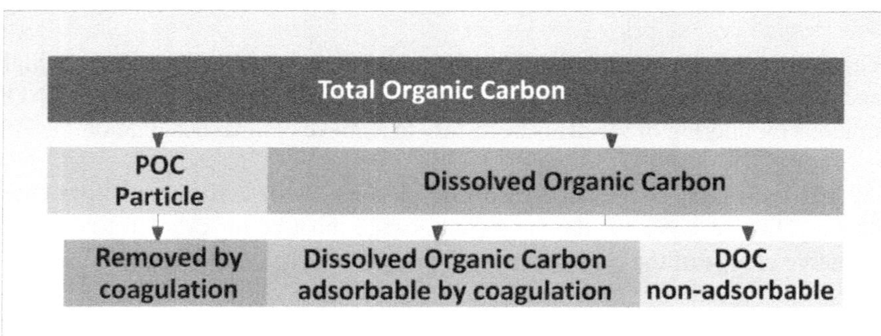

Figure 7.48. *Simplified diagram of organic carbon fractions removed by coagulation–flocculation–decantation (or flotation). For a color version of this figure, see www.iste.co.uk/gaid/watertreatment2.zip*

The particulate POC fraction consists of various particles: organic colloids, microorganisms, phytoplankton, algae, plant particles, etc. Most often, the ratio between TOC and DOC is estimated based on the turbidity of the water. In other cases, the presence of algae is taken into account to establish this ratio.

The DOC fraction consists primarily of non-HS (which include carbohydrates, proteins, peptides, fats, pigments and other low molecular weight compounds) and HS that form the bulk of the organic matter in water (which include hydrophilic and acidic colored complexes).

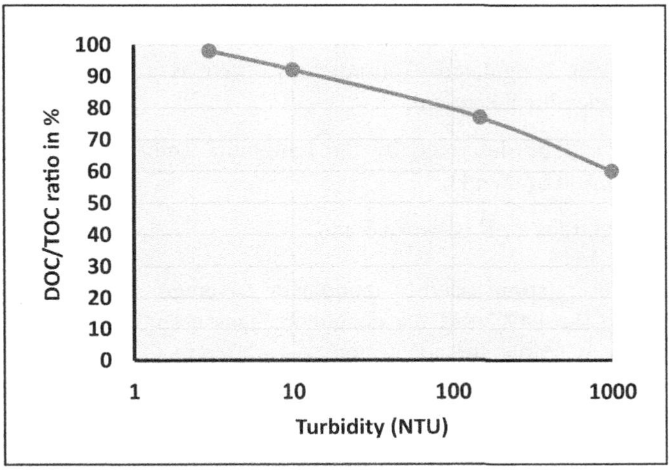

Figure 7.49. *Example of DOC/TOC ratio according to turbidity. For a color version of this figure, see www.iste.co.uk/gaid/watertreatment2.zip*

The DOC/TOC ratio is often estimated when field analyses cannot be performed. This ratio depends on turbidity and other water components.

The residual DOC after the addition of chemical reagents, followed by liquid–solid separation solid separation, is:

Residual DOC = non-adsorbable DOC + residual adsorbable DOC

The non-adsorbable DOC depends on the characterization of the organic substances, that is, the fractionation between high and medium molecular weight molecules and low molecular weight molecules, as well as various molecules not retained by the coagulation–flocculation mechanisms. It is linked to the SUVA of the water to be treated.

The residual adsorbable DOC depends on the operating conditions of the coagulation (pH, coagulant dosage, contact time, alkalinity, etc.). Two models are proposed.

7.4.1.10.1. First model

The first model is an empirical model based on the following assumptions:

– DOC removal is mainly due to adsorption on flocs of metal hydroxides formed during coagulation;

– part of the DOC is not adsorbable on the surface of the hydroxides and therefore cannot be coagulated. Consequently, there is a fraction that cannot be adsorbed by the metallic coagulant;

– adsorption can be described by the Langmuir isotherm, and the adsorption capacity is a function of the pH;

– it is only valid for a pH between 5 and 7.

The Langmuir relation assumes monolayer coverage, in which equilibrium is reached between the DOC and the coagulant. This assumes that there is a fixed number of sites available for adsorption. All sites are identical, and there is no interference for adsorption reactions. This relation is described by the following equation:

$$qr = \frac{qm.b\,.non\text{-}\,adsorbed\ DOC}{(1+\ b\,.non\text{-}\,adsorbed\ DOC)}$$

with:

– qr: amount of DOC removed relative to the weight of metallic coagulant added (mg DOC·mmole^{-1}Al^{3+} or Fe^{3+});

– b: Langmuir isotherm constant (L·mg^{-1});

– qm: maximum DOC retention capacity (mg·mmole^{-1} Al^{3+} or Fe^{3+});

– non-adsorbed DOC: concentration of non-adsorbed DOC remaining in solution and not adsorbed at equilibrium (mg·L^{-1}).

This approach makes it possible to initiate the following type of relation:

Non-adsorbed $(DOC)^2$ + non-adsorbed B (DOC) – C = 0

$$B = qm\,M + \frac{1}{b} - DOC\ raw\ water +\ non\ adsorbable\ DOC$$

$$C = \frac{DOC\ raw\ water - non\ adsorbable\ DOC}{b}$$

Non-adsorbable DOC: [– K_1 (SUVA) + K_2] DOC raw water

with:

– K1 and K2: empirical constants related to the metallic coagulant;

– M: coagulant dosage ($mmol \cdot L^{-1}$).

The qm parameter depends on the pH, according to the following relation:

$$qm = x_3\, pH^3 + x_2\, pH^2 + x_1\, pH$$

x_1, x_2 and x_3 are empirical constants related to the nature of the metallic coagulant:

x_1 (4.96), x_2 (–73.9) and x_3 (280)

The residual non-adsorbed DOC remaining in solution ($mg \cdot L^{-1}$) is then obtained by:

$$Residual\ non-adsorbed\ DOC = \frac{-B + \sqrt{(B^2 - 4\,C)}}{2}$$

The final DOC in solution is the sum of the non-adsorbable DOC and the residual non-adsorbed DOC remaining in solution. The removed DOC is the difference between the inlet DOC and the outlet DOC.

The various parameters applicable to this model (mean uncertainty 5–20%) are given in Table 7.9.

Parameters	**Fe^{3+}**	**Al^{3+}**
x_1	280	284
x_2	– 73.9	– 74.2
x_3	4.96	4.91
K_1 ($mg \cdot m \cdot L^{-1}$)	– 0.028	– 0.075
K_2	0.23	0.56
b ($L \cdot mg^{-1}$)	0.068	0.147

Table 7.9. *Parameters that are applicable to the model*

The results obtained at the Veolia sites are summarized in Table 7.10.

This comparison shows similar differences in the Delta between the DOC measured on-site and the calculated DOC for $FeCl_3$ and $Al_2(SO_4)_3$. Nevertheless, it can provide a first estimation on the coagulation performance with respect to DOC removal and be a guide for the choice of the best coagulant.

Parameters	Fe^{3+}					Al^{3+}			
$FeCl_3$ dosage, 41% ($g \cdot m^{-3}$)	105	80	80	150	70	105	50	80	45
DOC inlet ($mg \cdot L^{-1}$)	11	6.2	6.2	8.2	11	11	4.0	8.2	3.5
pH	5.9	6.4	5.8	6	6	5.9	7	6	6.3
SUVA ($L \cdot mg^{-1} \cdot m^{-1}$)	5.5	4.2	4.2	8	5.5	5.5	4.2	7.2	3.2
On-site DOC out ($mg \cdot L^{-1}$)	3.7	3.6	2.2	1.9	4	5.1	2.2	3.0	2.5
Calculated DOC ($mg \cdot L^{-1}$)	3.0	2.6	2.0	2	4	3.9	2.26	2.5	2.3
Non-adsorbed DOC ($mg \cdot L^{-1}$)	0.84	1.0	1.0	<0.1	0.84	1.62	1.01	<0.1	1.04
ΔDOC between site and calculation ($mg \cdot L^{-1}$)	0.7	1.0	0.2	0.1	0	1.3	0.06	0.5	0.2

Table 7.10. *Application of the first model to various Veolia sites*

7.4.1.10.2. Second model

The second model is also empirical and is based on relations that make it possible to determine the DOC out according to coagulation pH, coagulant dosage and DOC raw water.

According to a relation of the following form: DOC out = EXP (DOC raw water, pH, coagulant dosage).

The non-adsorbable fraction is obtained with the same previously used equation:

Non-adsorbable DOC: $[-K_1 (SUVA) + K_2]$. DOC raw water

Parameters K_1 and K_2 are identical to those used for the first model.

With aluminum sulfate as the coagulant, the final DOC after coagulation–flocculation and decantation (and filtration) can be obtained using the following modified relation, which is valid for pH values between 5 and 7:

$$(DOC)\ out = 0.62\ [Exp\ (-0.1639 + 1.159\ Ln\ (DOC\ inlet) - 0.4458\ Ln\ (AS) - 0.00689\ Ln\ (DOC\ inlet) \cdot Ln(AS) + 0.0566\ pH \cdot Ln(AS))]$$

with:

– DOC out: treated water DOC ($mg \cdot L^{-1}$);

– DOC inlet: DOC raw water ($mg \cdot L^{-1}$);

– pH: coagulation pH;

– AS: aluminum sulfate dosage ($g \cdot m^{-3}$).

The non-adsorbable DOC is given by:

$$[-0.075\ (SUVA) + 0.56]\ DOC\ inlet$$

With ferric chloride as a coagulant, the final DOC after coagulation–flocculation and decantation (and filtration) can be obtained using the following relation, which is valid for pH values between 5 and 7:

$$(DOC)\ out = 0.90\ [Exp\ (-0.06 + 1.12\ Ln\ (DOC\ inlet) - 0.63\ Ln\ (FeCl_3) - 0.078\ Ln(DOC\ inlet) \cdot Ln(FeCl_3) + 0.0938\ pH \cdot Ln(FeCl_3))]$$

The non-adsorbable DOC is given by:

$$[-0.028\ (SUVA) + 0.23]\ DOC\ raw\ water$$

Parameters	**Fe^{3+}**					**Al^{3+}**			
$FeCl_3$ dosage, 41% ($g.m^{-3}$)	105	80	80	150	70	105	50	80	45
DOC inlet ($mg \cdot L^{-1}$)	11	6.2	6.2	8.2	11	11	4.0	8.2	3.5
pH	5.9	6.4	5.8	6	6	6	7	6	6.3
SUVA ($L \cdot mg^{-1} \cdot m^{-1}$)	5.5	4.2	4.2	8	5.5	5.5	4.2	7.2	3.2
On-site DOC out ($mg \cdot L^{-1}$)	3.7	3.6	2.2	1.9	4	5.1	2.2	3.0	2.5
Calculated DOC ($mg \cdot L^{-1}$)	3.6	3.1	2.4	2.8	4.2	4.8	2.1	3.5	1.5
Non-adsorbed DOC ($mg \cdot L^{-1}$)	0.84	1.0	1.0	<0.1	0.84	1.62	1.01	<0.1	1.04
Δ DOC between site and calculation ($mg \cdot L^{-1}$)	0.1	0.5	0.2	0.9	0.2	0.3	0.1	0.5	1.0

Table 7.11. *Application of the second model at various Veolia sites*

The results obtained with the second model appear to be closer than those obtained with the first model. However, both models can be used to obtain a first estimate of DOC removal by coagulation.

7.5. Adsorption on activated carbon

As widely described in Chapter 9, which focuses on adsorption, activated carbon is an absorbent material that provides a surface on which molecules, that are present in raw water, can be retained.

7.5.1. *Mechanism of NOM removal by activated carbon*

When the NOM concentration is high, advanced coagulation does not achieve a final DOC concentration ≤ 2 mg·L^{-1}, additional treatment with activated carbon is one of the solutions.

The objective of activated carbon is therefore to remove residual organic substances after the advanced coagulation stage. After coagulation, the SUVA is most often less than 2 L·m^{-1}·mg^{-1}, and the molecules still present are low molecular weight molecules that are mostly non-humic and uncharged organic fractions, as shown in Figure 7.50.

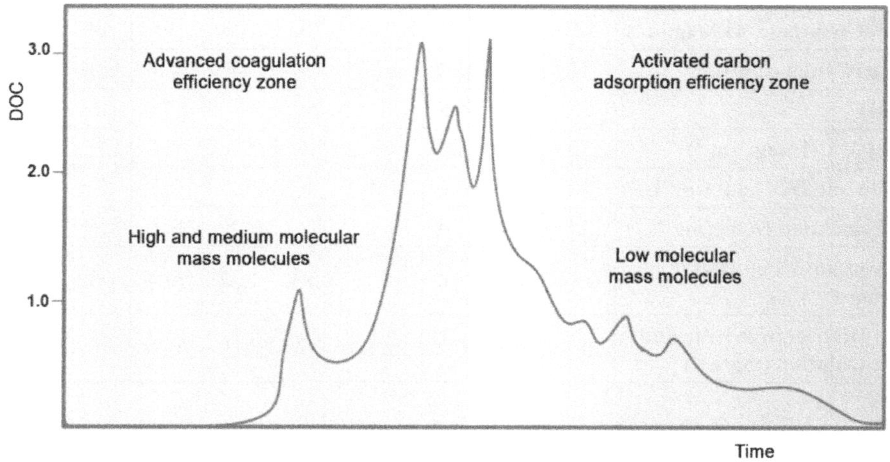

Figure 7.50. *Efficiency zone of activated carbon for DOC removal after advanced coagulation. For a color version of this figure, see www.iste.co.uk/gaid/watertreatment2.zip*

Let us recall that the carboxylic and phenolic groups of HS are protonated and neutral (-COOH) in an acidic medium. They therefore tend to collapse on themselves, leading to the formation of hydrogen bonds, electrostatic interactions and van der Waals interactions, generating a strong aggregation.

In a basic medium, the surface charge of HS is negative due to the ionization of carboxylic (-COOH) and hydroxyl (-OH) groups by the loss of hydrogen. This leads to a decrease in the number of hydrogen bonds. Consequently, the aggregation of HS decreases, and they start to have a stretched form.

Thus, depending on the pH, HS change their conformation, and this parameter can play a significant role in the adsorption mechanism of activated carbon.

The efficiency of activated carbon depends on the molecular structure of the substances, its porous structure and the significance of the surface functions that constitute its surface. It is the mesoporous structure (2–50 nm) that must allow the adsorption of these molecules, and the choice of activated carbon must be made in this respect.

The adsorption isotherm data are correlated with the two models detailed in the Chapter 9, which is represented by the Freundlich and Langmuir isotherms:

– Freundlich model:

$$qe = k\, Ceq^{\frac{1}{n}}$$

– Langmuir model:

$$qe = \frac{qm\, b\, Ceq}{1 + bCeq}$$

with:

– *Ceq*: equilibrium concentration of DOC ($mg \cdot L^{-1}$);

– *qe*: adsorption capacity ($mg \cdot mg^{-1}$);

– *k*: Freundlich constant related to the adsorption capacity ($L \cdot mg^{-1}$);

– *n*: parameter related to the adsorption efficiency;

– *Qm*: maximum adsorption capacity ($mg \cdot g^{-1}$);

– *b*: Langmuir adsorption constant related to the adsorption energy ($L \cdot mg^{-1}$).

As the adsorption efficiency increases with the reaction time (up to a maximum limit) and the reactors cannot be disproportionately large, the dosage of powdered activated carbon (PAC) or the volume of granular activated carbon (GAC) are increased to limit the reaction times to between 10 and 20 min.

Sorption occurs until the concentration of organic substances present in the water is in equilibrium with the concentration of organic substances retained on the internal surface of the activated carbon. The parameters of the Freundlich and Langmuir isotherms thus make it possible to describe this distribution.

For example, the higher the b parameter of the Langmuir isotherm is, the greater the affinity of the organic substance (expressed as DOC) for the activated carbon. When the tests are carried out using UV 254 nm analysis, the results show that the adsorption is better than that obtained by expressing the organic substances in DOC. This means that a large part of the substances are aromatic.

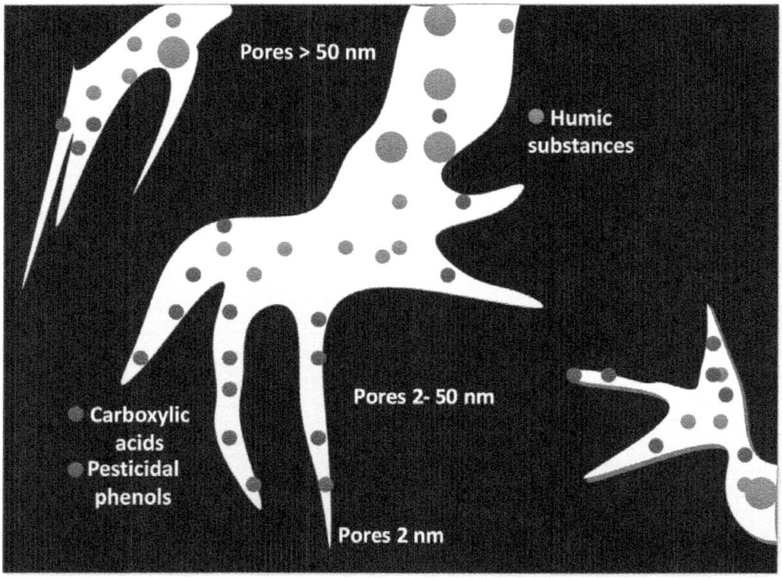

Figure 7.51. *Adsorption of various substances in the pores of activated carbon. For a color version of this figure, see www.iste.co.uk/gaid/watertreatment2.zip*

The conformation of humic acid, and therefore its size, influences its adsorption. Its size varies between a few nanometers (1.5–6 nm) and a few dozen nanometers (20–40 nm). As a result, the distribution and size of the pores of the activated carbon are two significant factors that affect the adsorption capacity of these substances;mesopores and macropores are the adsorption pores for HS or other substances that are greater than 2 nm in size. The micropores are only accessible to very small molecules such as micropollutants. For example, an activated carbon with a large mesoporous volume, and a pore diameter between 10 and 50 nm, has a good adsorption capacity for HS and other equivalent substances.

7.5.2. *Implementation of activated carbon adsorption for NOM removal*

Activated carbon can be applied in several ways:

– in the form of powdered activated carbon, upstream of a coagulation–flocculation–decantation (or flotation) stage, whether the coagulation is conventional or advanced;

– in a powdered activated carbon reactor, downstream of a coagulation–flocculation–decantation stage;

– in the form of micrograin activated carbon downstream of coagulation–flocculation–decantation;

– in the form of powdered or micrograin activated carbon, associated with an in situ ozone injection in the activated carbon reactor;

– in the form of granular activated carbon in a filter located downstream of a decanter, a flotation device or a sand filter;

– in the form of granular activated carbon in a filter located downstream or upstream of a microfiltration or ultrafiltration membrane.

7.5.2.1. *NOM removal processes using activated carbon as an adsorbent*

7.5.2.1.1. Powdered activated carbon

The use of PAC has the advantage of providing the necessary dosage during periods when the presence of various contaminants (pesticides, micropollutants, high organic matter, taste and odor) is reported.

Physical characteristics such as particle diameter, specific surface area and pore distribution are useful in estimating the potential for NOM adsorption. Figure 7.53 shows the results of DOC adsorption from river water (DOC: 3.3 $mg \cdot L^{-1}$; SUVA: 3.12 $L \cdot mg^{-1} \cdot m^{-1}$), with mesoporous carbon (Jacobi Aquasorb MP23) and microporous carbon (Jacobi Aquasorb). The adsorption on the mesoporous carbon is better due to the internal pore structure, which is more available for HS when compared to a microporous carbon with a smaller mesoporous pore volume. With 40 $mg \cdot L^{-1}$ of PAC, the residual DOC is 1.3 $mg \cdot L^{-1}$ with mesoporous activated carbon, while it is 2 $mg \cdot L^{-1}$ with microporous activated carbon. The ratios observed for these tests indicate an average of 0.50 mg DOC removed/10 mg mesoporous carbon, while it is approximately 0.3 mg DOC removed/10 mg microporous carbon.

The Freundlich parameters obtained with different activated carbons during the adsorption of HS are shown in Table 7.12.

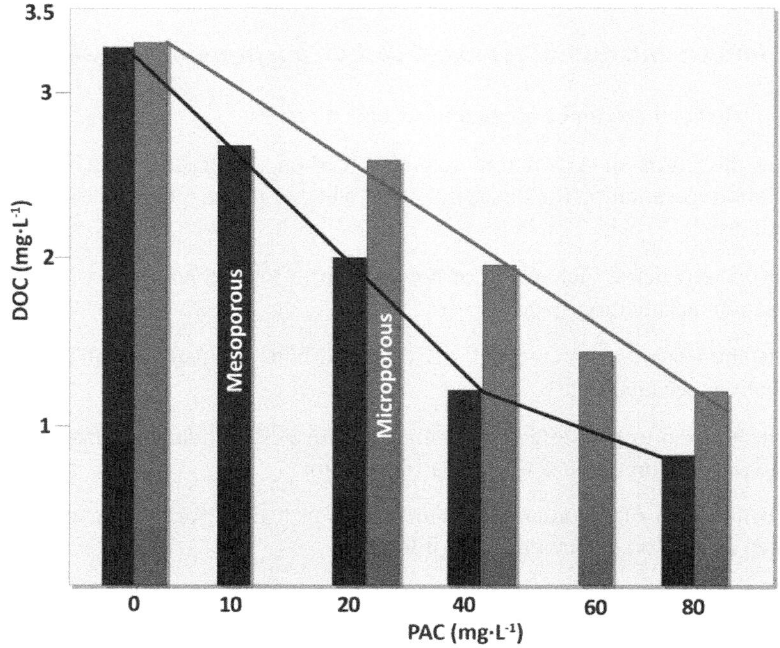

Figure 7.52. *Comparison of adsorption of DOC on two types of PAC (Veolia site). For a color version of this figure, see www.iste.co.uk/gaid/watertreatment2.zip*

The residual DOC concentration after powdered activated carbon injection can then be calculated.

Activated carbon	Freundlich isotherm parameters		
	K mg·L$^{1/n}$ L^{-1}·mg$^{-1/n}$	1/n	qe mg·mg^{-1}
Jacobi Aquasorb LP39	0.0102	2.23	0.0102 DOC $^{2.23}$
Jacobi Aquasorb MP22	0.025	1.089	0.025 DOC $^{1.089}$
Jacobi Aquasorb MP23	0.028	1.063	0.028 DOC $^{1.063}$
Chemviron Calgon Pulsorb C	0.0278	1.255	0.0278 DOC $^{1.255}$
Cabot Norit SA UF	0.032	1.21	0.032 DOC $^{1.21}$

Table 7.12. *Freundlich parameters for various PACs with respect to DOC removal*

For Aquasorb MP23 carbon, for example, we obtain:

$$qe = \frac{(DOC\ inlet - DOC\ out)V}{M} = k\ DOC\ out^{\frac{1}{n}}$$

or:

$$qe = \frac{(DOC\ inlet - DOC\ out)V}{M} = 0.028\ DOC\ out^{1{,}063}$$

The weight of carbon M is in mg·L$^{-1,}$ and the outlet DOC is in mg·L^{-1}.

Two modes of powdered activated carbon injection are possible:

– in the form of powdered activated carbon, upstream of a coagulation–flocculation–decantation (or flotation) stage, whether the coagulation is conventional or advanced;

– in a powdered activated carbon reactor located downstream of a coagulation–flocculation–decantation stage.

Injection before a coagulation stage

This solution is applied to a Multiflo® decanter, an Actiflo® or a Spidflow® flotation device during seasonal episodes that involve a DOC increase in the raw water. PAC dosages can vary between 10 and 100 mg·L^{-1}, depending on the DOC concentration. The existing type of coagulation can be conventional or advanced. However, it may not be sufficient when the DOC is occasionally too high. An additional treatment with PAC is necessary.

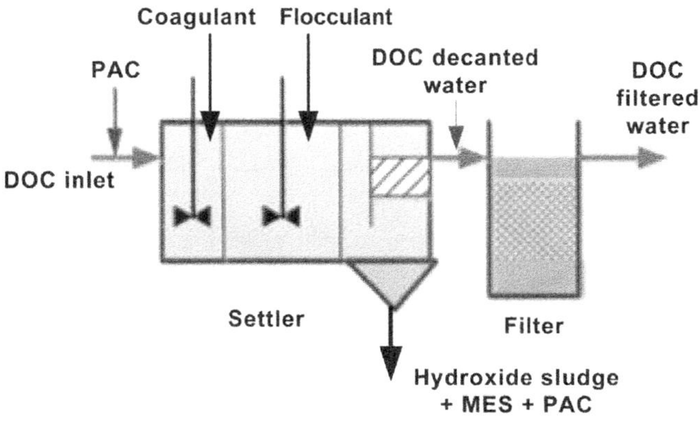

Figure 7.53. *Injection of PAC into raw water before coagulation. For a color version of this figure, see www.iste.co.uk/gaid/watertreatment2.zip*

Preliminary tests obtained in the laboratory on a dam water site with a DOC of 4.9 mg·L^{-1} and an SUVA of 4.1 mg·L^{-1}, are presented in Figure 7.54. The characteristics of the powdered activated carbons are also summarized in Table 7.13.

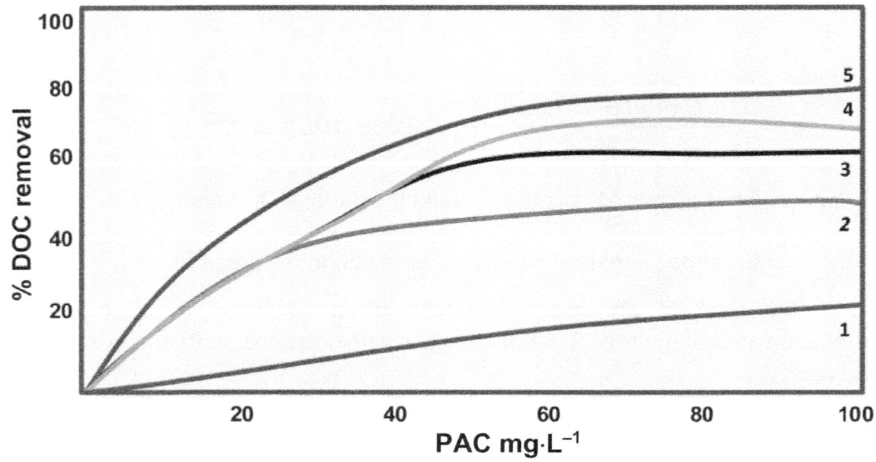

Figure 7.54. *Performance of various PACs with regard to DOC removal. For a color version of this figure, see www.iste.co.uk/gaid/watertreatment2.zip*

Numbers	Activated carbons	Types	Origins	Granulometry
1	Jacobi Aquasorb CP	Microporous	Coconut	15–35 µm
2	Jacobi Aquasorb MP22	Mesoporous	Lignite	15–35 µm
3	Jacobi Aquasorb MP23	Mesoporous	Lignite	15–35 µm
4	Chemviron Calgon Pulsorb C	Mesoporous	Coal	8–15 µm
5	Cabot Norit SA UF	Mesoporous	Bituminous coal	8 µm (d50)

Table 7.13. *Characteristics of tested powdered activated carbons*

The results show the importance of selecting a mesoporous carbon for DOC removal in water with SUVA >2 L·mg^{-1}·m^{-1}. A microporous carbon should therefore be avoided for this application.

The results obtained at the Veolia site in western France, on a raw water loaded with DOC between 7.5 and 17.1 mg·L^{-1} and a SUVA between 6.2 and 8.5 L·mg^{-1}·m^{-1}, show an advanced coagulation removal of 50–62% for a 41% variable dosage of $FeCl_3$ between 120 and 180 g·m^{-3} and a coagulation pH of 5.5.

The DOC of the decanted water varied between 3.7 and 6.5 $mg{\cdot}L^{-1}$. Injection of powdered activated carbon (Jacobi Aquasorb MP23) for a few days, during which the DOC concentration rose to more than 17 $mg{\cdot}L^{-1}$, was performed with a dosage of 32–36 $g{\cdot}m^{-3}$. This resulted in a decanted water DOC of approximately 4.2 $mg{\cdot}L^{-1}$, that is, a DOC reduction of 2.3 $mg{\cdot}L^{-1}$. To obtain a DOC of 2 $mg{\cdot}L^{-1}$, it would have been necessary to inject more than 80 $mg{\cdot}L^{-1}$ of PAC. Nevertheless, these results show that the solution of injecting PAC in the coagulation phase can be a palliative solution to improve the performance of DOC removal, provided that the PAC is well adapted to this type of organic substance.

The amount of DOC removed per weight of PAC is therefore:

$$\text{Ratio DOC removed/PAC} = (6.5 - 4.2)/36 = 0.063\ \text{mgDOC}{\cdot}\text{mg}^{-1}\text{PAC}$$

This gives a ratio of 0.63 mg of DOC removed per 10 mg of PAC.

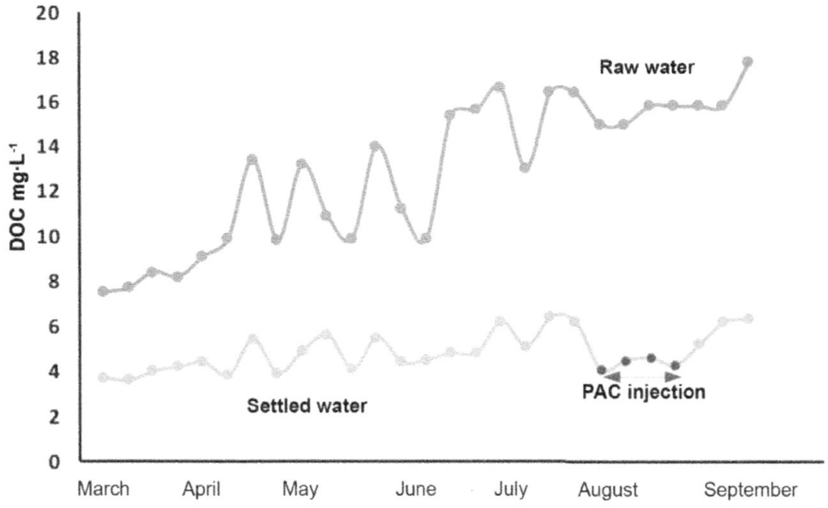

Figure 7.55. *Injection of PAC in raw water to improve DOC removal (Veolia site). For a color version of this figure, see www.iste.co.uk/gaid/watertreatment2.zip*

At some sites, it has been observed that the addition of low dosages of PAC to coagulated water with high dosages of coagulant presents possible competitive effects and reduces DOC removal. On the other hand, by increasing the dosage of PAC, clearer DOC removal is obtained, which can be explained by very fast adsorption kinetics due to more active sites.

A macroporous–mesoporous activated carbon is to be recommended for this type of water because the presence of macropores (>50 nm) allows fast kinetics and adsorption of metal complexes (HS-metal). This can also lead to a decrease in coagulant dosages.

Applying the previously described Freundlich parameters, the PAC weight required to achieve the desired final DOC concentration is:

$$\text{M CAP mg}\cdot\text{L}^{-1} = (\text{DOC inlet} - \text{DOC out})/(\text{k}\cdot\text{DOC out } 1/\text{n})$$

Without the addition of PAC, the residual DOC was 6.8 mg·L^{-1} (August period). After injection of CAP, the residual DOC was 4.1 mg·L^{-1}, that is, a DOC removal of 2.7 mg·L^{-1}.

$$\text{M PAC mg}\cdot\text{L}^{-1} = (6.8 - 4.1)/(0.028 \times 4.1^{1.063}) = 21.6 \text{ mg}\cdot\text{L}^{-1}$$

The weight of activated carbon thus obtained is lower than that applied on-site, which can be explained by the competition between the different substances present in the water, as well as between activated carbon and coagulation. The Freundlich parameters were obtained for raw water without the addition of coagulant.

Injection into a powdered activated carbon reactor located downstream of a coagulation–flocculation–decantation – mode of operation

PAC can be continuously injected into a reactor located downstream of a coagulation–flocculation–decantation stage. These PAC reactors are very valuable if they are optimized by reducing as much of the dosages of injected activated carbon as possible while obtaining the desired DOC residual.

A PAC reactor has a specific reaction tank in which the adsorption of organic matter by the PAC takes place. This reactor is usually installed downstream of a first coagulation–flocculation–decantation stage. Figure 7.56 shows the layout of the PAC reactor in the case of a treatment of surface water loaded with DOC. An example of Actiflo® is shown in this figure.

The first stage of Actiflo® allows the removal of turbidity and part of the NOM. The type of coagulation can be conventional or advanced, depending on the concentration of DOC in the raw water. The quality of the outlet decanted water of this first stage is made up of a turbidity <2 NTU and a residual DOC corresponding to 30–60% of the DOC of the raw water.

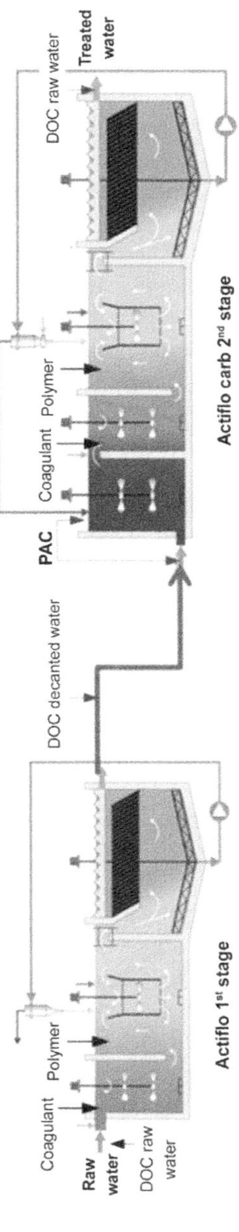

Figure 7.56. *DOC removal in two decanters in a row: Actiflo® in the first stage and Actiflo® Carb in the second stage. For a color version of this figure, see www.iste.co.uk/gaid/watertreatment2.zip*

The second stage of the Actiflo®, which is an Actiflo® Carb with a continuous dosage and recirculation of the PAC, allows the removal of approximately 50% of the residual DOC of the first stage. This removal can be higher, depending on the increase in the PAC dosage. This overall solution thus removes more than 70–80% of the DOC from the raw water to achieve the desired final DOC concentration. It should usually be lower than 2 $mg \cdot L^{-1}$ and sometimes lower than 1.5 $mg \cdot L^{-1}$, depending on the final desired THM concentration. The outlet turbidity of the second stage is <1 NTU. The contact time in the PAC tank is 10 min. The mirror velocity in Actiflo® is 50–60 $m \cdot h^{-1}$ and is equal to 40 $m \cdot h^{-1}$ in Actiflo® Carb.

This configuration with PAC in the second stage can be replaced by other Veolia technologies, such as Multiflo® Carb or Opacarb® MF (see Chapter 9). The first stage can be formed by a Spidflow® (flotation) as an alternative to a decantation.

When turbidity is low (<5 NTU) and the DOC concentration is ≤4 $mg \cdot L^{-1}$, the PAC reactor can be installed directly in the first stage.

The advantage of a PAC reactor operating continuously, over a solution where it is injected in the first stage (at the same time as coagulation), is that it has a high concentration of activated carbon (1–5 $g \cdot L^{-1}$) in the reaction tank, which are all sites that are available for DOC adsorption. The adsorption capacity is always available and can adsorb the DOC spikes. It also allows a reduction in the quantity of PAC injected compared to the PAC solution dosed in the first stage.

Figure 7.57 shows the influence of PAC contact time with respect to turbidity and UV 254 removal.

A contact time of 10 min is sufficient to obtain a significant removal of UV 254 and thus of DOC.

Figure 7.58 shows the LC-DOC chromatogram of the different organics in the raw water, which is characterized by a DOC of 4.2–6.3 $mg \cdot L^{-1}$ and a SUVA of 3.1–4.2 $L \cdot m^{-1} \cdot mg^{-1}$. The inlet turbidity is between 8 and 20 NTU.

Preozonated water has a slightly higher organic carbon than raw water, which is often observed after this type of chemical preoxidation. This is due to the reaction of ozone on particulate organic carbon, and the additional dissolved DOC is accounted for in the dissolved organic matter fractions.

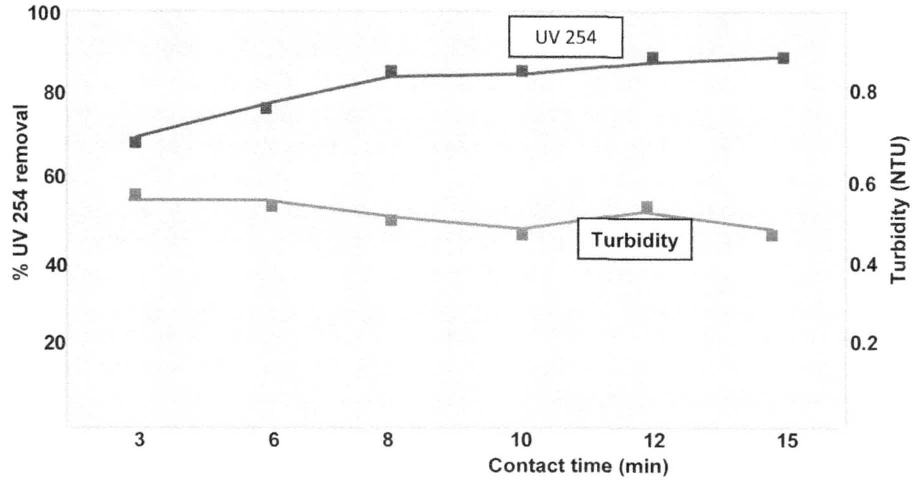

Figure 7.57. *Influence of the PAC contact time on the reactor performance. For a color version of this figure, see www.iste.co.uk/gaid/watertreatment2.zip*

Qualitatively, we also observe a transformation of the organic matter:

– at the global level, with a significant decrease in the UVD response and a decrease in the SUVA (UV/OCD ratio), which confirms the priority degradation of aromatic substances by the reaction of ozone on the double bonds;

– at the level of HS + BB fractions, with a shift toward the lowest molecular weights of HS observed in the OCD response (shoulder of the BB peak and increase of the acid peak) and a significant decrease of the UVD signal for this fraction, which is also shown through a decrease in the aromaticity of this fraction;

– a slight increase in the acid and neutral fractions, which remains coherent with the ozone reaction to break the molecules.

In terms of the treatment by Actiflo® and Actiflo® Carb, a global removal of DOC is observed, with a significant decrease in the aromatic fractions. The type of coagulation is advanced, operating at a pH of 6.8–7 with a polymerized organic coagulant based on aluminum (PAX) with a dosage of 80 $g \cdot m^{-3}$. SUVA at 3.1–4.2 $L \cdot m^{-1} \cdot mg^{-1}$ was predicted to have good organic removal.

A DOC removal efficiency of 35–40% is obtained with high PAX dosages, as peaks of 120 $g \cdot m^{-3}$ have sometimes been injected. The DOC of the decanted water varied between 2.7 and 3.8 $mg \cdot L^{-1}$.

At the qualitative level, the chromatogram shows an interesting removal of polysaccharides, HS and building blocks. As expected, low molecular weight molecules are rarely removed in the Actiflo® first stage. The process of the Actiflo® treatment, like most advanced coagulation–flocculation treatments, usually removes 30–50% of the DOC, depending on the coagulant dosages and the operating pH, and mainly acts on the HS fraction and to a lesser extent on the P fraction. The chromatogram shows that the difference between the two Actiflo® treatments is not localized to a single fraction but is distributed over the entire molecular weight range, as if the raw water matrix was different for the two decanted water samples.

The passage of water in a PAC reactor (20 g·m^{-3}) allows a decrease in low molecular weight molecules and some residual HS. The outlet DOC of the Actiflo® Carb is always lower than 2 mg·L^{-1}.

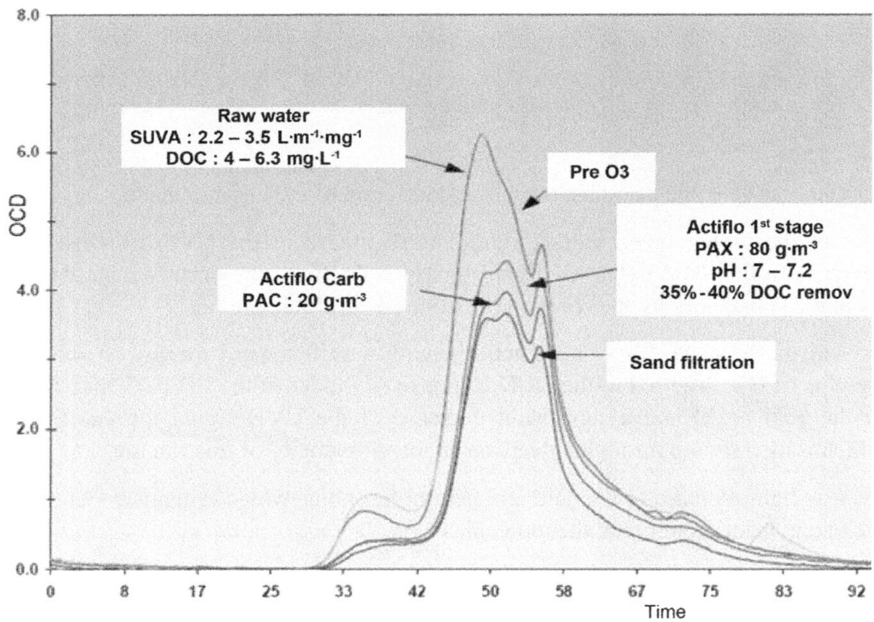

Figure 7.58. *LC-DOC characterizing the different organic substances removed by Actiflo® (coagulation) and Actiflo® Carb (activated carbon). For a color version of this figure, see www.iste.co.uk/gaid/watertreatment2.zip*

A continuous measurement of UV 254 on both the raw water and the decanted water of the first stage assures the dosages of coagulant and PAC, depending on the variation of the quality of the raw water. A good correlation between the DOC and UV 254 makes it possible to deduce the residual DOC values.

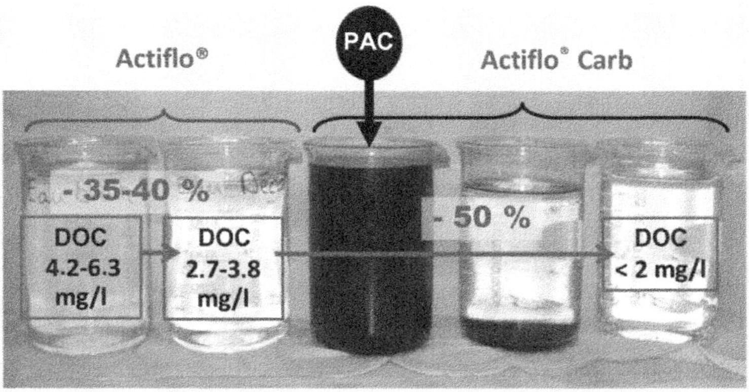

Figure 7.59. *Illustration of DOC removal with the two stages: coagulation and PAC reactor. For a color version of this figure, see www.iste.co.uk/gaid/watertreatment2.zip*

Results on Veolia sites

The results obtained on various sites in operation show the performance of this combination of decanters in series. For example, on a Veolia site where two Actiflo® and Actiflo® Carb lines are operating in parallel, the continuous recordings of organic substance removal (UV 254) show average efficiencies of 50–55%. This corresponds to an average DOC removal of 45–50%. A constant 20 $g \cdot m^{-3}$ dosage of PAC (Jacobi Aquasorb MP23) is applied.

The variation of the PAC dosage (Jacobi Aquasorb MP21), according to the DOC of the inlet water to the Actiflo® Carb, indicates an average value of 25 $g \cdot m^{-3}$ for the removal of 1 $mg \cdot L^{-1}$ of DOC.

Ratio of mg of DOC removed per mg of PAC injected

The ratio of DOC removed/10 mg. The injected PAC obtained after 10 analysis campaigns on various Veolia sites is, on average, equal to 0.5–0.6 mg DOC/10 mg PAC (Jacobi Aquasorb MP23). This value confirms the value obtained during the tests with PAC injected before coagulation.

Other results obtained with Cabot Norit W35 PAC indicate a ratio equal to 0.62 mg DOC for 10 mg PAC, and with Aquasorb MP21 PAC, the ratio drops to 0.37 mg DOC/10 mg PAC.

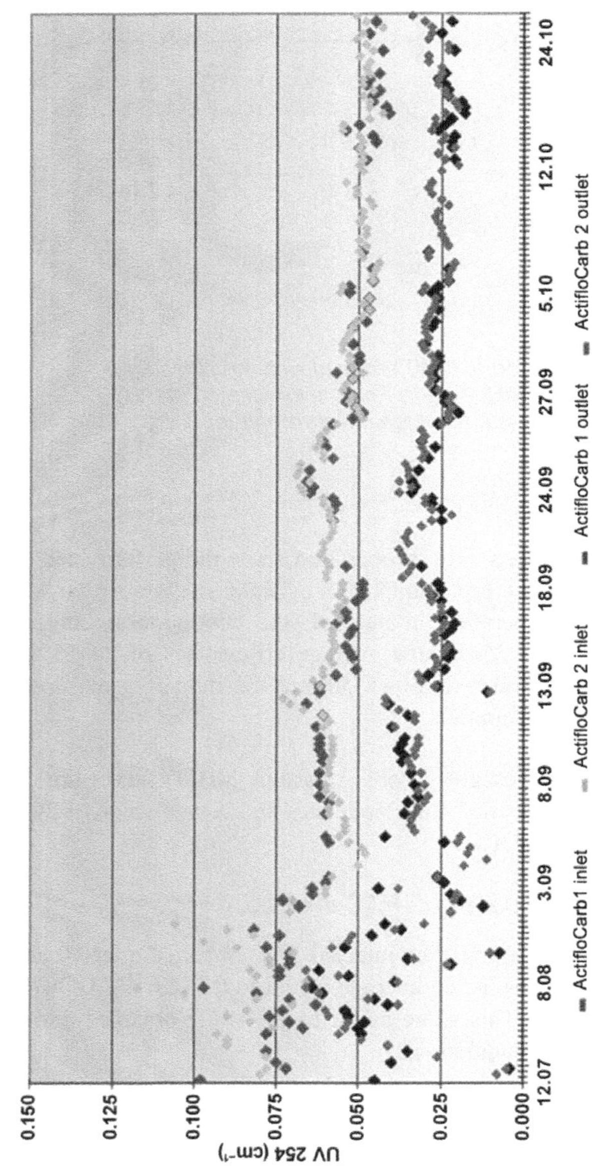

Figure 7.60. *Removal of organic substances (UV 254) on two Actiflo® Carb operating in parallel. For a color version of this figure, see www.iste.co.uk/gaid/watertreatment2.zip*

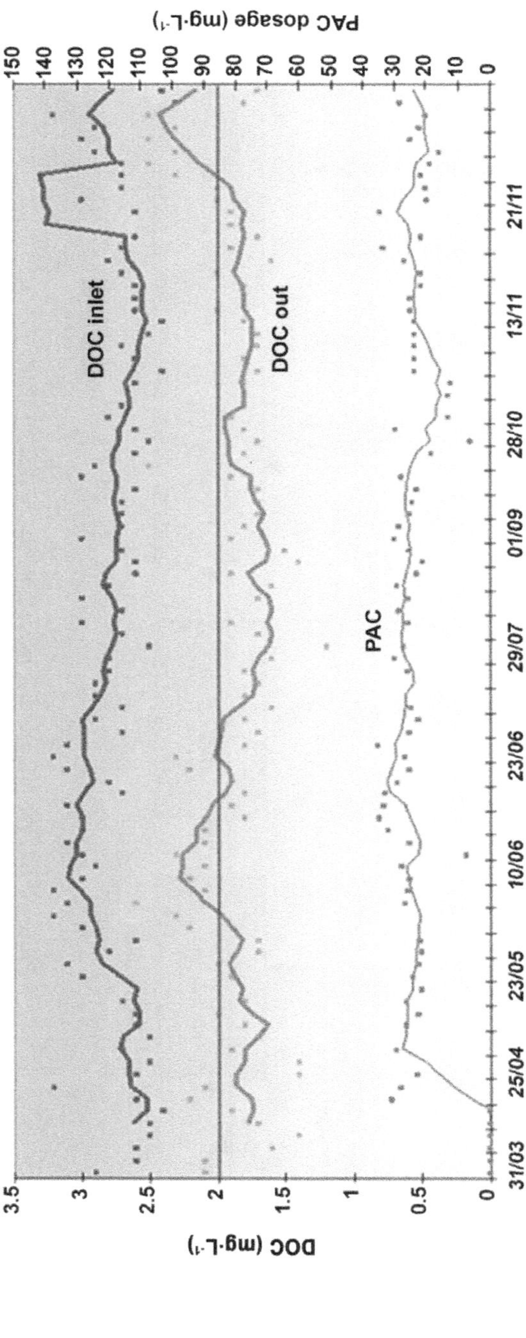

Figure 7.61. *DOC removal according to PAC dosage. For a color version of this figure, see www.iste.co.uk/gaid/watertreatment2.zip*

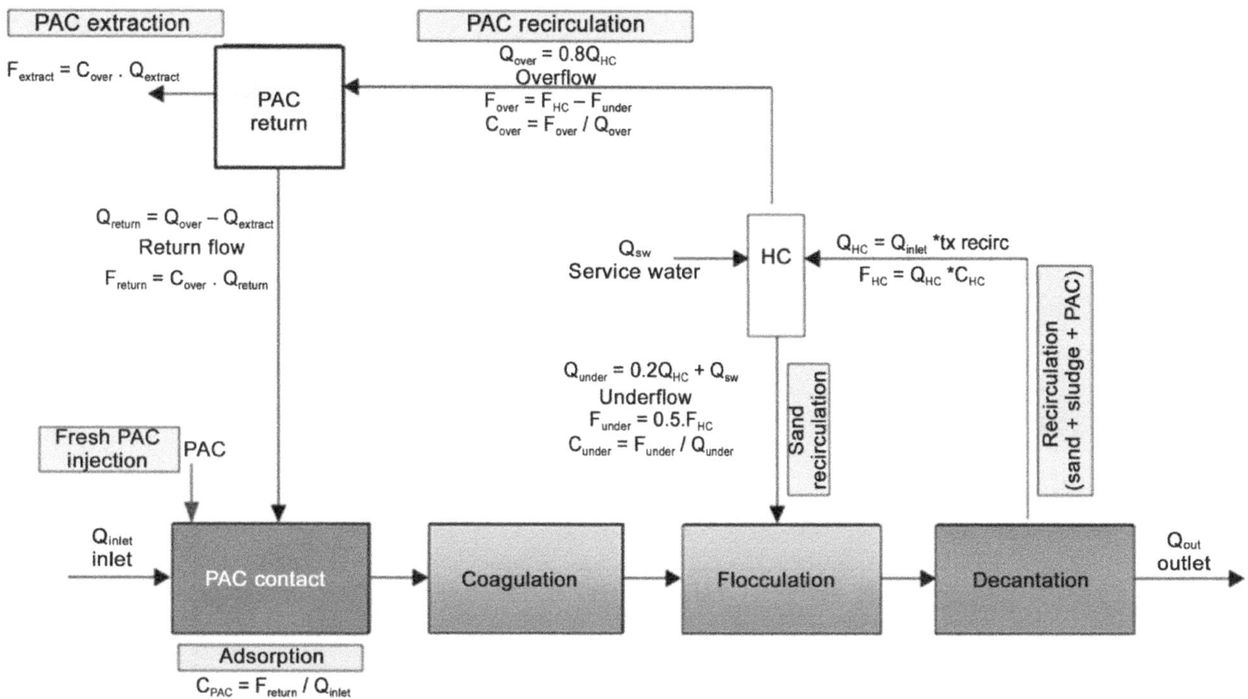

Figure 7.62. *Weight and hydraulic balance of Actiflo® Carb. For a color version of this figure, see www.iste.co.uk/gaid/watertreatment2.zip*

	Raw water		Actiflo® Carb DOC out		PAC dosage	PAC concentration in contact tank
	Turbidity NTU	DOC $mg{\cdot}L^{-1}$	Turbidity NTU	DOC $mg{\cdot}L^{-1}$	$mg{\cdot}L^{-1}$	$g{\cdot}L^{-1}$
Average	4.95	2.10	0.72	1.30	15.6	1.45
Maximum	6.40	2.20	1.10	1.40	19.8	1.60
Minimum	3.10	2.00	0.48	1.30	14.1	1.30
30%ile	4.63	2.09	0.55	1.30	14.4	1.40
60%ile	5.36	2.10	0.69	1.30	16.6	1.50
95%ile	6.27	2.19	1.04	1.39	15.7	1.54

Table 7.14. *Performance results on DOC removal according to PAC dosage (Aquasorb MP23)*

Weight and hydraulic balance

The sizing of the equipment related to the recirculation of PAC, its extraction and injection can be estimated from the weight and hydraulic balance shown in Figure 7.62.

7.5.2.1.2. Combination of PAC and in situ ozonation

As detailed in the adsorption chapter, activated carbons have both hydrophilic surface functional groups, which act as centers. In general, a larger specific surface and the increased presence of surface functions imply a greater adsorption capacity of the activated carbon in relation to organic pollutants.

These various surface functions play a very important role during adsorption, particularly during chemisorption, because, according to their nature and their number, they determine the qualitative and quantitative aspects of adsorption. They can be of four types: acid, basic, neutral or polar.

Depending on the predominance of their structure, these functional groups (Figure 7.63) lead to acidic (oxygen-bearing functions) or basic (nitrogen-bearing functions) carbons. The functional groups with O_2 increase the polarity of the carbon surface. Activated carbons with acidic surfaces are called L-type activated carbons. These carbons have mostly phenolic, n-lactone, f-lactone and carboxylic acid groups. The polarity of activated carbons is explained by the presence of very polar carbon–oxygen complexes on their surfaces, the carbons having free valencies (non-tetravalent carbons).

It therefore appears that not only are the porosity and the specific surface important characteristics of activated carbons, but the surface properties must also be taken into account. From this perspective, the adsorption of an organic substance, for example, does not only consist of the introduction of the adsorbent into the pores, but these pores are also the site of a chemical reaction between this substance and the surface groups of the activated carbon. The reaction mechanism can therefore be studied through the formation of chemical bonds between the organic substance and the surface groups of the activated carbon. The adsorption of organic acids increases with the presence of OH and NH_2 groups present on the surface of the activated carbons.

The surface functional groups therefore participate in the adsorption process. The introduction of new surface functional groups would lead to an increased adsorption capacity because they strengthen the electrostatic and dispersion interactions. Carboxylic groups, as well as basic groups containing oxygen (such as the chromene and pyrone type), would promote adsorption.

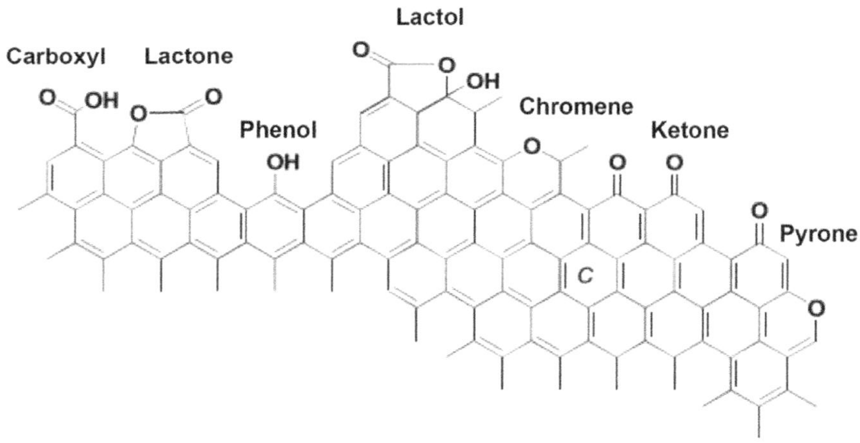

Figure 7.63. *Functional groups on the inner surface of the activated carbon*

The use of a strong oxidizing agent such as ozone is involved in the increase in these functional groups and, in particular, in the carbonylated compounds. They are, indeed, derived from the oxidation of the substituted aromatic rings after theformation of phenols, diphenols or quinones and the opening of the aromatic ring. Figure 7.64 shows an example of surface functions on the internal surface of the micropores. These surface functions are also present on the inner surface of other pore types.

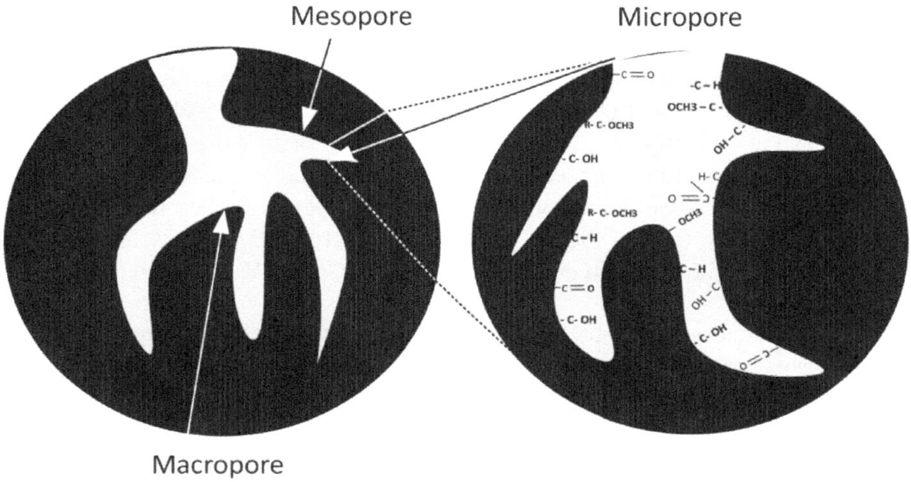

Figure 7.64. *Example of surface functions on the internal surface of micropores*

Determination of surface functions on activated carbon

Surface functions are obtained according to the Boehm method. The PZC (zero point of charge) pH of the water containing the activated carbon is 8.05 when new and 8.28 after 10 min of ozonation. It is known that the surface of the activated carbon is negatively charged when the pH is higher than the PZC pH, while the surface of the activated carbon is positively charged when the pH is lower than the PZC pH. In the latter case, the activated carbon will have a high affinity for anionic compounds. The pH of the water containing the tested activated carbon is between 7.4 and 7.9, meaning that it is suitable for negatively charged organic substances. Laboratory tests of direct ozonation (2 $g \cdot m^{-3}$) in a contact tank containing PAC (concentration at 1.2 $g \cdot L^{-1}$) for 10 min lead to the results shown in Table 7.15.

Dosage in O_3 $g.m^{-3}$	**Group I $meq \cdot g^{-1}$**	**Group II $meq \cdot g^{-1}$**	**Group III $meq \cdot g^{-1}$**	**Group IV $meq \cdot g^{-1}$**	**Total groups $meq \cdot g^{-1}$**	**Iodine index**	**pHPZC**
0	2.58	0.66	2.36	3.24	8.84	950	8.05
2.0	4.30	0	0	9.28	13.58	–	8.28

Table 7.15. *Characterization of surface functions with and without in situ ozonation*

It appears (Table 7.15) that, in its initial activated state, Chemviron's Pulsorb RD 90 activated carbon presents fairly interesting surface functions, with strong carboxyl functions (Group I) at 2.58 $meq \cdot g^{-1}$, hydroxyl and phenol functions (Group III) at 2.36 $meq \cdot g^{-1}$, and especially functions of the carbonyl and quinone groups (Group IV) at 3.24 $meq \cdot g^{-1}$, which are largely superior to those usually observed. These high values are associated with the activation and manufacturing conditions of the activated carbon. In many cases, the values of the surface function groups are approximately 0.3–0.6 $meq \cdot g^{-1}$ for Group I, 0.10–0.35 $meq \cdot g^{-1}$ for Group II, 0.5–1.5 $meq \cdot g^{-1}$ for Group III and <0.2 $meq \cdot g^{-1}$ for Group IV.

The high values of the surface functions tend to increase the adsorption capacity of activated carbons, particularly with respect to humic and fulvic organic substances. The in situ ozonation of activated carbon acts directly on the surface of the material by creating or increasing the surface groups. Thus, according to the results of direct ozonation of activated carbon within the reactor, Table 7.15 shows an increase in surface groupings. Group I increases from 2.58 to 4.30 $meq \cdot g^{-1}$, signifying a clear increase in strong carboxyl groups. Group III decreases from 2.36 to 0 $meq \cdot g^{-1}$, which is explained by an oxidation of the hydroxyl and phenol groups into carbonyls and carboxyls, and Group IV increases from 3.24 to 9.28 $meq \cdot g^{-1}$. Groups I and IV are involved in the adsorption of organic substances. In particular, interactions of CO and OH (organic substance–surface functions) occur. We therefore note an increase in the various functional groups, which should ensure a better adsorption capacity of the activated carbon.

The reactions of O_3 on PAC give:

$$O_3 + PAC \rightarrow \text{Oxidized PAC} + OH\bullet$$

In the O_3-PAC system, ozone is partially transformed into OH• radicals, and the generated OH• radicals are trapped on the activated carbon surface.

Adsorption isotherms with PAC and in situ ozonated PAC give curves showing the increase in adsorption capacity of in situ ozonated PAC. Three distinct zones are obtained: a low adsorbable zone, a moderate to good adsorbable zone and a strongly adsorbable zone. The isotherm data selected are chosen from the third zone and sometimes, to a lesser degree, between the second and third zones. The least adsorbed compounds have a slope of 1.0 at highequilibrium concentrations. Compounds that are more strongly adsorbed have a slope of 1.0 at much lower concentrations.

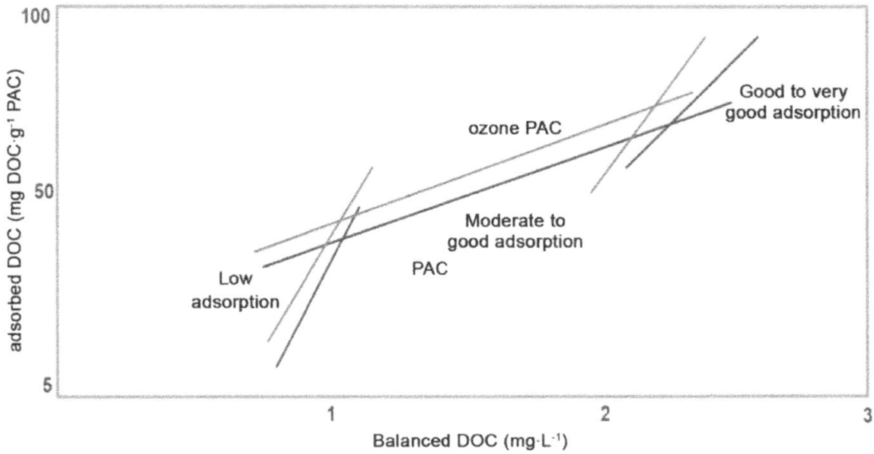

Figure 7.65. *Adsorption isotherm results with PAC and ozonated PAC in situ. For a color version of this figure, see www.iste.co.uk/gaid/watertreatment2.zip*

Efficiency of Actiflo® Carb with ozone injection in the carbon contact tank

The Actiflo® Carb process, using an injection of O_3 (1–2.5 $g{\cdot}m^{-3}$) in the PAC contact tank, was tested on a pilot project. The PAC contact time in the reaction tank was 10 min, and the PAC dosages varied between 10 and 20 $mg{\cdot}L^{-1}$. The tests were conducted at clarification rates of 40 $m{\cdot}h^{-1}$.

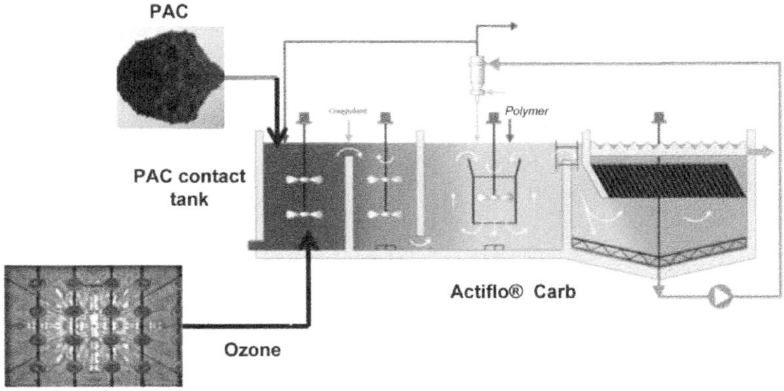

Figure 7.66. *Operating diagram of the combination O_3 in situ + PAC reactor. For a color version of this figure, see www.iste.co.uk/gaid/watertreatment2.zip*

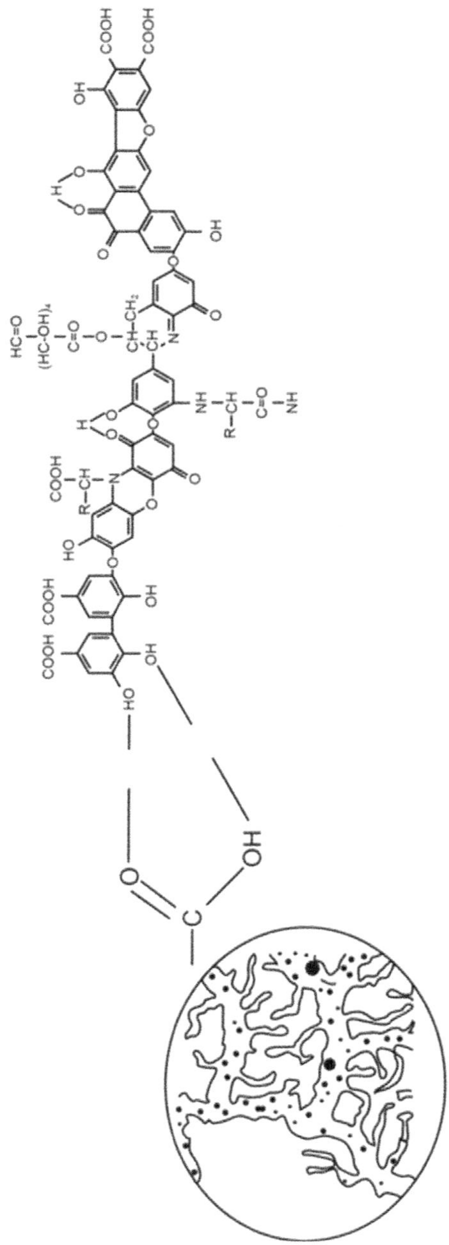

Figure 7.67. *Example of the link between humic substances and surface functions*

The results obtained illustrate the efficiency of chemical oxidation with ozone with respect to DOC removal by ozonated activated carbon.

Thus, on the first site, a gain of approximately 10–15% in average DOC removal performance was observed, bringing the removal efficiency to 55–60%. A better DOC removal is thus obtained with this in situ O_3 and PAC reactor combination, without releasing any by-products in the produced water. Moreover, ozone is injected directly into the PAC contact tank, whose concentration is between 1.5 and 2 $g \cdot L^{-1}$, there is a direct reaction between the oxidant (O_3) and the reducer (PAC) to the detriment of the organic matter. This is the reason why the ozonation reaction does not modify the DOC because the O_3-PAC reaction kinetics are much faster. Furthermore, the residual ozone was always found to be equal to 0 $mg \cdot L^{-1}$ in the interstitial water. The SEC-UVD chromatogram (Figure 7.68) shows the evolution curves of the different types of organic substances obtained with PAC alone and with ozonated PAC. It can be seen that the injection of O_3 results in a systematic reduction of HS peaks compared to those obtained with Actiflo® Carb (without O_3). The sample thus treated contains a lower concentration of organic carbon. This result can be explained by the better adsorption efficiency of PAC oxidized in situ by ozone, without generating the appearance of new peaks, which would have been attributed to oxidation by-products.

The size exclusion chromatogram (Figure 7.68) shows a small peak for biopolymers (retention time of 28 min), followed by a much larger peak related to HS and building blocks (retention time of 42–48 min), which constitute more than 50% of the DOC.

Low molecular weight substances (retention time of 51 min) are composed of acids and HS that elute at the same time. Neutral substances (retention time > 55 min) are neutral or amphiphilic molecules that interact with the chromatography column matter and have a longer retention time than water (55 min).

After passage through the Actiflo® Carb (without addition of O_3), we observed the following (brown curve):

– the biopolymer peak is absent;

– the HS and building block peaks have decreased;

– the low molecular weight substances peak has also decreased;

– the neutral and amphiphilic substance peaks are higher, probably due to desorption of this type of substance.

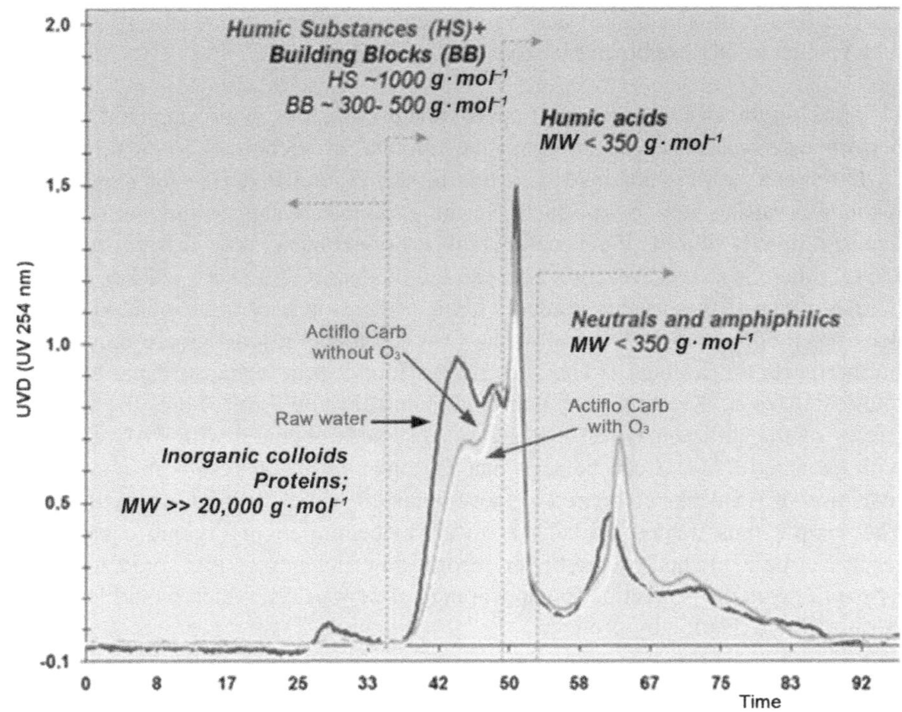

Figure 7.68. *LC-DOC of samples treated with Actiflo® Carb with and without activated carbon ozonation. For a color version of this figure, see www.iste.co.uk/gaid/watertreatment2.zip*

The same chromatogram, on the results of the addition of O_3 in situ in the PAC reactor, shows (yellow curve) an additional decrease in the peaks observed after Actiflo® Carb. Even the neutral and amphiphilic substances that increased after Actiflo® Carb treatment were reduced by O_3.

Bromate formation?

It is well known that the ozonation of water containing bromides leads to the formation of bromates under certain operating conditions. To determine if this is the case in an ozonation configuration within the activated carbon reactor, tests were carried out on a tertiary wastewater containing approximately 6.5–7 mg·L^{-1} of bromides. The results obtained are summarized in Table 7.16. Bromate concentrations <10 μg·L^{-1} were analyzed. In contrast, the same clarified water, which was directly ozonated, led to the formation of 137 μg·L^{-1} bromates for an O_3 dosage of mg·L^{-1} and a contact time of 10 min.

Parameters	Actiflo® Carb (without O_3)	O_3 in situ in Actiflo® Carb
Bromates $\mu g \cdot L^{-1}$	<10	<10

Table 7.16. *Bromate formation in various ozonation tests*

This confirms that the traditional mechanisms of ozonation are not implemented in such a case because the presence of activated carbon, which is in direct contact with the oxidant, does not allow any residual concentration of O_3 that could give rise to secondary reactions and prevents the reaction of O_3 with the organic substances present in the water, as well as bromides.

The process combining a PAC reactor with ozonation injected in situ in the contact tank leads to an increase in DOC removal. Implemented in the Actiflo® Carb process, this PAC reactor combines the flocculation and fast decantation performances of the Actiflo® with the adsorption capacities of PAC and an advanced ozone oxidation. DOC is fixed on the PAC adsorption sites, as well as on the surface functions developed on the activated carbon surface. Modification of the activated carbon surface results in an increase in the surface functions that participate in a better adsorption capacity. This results in much better DOC removal for the same amount of activated carbon used. Equally, it is possible to significantly reduce the PAC dosages because of the injection of O_3, all while maintaining the performances of the process at the same level.

7.5.2.1.3. Granular activated carbon

Granular activated carbon (GAC) is usually installed after the coagulation–flocculation–decantation (or flotation) sand or bilayer filtration steps. It can also be installed directly after a decanter or a flotation device and plays, in this case, the role of an adsorbent filter for organic substances and a mechanical filter for residual flocs.

The efficiency of GAC filters depends on the characteristics of the GAC (type, granulometry, reactivation mode) and the operating conditions (filter speed, contact time in empty drums, backwash conditions, GAC life, etc.) and quality of the feed water.

Its high specific surface and porous structure suggest that GAC has high adsorption capacities in relation to the DOC present in surface waters. In reality, GAC is often used because of its ease of implementation, although its adsorption capacity in relation to DOC tends to be rapidly depleted. For example, raw water with a DOC of 6 $mg \cdot L^{-1}$ leads to GAC life spans between 1 and 4 months for contact times of 10 and 30 min, respectively. This is one of the reasons why the GAC filter is always placed in the first or second filtration stage after a coagulation–flocculation–decantation or flotation stage, which removes between 30 and 60% of the DOC (see advanced coagulation). This reduces the flow of DOC supplying the GAC filter and allows it to have a longer life span.

Indeed, when the GAC filter is supplied directly from the raw water, the regular variation in DOC leads to desorption of the adsorbed substances during periods of high concentration due to the rebalance with a lower DOC concentration.

The mechanism of NOM adsorption on GAC involves competition between the various organic substances present in the water, even if the upstream conventional or advanced coagulation has removed some of the high and medium molecular weight substances.

As explained with PAC, the effectiveness of GAC in removing NOM partly depends on the volume and type of pores, in that they must match the size of the molecules to be removed. This is because NOM adsorption is largely controlled by the relation between the molecular size distribution of NOM and the pore size distribution of the activated carbon (Figure 7.69). The adsorption capacity theoretically increases with the available volume of macropores and mesopores, and a strong correlation is obtained for the removal of HS. However, there may be a size exclusion effect for the very high molecular weight fraction (>15,000 Da) of NOM.

However, this NOM fraction is generally well removed by coagulation and is only frequently present in low concentrations at the inlet of the GAC filter.

After the coagulation–flocculation–decantation stage, passage through a GAC filter shows that fractions of intermediate molecular weight (500–1,000 Da) are well removed (50–60%) by the GAC because these substances have access to a large percentage of the pore volume. Even micropores with a width of 1.5–2 nm can participate. The removal of high molecular weight substances (>4,000 $g \cdot mol^{-1}$) is up to 25–30%, while low molecular weight substances (<500 $g \cdot mol^{-1}$) are less well removed (<20%). A small increase in these molecules can be observed, but this has very little impact on the DOC of the treated water.

The efficiency of a GAC that has regenerated more than three times is reduced in relation to neutral and amphiphilic molecules.

In contrast, the adsorption capacity of non-aromatic substances from organic matter is related not only to the pore size but also to other characteristics, such as the surface chemistry of activated carbons. Due to the presence of carboxylic acids and phenolic groups in the structure of NOM, they carry negative charges at surface water pH values. Depending on the pH, the GAC base material and its activation mode, the GAC pore surface can be positive, negative or neutral.

As a result, electrostatic interactions may affect NOM adsorption. However, the surface and total pore volumes are not sufficient to select GAC as a means of NOM removal, as these parameters do not provide information on the distribution of accessible pores.

In some specific cases, direct adsorption on GAC is generally not recommended because the adsorption capacity can be rapidly reduced by blocking the pores with large molecules of HS. The adsorption isotherms with GAC convey the equilibrium relation between the amount of organic compound that remains in solution (equilibrium concentration, Ce) and the amount of compound that is on the surface of the activated carbon (surface concentration, q).

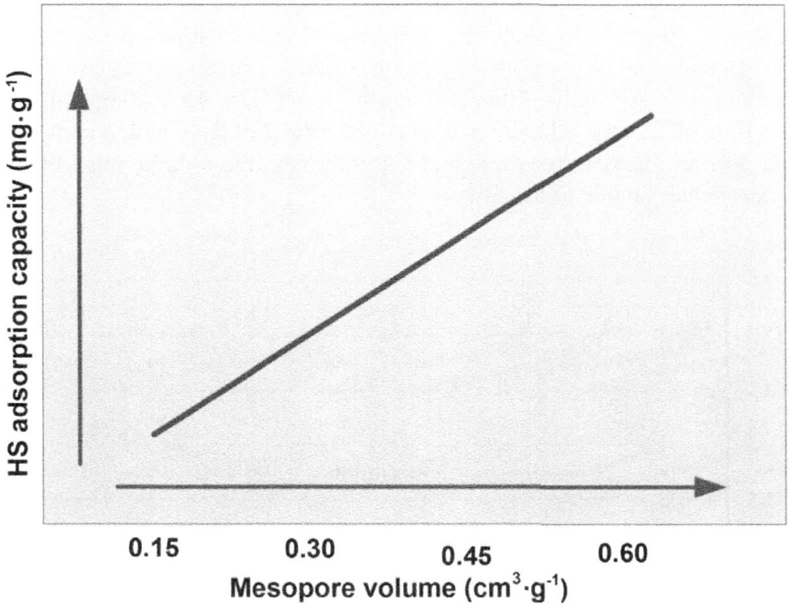

Figure 7.69. *Mesopore volume versus adsorption capacity for NOM. For a color version of this figure, see www.iste.co.uk/gaid/watertreatment2.zip*

There are often significant differences between the adsorption characteristics of different types of GAC in relation to NOM, whether they are based on bituminous coal, coconut or wood (Table 7.17).

Activated carbons	Freundlich parameters		
GAC	K	1/n	qe $mg \cdot mg^{-1}$
Jacobi Aquasorb 2000	0.0248	0.64	$0.0248\ C^{0.64}$
Cabot Norit 4000	0.0225	0.71	$0.0225\ C^{0.71}$
Cabot Norit 830P	0.0236	0.68	$0.0236\ C^{0.68}$

Table 7.17. *Freundlich parameters for DOC for some GACs*

Life time of GAC before its renewal

The life time of GAC depends on various parameters, such as DOC concentration, contact time, characteristics of the activated carbon and characteristics of the organic substances.

Figures 7.70 and 7.71 show the influence of contact time on two GAC filters with identical carbon supplied with the same decanted water. The DOC concentration at the outlet filter is 2 $mg \cdot L^{-1}$ after 120 days of operation with a contact time of 12 min, whereas it is obtained after 250 days with a contact time of 30 min. For an identical feed rate and filtration rate, the volume must therefore be 2.5 times greater on one of the filters.

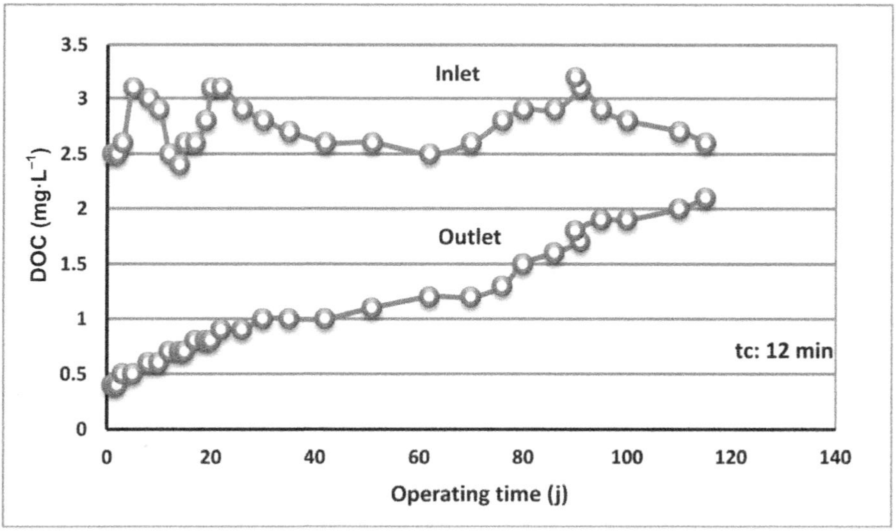

Figure 7.70. *Evolution of DOC at outlet of GAC filter (Aquasorb 2000) for a contact time of 12 min. For a color version of this figure, see www.iste.co.uk/gaid/watertreatment2.zip*

There are many empirical equations for calculating the life span of a GAC. Many of them are based on Freundlich or Langmuir parameters, which are related to the type of AC and its characteristics.

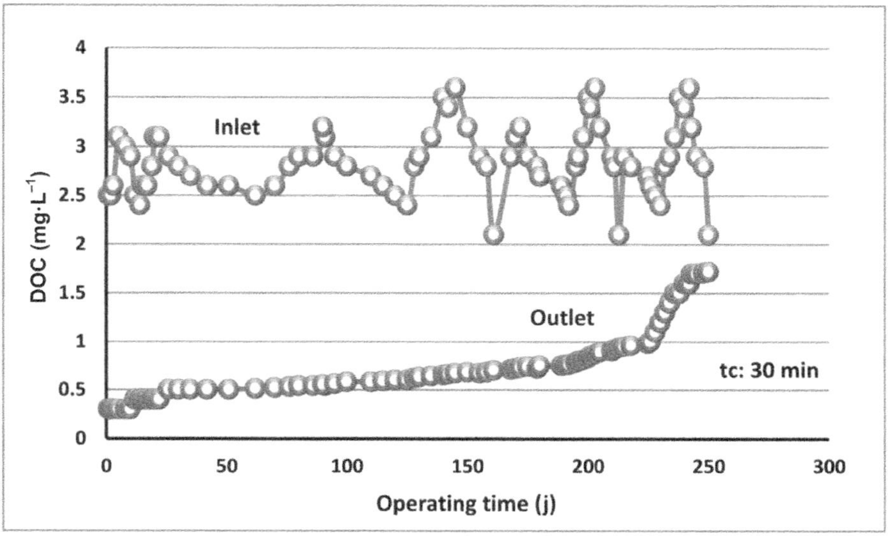

Figure 7.71. *Evolution of DOC at a GAC filter outlet (Aquasorb 2000) for a contact time of 30 min. For a color version of this figure, see www.iste.co.uk/gaid/watertreatment2.zip*

The calculation can be done as follows:

– amount of DOC to be removed ($kg \cdot d^{-1}$) = (DOC inlet – DOC out) × flow rate;

– total volume of GAC in the filter (m^3) = filter surface × GAC height;

– GAC weight (kg): volume × density ($kg \cdot m^{-3}$);

– contact time of the water in the filter (min) = height (m) × 60/passage velocity ($m \cdot h^{-1}$);

– qe = adsorption capacity (mg $DOC \cdot g^{-1}$ GAC);

with:

– ρGAC: density of GAC;

– life time (d) of GAC = $V_{CAG} \cdot qe \cdot \rho GAC$)/ Q(C inlet – C outlet);

– Q: flow to be treated/day ($m^3 \cdot d^{-1}$).

For example, for a flow of water to be treated from 1,000 $m^3 \cdot d^{-1}$ of settled water containing 3 $mg \cdot L^{-1}$ of DOC on a GAC filter with the following characteristics: a filtration speed of 6.94 $m \cdot h^{-1}$ in a GAC volume of 9 m^3, a surface area of 6 m^2 and a contact time of 12 min. The density of the GAC is 450 $kg \cdot m^{-3}$.

The adsorption capacity (qe) deduced from the Freundlich parameters is 38.6 mg $DOC \cdot mg^{-1}$ GAC.

The desired DOC concentration was 2 $mg \cdot L^{-1}$.

We deduce that:

Life time of GAC (d) = 9. 450 . 38.6/((3 –2)1000) = 156 days.

7.5.2.1.4. Micrograin activated carbon (GAC)

Whether with Opacarb® FL or Filtraflo® Carb, the adsorption mechanism of NOM with micrograin activated carbon is identical to that of PAC or GAC.

Usually, a GAC filter is scaled with a contact time of 10 minto reduce the organic contamination of the water to be treated. This contact time is determined from laboratory kinetics and has been used to deduce Freundlich and/or Langmuir adsorption parameters. Nevertheless, the weight of GAC present in a filter at the beginning of the life of the carbon is very important, and in this case, the contact time of 10 min is excessive compared to the kinetics, which are related to the initial concentration of pollutants, as well as to the weight of carbon present. If we use the C·t/M parameter (concentration of pollutants removed × contact time and per weight of activated carbon), the decontamination in a tank of water containing 3 $mg \cdot L^{-1}$ of DOC for a DOC outlet of 1.5 $mg \cdot L^{-1}$ requires approximately 20 $mg \cdot L^{-1}$ of PAC for a contact time of 10 min. This leads to a C removal/Mtratio that is equal to 0.0075 $mgDOC \cdot mgPAC^{-1} \cdot min^{-1}$. In a well-mixed micrograin activated carbon reactor, the Ct/M parameter becomes very low due to the weight of carbon present (250–350 $g \cdot L^{-1}$), and a 10-min contact time is no longer necessary. For the same DOC concentration to be removed, the ratio becomes equal to 0.0035 $mgDOC \cdot mgCAP^{-1} \cdot min^{-1}$. Half the contact time would be quite acceptable with the same performance. This was verified by establishing DOC removal tests with different GAC µgrain concentrations and for different contact times (tc).

Activated carbons	Freundlich parameters		
µgrain	K	1/n	qe ($mg \cdot g^{-1}$)
Chemviron Microsorb 400R	0.0243	0.96	0.0243 $C^{0.96}$

Table 7.18. *Freundlich parameters for DOC with micrograin activated carbon*

The adsorption capacity of micrograin activated carbon is 31 $mg \cdot g^{-1}$micrograin. This means that obtaining 1.5 $mg \cdot L^{-1}$ of DOC in treated water for a DOC inlet of 3 $mg \cdot L^{-1}$ requires 42 $g \cdot m^{-3}$ of micrograin activated carbon.

7.6. Ozonation

7.6.1. *Preozonation*

In the treatment of drinking water, preozonation can have direct effects, such as the oxidation of iron and manganese, presence in the raw water, treatment of part of the color, oxidation of the molecules responsible for tastes and odors, and algal lysis, and it can promote indirect effects, such as a better removal of NOM by coagulation.

Preozonation is placed directly on the raw water, with an injection of ozone in a reaction tank with a contact time of 2 min. The ozone is injected via a static mixer or is mixed directly in the tank.

The figure shows the chromatogram of a treatment process, including a preozonation of O_3 at 0.5 $mg \cdot L^{-1}$ on average and 1 $mg \cdot L^{-1}$ at peak. Compared to the average DOC concentration of the raw water, which is 3.9 $mg \cdot L^{-1}$, the ratio applied in pre O_3 is 0.13 $mgO_3 \cdot mgDOC^{-1}$.

The UVD chromatogram of the raw water shows a standard organic matter profile for natural surface water, with the predominance of aromatic HS (60% of DOC). After the preozonation stage (Figure 7.72), a clear decrease in the peaks of high molecular weight substances is observed, as is the case for medium and low molecular weight substances. Thus, O_3 acts significantly on the transformation of organic matter, with a decrease in the aromaticity of the predominant fraction of HS.

Preozonated water has a DOC that is equivalent to that of raw water. Other chromatograms showed a slight increase in DOC after the preozonation stage, which can be explained by the fact that ozone acts on particulate organic matter, which is then counted in the dissolved organic matter fractions.

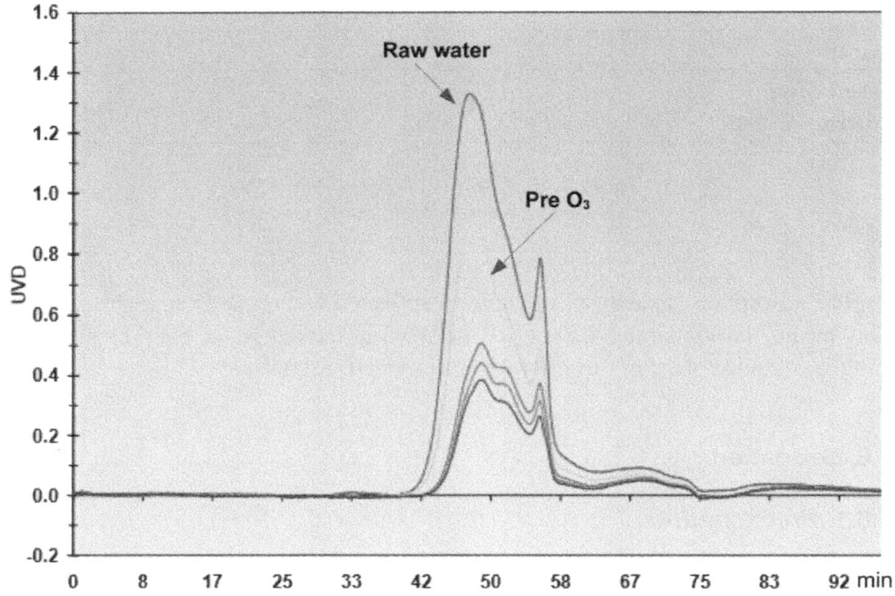

Figure 7.72. *UV 254 chromatogram on raw water and preozonated water. For a color version of this figure, see www.iste.co.uk/gaid/watertreatment2.zip*

Qualitatively, Figures 7.72 and 7.73 show:

– a significant decrease in the UVD response and a decrease in the SUVA (UV/OCD ratio), which confirms the priority degradation of aromatic substances by the ozone reaction on the double bonds;

– at the DOC level, a weak transformation of the organic matter is observed, which can be explained by the dosage applied (0.13 mg $O_3 \cdot mg^{-1}$ DOC);

– at the HS + BB fractions level, a shift toward the lowest molecular weights of HS, observed in OCD response (shoulder of the BB peak and increase inthe acid peak), and a significant decrease of the UVD signal for this fraction, which is also conveyed by a decrease in aromaticity for this fraction;

– the organic matter distribution diagram shows the mutual compensation of the concentrations of the HS and BB fractions;

– at the level of the acid and neutral fractions, a slight increase in these fractions is observed, which remains coherent with the ozone reaction to break the molecules.

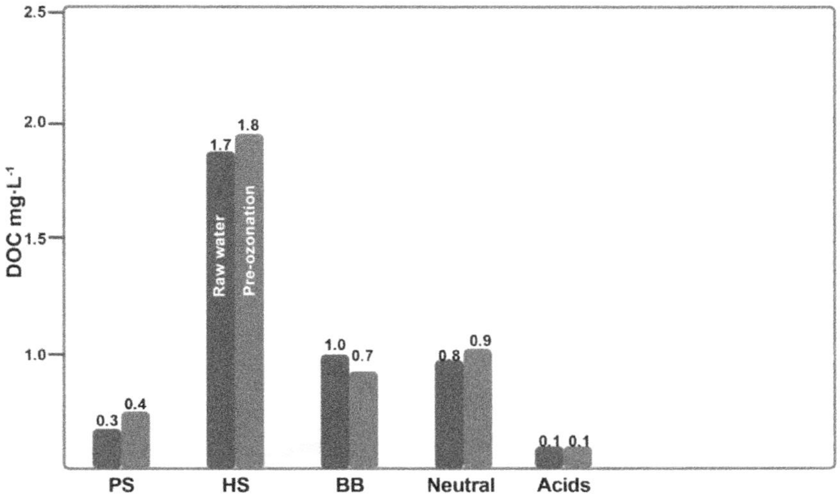

Figure 7.73. *Characterization of NOM fractions in raw water and after ozonation. For a color version of this figure, see www.iste.co.uk/gaid/watertreatment2.zip*

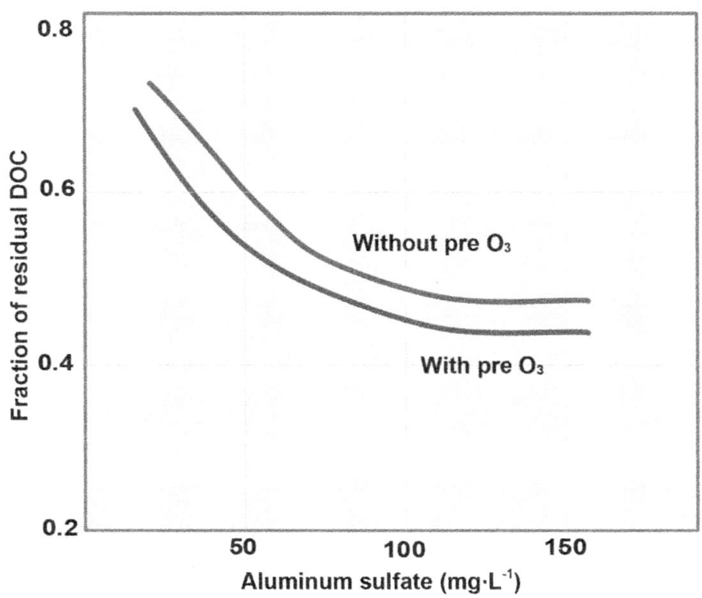

Figure 7.74. *Effect of preozonation at alow mg $O_3 \cdot mg^{-1}$ DOC ratio on the coagulation efficiency of NOM. For a color version of this figure, see www.iste.co.uk/gaid/watertreatment2.zip*

Concerning the polysaccharide fraction, usually, the ozone reaction only results in a transformation of the OM of this fraction (peak shift toward the lowest retention times). Here, we observed a significant increase in the concentration of this fraction (visible in OCD response).

At many Veolia sites, it has been observed that preozonation improves the efficiency of the coagulation–flocculation–clarification stage, both in terms of the residual turbidity of the decanted water and the residual DOC (Figure 7.74).

Ozone, applied to raw water at high dosages (>0.4 mg O_3·mg DOC^{-1}), may preferentially react with the humic fraction of NOM, resulting in a decrease in coagulation efficiency that is located downstream. At lower dosages (<0.2 mg O_3·mg^{-1} DOC), the effect of pre O_3 can improve coagulation efficiency.

7.6.2. *Interozonation*

In interozonation, the injection of O_3 dosages is usually carried out between 0.5 and 2.5 mg·L^{-1}, depending on the variation in DOC and the water temperature.

The interozonation of NOM is complex due to the variety of substances present. Generally speaking, high molecular weight substances (HPM > 4,000 Da) are affected by O_3 by being transformed into lower molecular weight molecules. O_3 also acts on molecules (1,000–4,000 Da) but with less efficiency. Interozonation does not lead to NOM removal, but it does change the structure of high molecular weight substances into smaller substances and turn hydrophobic molecules into more hydrophilic ones. Thus, the TOC value remains the same as that of raw water because there is a slight decrease in the fraction of high molecular weight compounds (HMWs), while the fraction of low molecular weight molecules (LMWs) increases. In contrast, a good reduction in color and UV 254 is obtained.

When ozone directly (O_3) or indirectly (OH•) reacts with NOM by converting a fraction of the DOC into BDOC, hydrophobic organic carbon is converted to hydrophilic organic carbon without a significant decrease in DOC.

During ozonation, the increase in BDOC mainly occurs for ozone dosages that are at or below 0.5 mg O_3·mg^{-1} DOC, while a dosage above 1.0 mg O_3·mg^{-1} DOC does not induce a significant increase in DOC (Figure 7.75).

Thus, in water containing both humic and non-HS, the beneficial reactions of ozonation with respect to NOM removal are only obtained if ozone reacts exclusively with the non-humic NOM substances.

This is possible if coagulation is optimal and has removed high molecular weight substances. This strategy allows O_3 to target the substances that are most sensitive to its reactions.

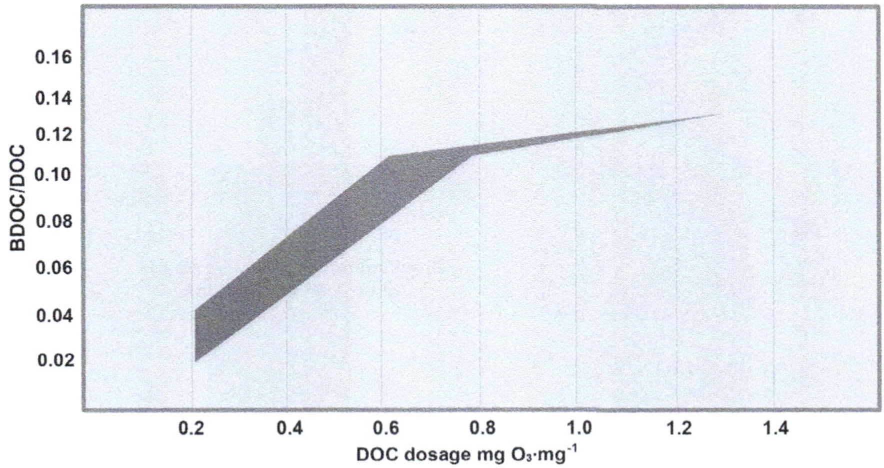

Figure 7.75. *BDOC/DOC ratio according to ozone dosage. For a color version of this figure, see www.iste.co.uk/gaid/watertreatment2.zip*

Thus, the interozonation of NOM leads to the formation of several by-products, including aldehydes and carboxylic acids, that contribute to the amount of BDOC and assimilable organic carbon (AOC). For this reason, interozonation is usually followed by biofiltrationto remove BDOC and AOC from the water (see section 7.7).

In general, when ozone is injected into water, it is consumed in two stages: an initial stage of rapid ozone consumption, followed by a stage of slow decay. The rapid consumption of ozone is mainly due to reactions with NOM.

Ozone consumption is influenced by organic carbon structures and functionality. For example, unsaturated bonds can consume ozone directly, alkyl compounds can inhibit ozone consumption and aryl compounds can promote or inhibit ozone consumption. Given its heterogeneous nature, NOM can be involved in some or all of these mechanisms (Figure 7.76). Due to the direct consumption of ozone by unsaturated bonds, larger organic components are oxidized into smaller organic components, and consequently, a decrease in UVA 254 occurs during ozonation.

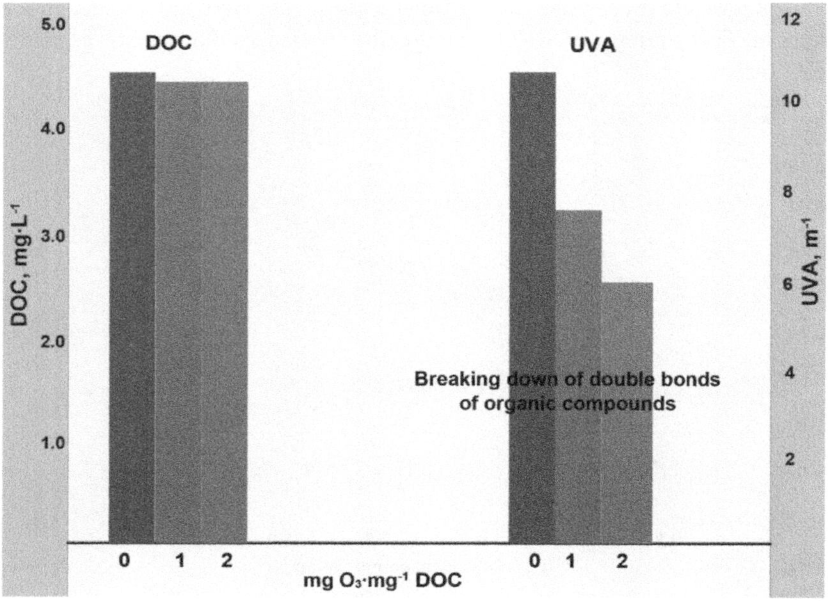

Figure 7.76. *Influence of interozonation on DOC and UVA. For a color version of this figure, see www.iste.co.uk/gaid/watertreatment2.zip*

7.6.2.1. *Effect of NOM nature and removal on AOC formation*

Part of the oxidized HS and building blocks are transformed into AOC. Low molecular weight acids are represented by small molecules of carboxylic acids, ketones and aldehydes. These small acid molecules are biodegradable organic compounds, such as formic acid and acetic acid.

Bromate formation during ozonation is higher for NOM consisting of a higher fraction of HS compared to that of building blocks.

The relations between the structure of organic matter and its reactivity are useful for predicting potential impacts on interozonation dynamics associated with variations in the nature and characteristics of NOM. These often depend on hydrological and seasonal factors such as agricultural runoff events, reservoir turnover and increased algal growth. These variations can then significantly affect the performance of interozonation.

7.7. Biological treatment

The presence in water of a biodegradable fraction of DOC favors bacterial regrowth processes in distribution networks. This requires the application of high chlorination rates, which leads to the formation of unwanted halo-organic compounds, as well as taste and odor problems. To avoid these problems, residual DOC should be removed as much as possible before disinfection and distribution of the water. This significantly reduces the risks of chlorination by reducing the demand for chlorine and, consequently, the formation of halo-organic compounds.

Two biological treatment technologies are presented:

– biological treatment with PAC with or without ozone and then separated by a microfiltration or ultrafiltration membrane;

– biological treatment with GAC placed in a filter and with ozone.

7.7.1. *Biological treatment with PAC: operating principle of the Opaline® B process (Veolia)*

The Opaline® B process combines PAC with ultrafiltration membranes and a PAC suspension.

The Opaline® B process is a hybrid process that combines suspended activated carbon and ultrafiltration membrane separation technology. This process is built around a PAC reaction tank and a membrane system. The ultrafiltration membranes (made of polyethersulfone with a cutoff of 0.045 μm) are made of high-strength hollow fibers and are directly immersed in the reactor.

The filtration is done in the external–internal direction: the treated water is absorbed by a slight depression (<0.4 bar) inside the fibers and is then collected at their end in a collector. An air injection at the base of the fibers ensures the renewal of the suspension near them and limits their clogging. Figure 7.77 shows the treatment process in which the Opaline® B process is implemented.

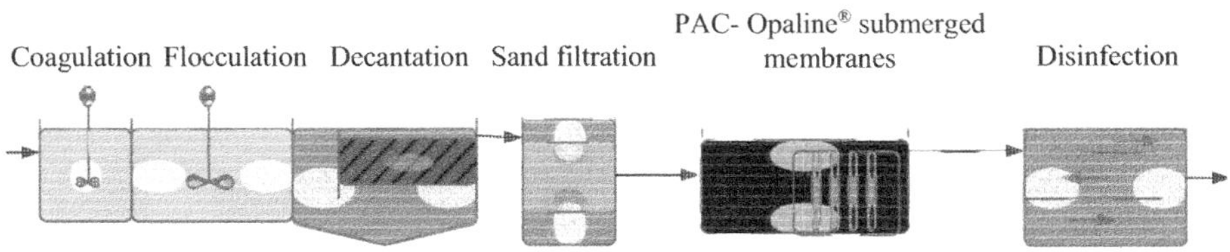

Figure 7.77. *Treatment process including the Opaline® B process. For a color version of this figure, see www.iste.co.uk/gaid/watertreatment2.zip*

All substances larger than the membrane's cutoff point accumulate around the fibers to form a "sludge", whose suspended matter concentration can be easily controlled by simple purging.

During operation, microorganisms grow by consuming bio-assimilable organic matter or ammonia and accumulate inside the reactor. The development of biomass improves the overall treatment capacity of the process by removing dissolved species such as DOC that is not retained by the ultrafiltration membrane.

The addition of PAC in the reactor allows a complete refining of the water by removing the micropollutants, the non-bio-assimilable fraction of the organic matter and the molecules generating tastes and odors by adsorption.

The introduction of a large weight of PAC inside the reactor (several $g \cdot L^{-1}$) allows the creation of a buffer adsorption capacity capable of stopping the peaks of micropollutants and organic matter. The maintenance of this adsorption capacity is achieved by renewal of the PAC according to a method, which makes it possible to optimize its use and to limit the production of sludge. This management method consists of purging a small volume of the reactor contents each day and compensating for the loss of activated carbon with an equivalent amount of new product. The complete renewal of the PAC held in the reactor is obtained after an operating time equal to the ratio of the volume of the reactor (m^3) to the volume of the purge ($m^3 \cdot d^{-1}$).

With this type of technology, the principle of managing PAC consists of renewing it from time to time so that it maintains an adsorption capacity. When saturation is reached, the activated carbon is purged, and a new PAC load is introduced into the reactor.

The complete renewal time of the suspension is called the PAC age. The weight of fresh activated carbon that is injected to compensate for the purge is equal to the weight of purged PAC; by reducing this value to the daily volume of water treated, we obtain a dosage of PAC that represents the treatment rate or PAC consumption in the Opaline® B process.

This management method makes it possible to work with high PAC ages (thus low PAC extraction) and to accommodate the two functions that are adsorption and biological activity.

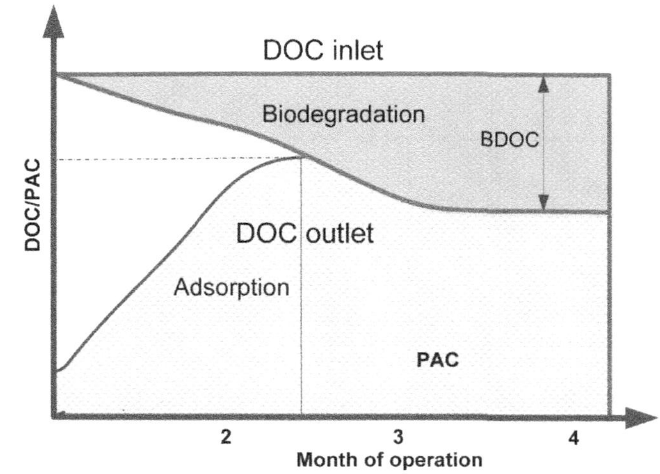

Figure 7.78. *Simplified diagram of the adsorption–biodegradation passage. For a color version of this figure, see www.iste.co.uk/gaid/watertreatment2.zip*

This combination ensures both DOC refining and clarification. An aeration system located at the base of the module limits the clogging of the membranes by agitating them. It also makes it possible to suspend the activated carbon and to prevent its decantation in the reactor because of an air-lift movement.

Table 7.19 summarizes the operating parameters for the PAC section.

Parameters	Activated carbon
Type	Aquasorb LP39 and Aquasorb MP23
Origin	Wood
Granulometry, µm	15–25
PAC concentration, $g \cdot L^{-1}$	6–12
Nominal age, j	10–80
Treatment rate, $g \cdot m^{-3}$	4–10
Purge rate, %	0.15

Table 7.19. *Operating PAC parameters in the Opaline® B process*

The results obtained over more than a year of operation (Figure 7.79) show that DOC removal by adsorption and biodegradation is significant in warm periods. In cold periods, biological activity is strongly slowed down, and DOC removal is obtained by a more frequent renewal of the PAC. BDOC is removed at 0.1–0.6 $mg \cdot L^{-1}$, depending on the temperature.

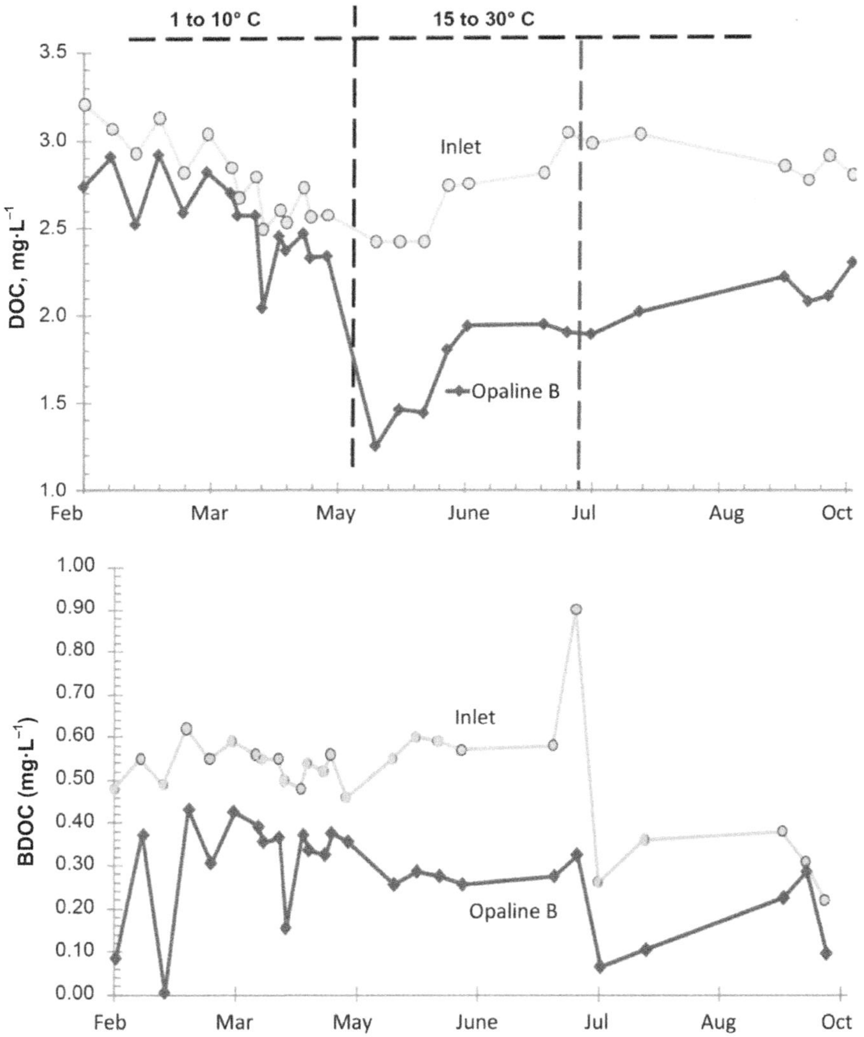

Figure 7.79. *DOC and BDOC removal with the Opaline® B process. For a color version of this figure, see www.iste.co.uk/gaid/watertreatment2.zip*

The operating parameters of the submerged membrane are summarized in Table 7.20 and Figure 7.80. The permeability ($L \cdot m^{-2} \cdot h^{-1} \cdot b^{-1}$) decreases from 780 $L \cdot m^{-2} \cdot h^{-1} \cdot b^{-1}$

to 400–450 $L \cdot m^{-2} \cdot h^{-1} \cdot b^{-1}$ in 80 days of operation in the first phase and then, after a CIP, decreases from 900 to 300 $L \cdot m^{-2} \cdot h^{-1} \cdot b^{-1}$ after more than 200 days of operation.

The influence of a preozonation upstream of the treatment plant makes it possible to significantly increase the BDOC content of the decanted water and thus to favor the biodegradation section in the Opaline® B process. With a PAC age of 40 days and a PAC concentration in the reactor of 12 $g \cdot L^{-1}$, the results of BDOC removal with and without pre O_3 are presented in Figure 7.81.

In fact, it appears that preozonation provides an additional 0.1–0.2 $mg \cdot L^{-1}$ of BDOC removal.

Parameters	Membranes
Type	Ultrafiltration
Critical flux, $L \cdot m^{-2} \cdot h^{-1}$	25 at 7°C and 35 at 20°C
Operating flux, $L \cdot m^{-2} \cdot h^{-1}$	15–25
Backwash flux, $L \cdot m^{-2} \cdot h^{-1}$	300
Filtration direction	External–internal
Configuration	Submerged
Filtration time, min	5 min filtration then 1 min BW
Hydraulic time, min	92 min at 15 lmh 55 min at 25 lmh

Table 7.20. *Operating parameters of submerged membranes*

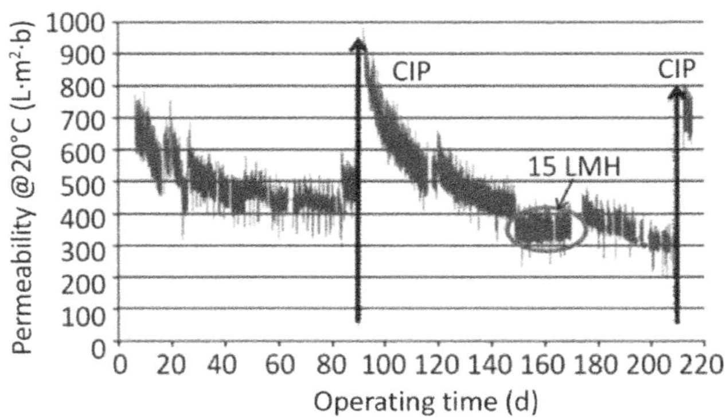

Figure 7.80. *Evolution of the permeability according to the operating time*

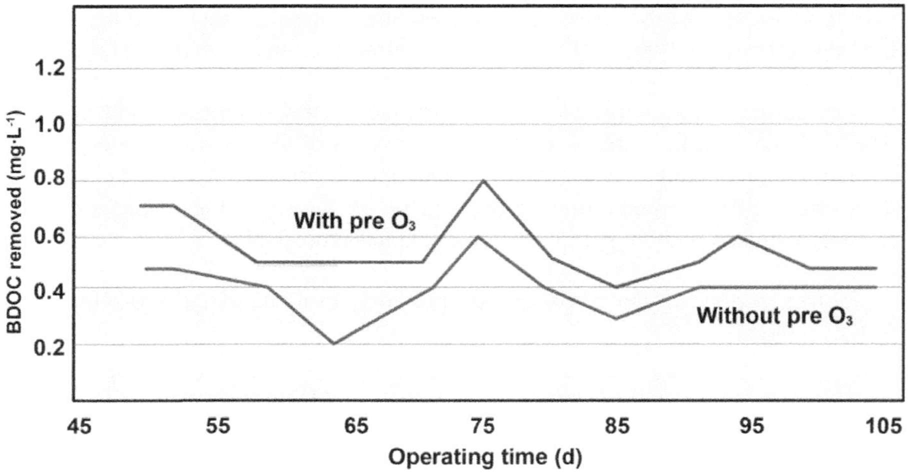

Figure 7.81. *Preozonation performance in relation to BDOC removal in Opaline® B. For a color version of this figure, see www.iste.co.uk/gaid/watertreatment2.zip*

The Opaline® B process, which is a combination of a PAC and an ultrafiltration membrane that is submerged in the PAC reactor, shows that the DOC can be adequately removed under certain operating conditions, such as the PAC age, renewal, dosage of fresh carbon and purge rate. The characteristics of PAC are also an important parameter with regard to the adsorption capacity of DOC.

7.7.2. *Biological treatment combining ozone and GAC*

In the field of drinking water, biofiltration is a filtration process on a granular base where biological activity develops, which allows the removal of biologically oxidizable compounds such as ammonia, biologically reducible compounds such as nitrates, or organic compounds such as part of the organic carbon present in water. When biological activity alone is sought, materials such as expanded clays or sand can be used, or materials such as activated carbon, for which adsorption and biological activity are complementary.

Biological filtration through an O_3-GAC combination has several advantages: the formation of AOC and BDOC through ozonation. These compounds are then responsible for bacterial growth on GAC medium. Another advantage is the removal of ozonation by-products such as short chain molecules (formaldehyde, acetaldehyde, glyoxal, methyl-glyoxal, etc.).

7.7.2.1. *Mechanism of the O_3-GAC combination*

O_3-GAC coupling is based on the interaction between ozone (oxidant) and GAC (adsorbent material).

Much of the effectiveness of this combination is attributed to the ozonation stage, as it is responsible for the reduction of UV-absorbing substances and the conversion of DOC to BDOC.

Indeed, for the dosages used in interozonation in drinking water plants (1–3 $mg \cdot L^{-1}$) for 6–8 min, the residual DOC present in sand filtered water is partially oxidized by ozone.

Ozone reacts with DOC in the molecular form O_3 (direct oxidation) or in the radical form OH• (indirect oxidation). Direct oxidation is a fast and rather selective reaction in which the O_3 molecule reacts with aromatic compounds, alkenes and amines. The O_3 molecule reacts faster with dissolved and ionized organic compounds than with neutralcompounds. Organic compounds that react slowly with ozone (aliphatic acids, aldehydes, ketones, some pesticides such as atrazine, etc.) are more likely to react by indirect oxidation with the OH• radical, which is non-selective.

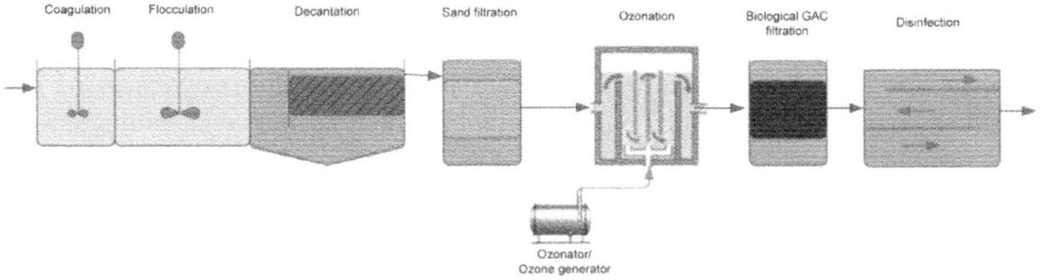

Figure 7.82. *Treatment process including an O_3-GAC stage. For a color version of this figure, see www.iste.co.uk/gaid/watertreatment2.zip*

HS reacts very quickly with molecular ozone O_3 and with hydroxyl radicals OH•. Ozone modifies the structure of HS. This results in a decrease in their molecular weight, an increase in the -COOH groups, a reduction in the color and a decrease in the UV absorbance.The decomposition products of DOC in the presence of ozone include aldehydes, ketones and carboxylic acids. As the oxidation products are more easily and quickly biodegraded than the NOM that is originally present in the raw water, an increase in the biodegradable fraction is obtained.

7.7.2.1.1. Operation of a GAC biofilter

The properties of a medium such as GAC are:

– an external specific surface that is available for the development of biofilm on the surface of the grains;

– a capacity for attachment of a biofilm and thus the ease of detachment of the biofilm during the washing of the filter;

– an internal adsorption capacity of the organic substances.

At the inlet of the second-stage filter, the DOC consists of various fractions that can be classified into four categories:

– adsorbable, biodegradable (BDOC);

– adsorbable, non-biodegradable (NBDOC);

– non-adsorbable, biodegradable;

– refractory (non-adsorbable, non-biodegradable).

Biodegradable organic substances include those that degrade rapidly and slowly. Organic matter that is most susceptible to biological attack falls into the non-adsorbable and biodegradable category. During GAC treatment, this organic matter could be removed by the bacteria on the GAC without affecting the adsorption process. The adsorbable and biodegradable organic matter category is of more interest because of the anticipated strong interaction between biological and adsorption processes. This distinguishes GAC from sand or other inert media. Biological removal of this class of organic matter, rather than adsorption, increases the ability of the GAC filter to remove adsorbable and non-biodegradable organic matter.

Molecules with molecular weights greater than 10,000 daltons have negligible BDOC, and it is important that they are removed upstream during the coagulation step. On the other hand, most of the molecules with molecular weights below 10,000 daltons can be found in the BDOC of ozonated water. Molecules below 1,000 daltons represent 61% of the total BDOC, and molecules with molecular weights between 500 and 1,000 contain the largest fraction of BDOC.

Bacteria and other microorganisms also actively interact with the surface of GAC in contact with the water. These interactions lead to the attachment, retention and, consequently, growth and maintenance of bacteria in GAC filters. Thus, at the beginning of GAC biofilter operation, adsorption is the primary mechanism for DOC removal because there is no bacterial growth yet. It is therefore used as an adsorbent for organic compounds. During operation, the adsorption capacity is gradually depleted, but biological activity takes over (Figure 7.83) by feeding on BDOC. Adsorption and biodegradation then become complementary. Biodegradation occurs gradually over a period of 1–2 months, depending on the inlet DOC concentration of the GAC filter and the temperature. Nevertheless, after a certain period of operation, the biofiltration mechanism is predominant because the adsorption reaches its saturation limits.

GAC biofilters contain five to eight times more biomass than a sand or anthracite filter, which explains the efficiency of activated carbon compared to inert materials.

7.7.2.2. *DOC removal efficiency*

Compared to filtration followed by ozonation, biological filtration after ozonation reduces the concentration of assimilable organic matter in the treated water.

During the adsorption period, the adsorbable DOC, including BDOC and NBDOC, is retained.

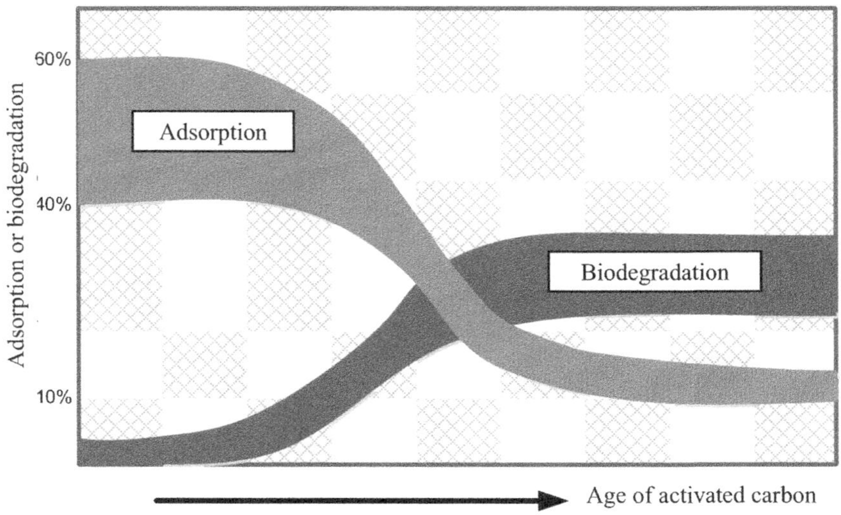

Figure 7.83. *Principle of the primary mechanism depending on the age of the activated carbon. For a color version of this figure, see www.iste.co.uk/gaid/watertreatment2.zip*

During this period, the percentage of BDOC removal, which is solely the result of the adsorption process, decreases rapidly and often reaches less than 15%. This highlights the rapid decrease in the adsorption capacity of activated carbon in relation to dissolved organic matter. Then, the percentage of BDOC removed increases again when bacterial colonization of the carbon has developed (Figure 7.84).

During the colonization phase, the biodegradation process progressively overtakes the adsorption process, removing the dissolved organic matter during filtration. Due to the adsorption process alone, the removal of non-biodegradable DOC goes from more than 60% at the beginning of the colonization period to less than 25% after a few months.

DOC removal in a biological filter is variable and depends on thewater temperature and contact time in the filter. Raw water DOC varies throughout the year and induces significant variations in the biological removal of organic matter due to large fluctuations in water temperature (3–25°C). The low water temperature during winter is a serious problem for DOC removal. This is the major disadvantage of GAC filtration because during this period, which can be quite long, the loss of efficiency has to be compensated by injecting PAC into the coagulation. The adsorption mechanism removes less than 10% of the total DOC.

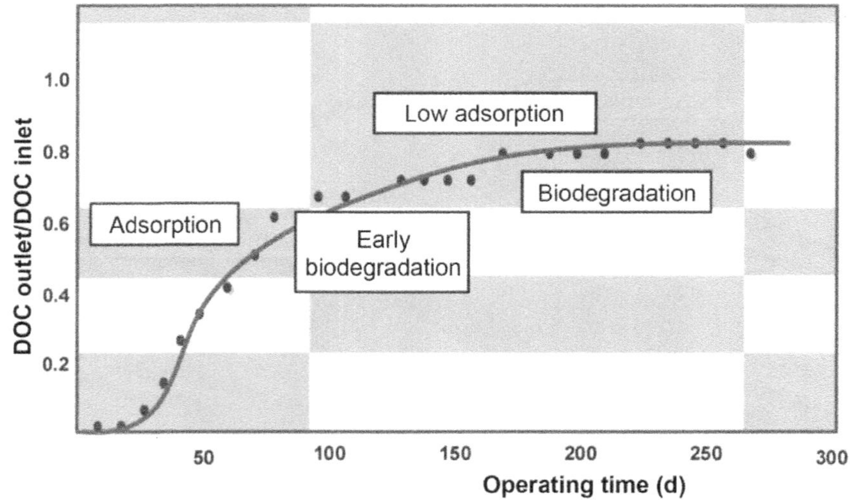

Figure 7.84. *Example of the operation of an O_3-GAC association (Veolia site, Australia). For a color version of this figure, see www.iste.co.uk/gaid/watertreatment2.zip*

Generally speaking, 20–35% of the DOC fraction can be converted to BDOC by ozonation. Therefore, biological filtration on GAC can achieve BDOC removal efficiencies of up to 90% at 20°C but no more than 30% at 8°C and less than 5% at 5°C.

GAC filter performances fluctuate significantly throughout the year. For example, during the cold season (<5°C), the BDOC of ozonated water is often less than 0.4 $mg \cdot L^{-1,}$ and the biological filtration removes very little.

During the intermediate period (5–12°C), BDOC can reach 0.6 $mg \cdot L^{-1,}$ and biological filtration removed 0.4 $mg \cdot L^{-1}$ of it. At higher temperatures, BDOC can reach 2.0 $mg \cdot L^{-1}$, and biological filtration typically provides 85–90% removal, with contact times of 20 min. For contact times of 10 min, removal in the warm period is approximately 65% but can be higher depending on the characteristics of the BDOC. Figure 7.85, obtained on surface water (in France) at 10°C, shows the influence of contact time on DOC and BDOC removal with Jacobi Picabiol activated carbon.

Unlike GAC filters operating solely on adsorption, which have a life between 3 and 12 months depending on the DOC concentration, biological GAC filters operate for several years. Figure 7.86 shows the evolution of the outlet BDOC after the ozonation and biological filtration stages on GAC. During cold periods, BDOC removal is slowed down.

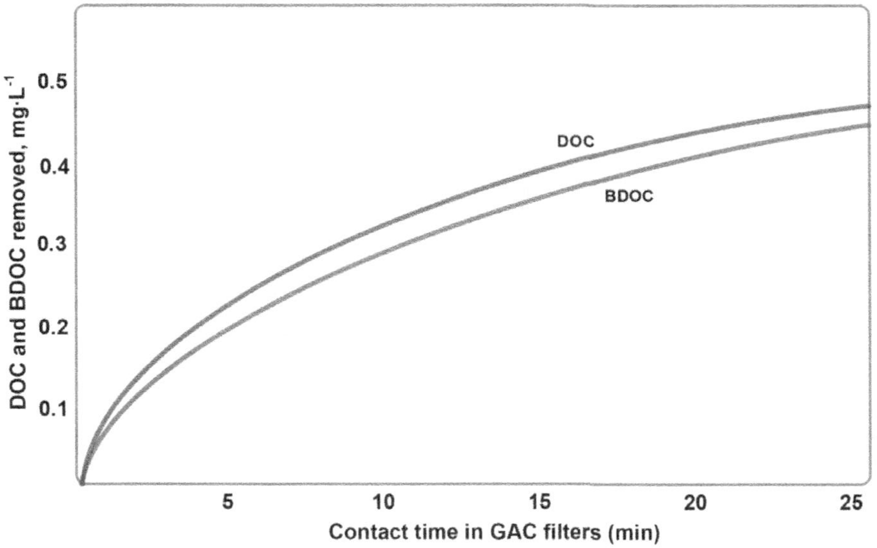

Figure 7.85. *Outlet DOC after biological GAC filter versus different contact times (Veolia site, France). For a color version of this figure, see www.iste.co.uk/gaid/watertreatment2.zip*

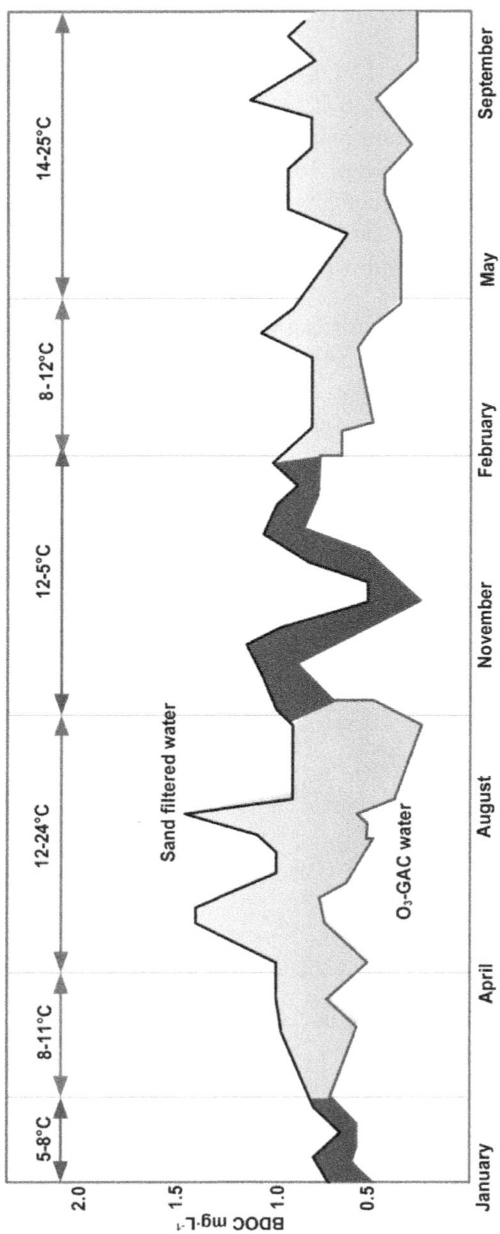

Figure 7.86. *Evolution of the BDOC at the O_3-GAC water outlet during one year of operation, with a contact time of 10 minutes on the GAC filter (Veolia site, France). For a color version of this figure, see www.iste.co.uk/gaid/watertreatment2.zip*

7.7.2.3. *Attachment of microorganisms and macroorganisms*

Only carbon pores with a radius greater than 500 nm are potentially available for bacterial attachment. The fraction of the GAC surface that is made up of these pores is only a small fraction of the total surface area, but due to the roughness and porosity of the material, this surface area is much larger than that of other materials, such as sand or anthracite. In addition, the hydrophobic nature and presence of functional groups strengthen the attachment of bacteria to the GAC surface.

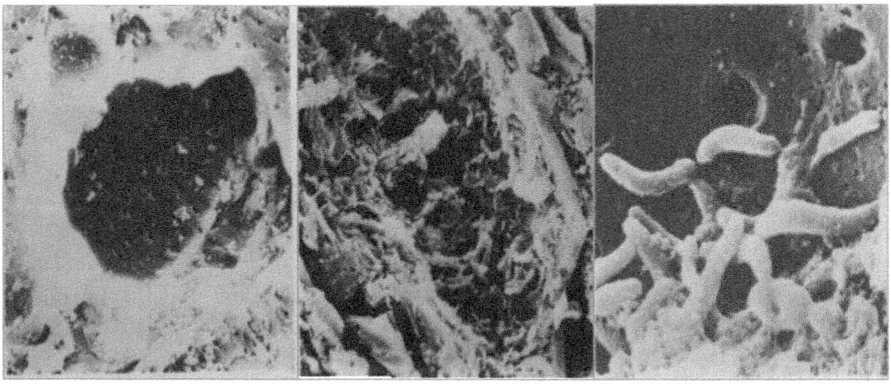

Figure 7.87. *Bacteria grown on a biological GAC filter*

Figure 7.87 shows electron microscope photographs of a pristine activated carbon (left photo) and a seeded activated carbon. The bacteria mainly grow in the macropores to which they have access. Various families of bacteria can be analyzed on biological filters, such as *Polaromonas, Lepadella, Achromobacter, Arthrobacter, Alcaligenes, Acinetobacter, Aeromonas, Bacillus, Caulobacter, Chromobacterium, Corynebacterium, Flavobacte-rium, Morax ella, Micrococcus, Microcyclus, Nitrosomonas, Nitrobacter*, and so on.

Macroorganisms may also grow (small worms, for example) on GAC filters. Ciliate and rotifer protozoa, copepods, oligochaeta and tardigrades are often detected in the liquid phase of the biological GAC filter and are seen as good indicators of significant organic carbon removal by bacteria. Nevertheless, they are not easily removed by backwashing, and it is important to carefully control and even optimize the cleaning conditions.

Moreover, since it is possible to find them in the filtered water after cleaning or even during normal operation, it is important to pay particular attention to them.

7.7.2.3.1. Bioregeneration

The long life of biological GAC filters is explained by a bioregeneration phenomenon of the internal adsorption capacity of the pores of the material. Two mechanisms are proposed:

– the first mechanism involves biodegradation following desorption of organic compounds. The desorption is caused by the concentration gradient between the surface of the GAC and the liquid phase of the filter;

– the second mechanism involves the excretion of exoenzymes by the microorganisms and their reactions with the substrates adsorbed within the internal porosity of the material.

7.7.2.3.2. Putting it into operation

When a biological GAC filter is put into operation, it takes several weeks for the biomass to develop. This development time for biomass depends on the water temperature and the concentration of BDOC. A GAC filter that is 2 m in height and operating at a filtration speed of 10 $m \cdot h^{-1}$ (i.e. a contact time of 12 minutes), and having a surface area of 10 m^2, requires between 5,000 and 6,500 m^3 of water$\cdot m^{-2}$ of filtering surface for an average biological colonization of the filter.

7.7.2.3.3. Cleaning the filters

The interozonation stage upstream of the filter induces a dissolved O_2 concentration that is always very high (supersaturation conditions). This implies that this parameter is never limiting for the aerobic bacteria present on the medium. For this reason, the design of high-water level filters (MD OTV Veolia type) is recommended to avoid embolisms due to the O_2 that can be trapped within the media.

Cleaning is performed to renew the mechanical retention capacity of the particles, to control the biomass and to avoid the proliferation of higher organisms on the surface of the filters, such as protozoa. The frequency of washes generally varies between once a day (warm periods), once or twice a week (intermediate periods) and once every 15–30 days in cold periods.

The biofilters are cleaned in an ascending flow according to the three-phase protocol (air, air + water, and water alone) in hot and intermediate periods. In cold periods, two phases (air, then water alone) can be sufficient. The loss of biomass generated by cleaning does not affect the biological efficiency of the filter because there is enough biomass left to restore efficiency as soon as the GAC filter is returned to service.

7.7.2.3.4. Fine activated carbon particles

Given that GAC is brittle, fine particles may be released. The amount of fine activated carbon particles released depends on the characteristics of the activated carbon. These particles can end up in the treated water when a filter is reimplemented after a wash or even after stopping the operation. Cleaning influences the attrition under the abrasive mechanical effect and implies a loss of material. It is estimated that between 2 and 5% of the activated carbon is lost depending on the GAC characteristics and the conditions and frequency of cleaning.

7.7.2.3.5. Control of the efficiency of biological GAC filters

Three monitoring parameters can be used.

DOC can be analyzed continuously and directly indicate the performance of the O_3-GAC stage. This makes it possible to anticipate the value of disinfection by-products. Regular BDOC analyses can also be performed in the laboratory to estimate the involvement of this parameter in the distribution system.

The UV 254 absorbance before and after ozonation is an indicator of the oxidation of HS (aromatic molecules). Its analysis at the GAC filter outlet is a good indicator of water quality after disinfection.

Using UV absorbance and DOC measurements, SUVA can be calculated.

7.7.2.4. *Sizing parameters*

The main design parameters are summarized in Table 7.21.

Ozonation
Number of reaction tanks: 3 Contact time: 6–8 min Dosage: 1–3 $g \cdot m^{-3}$
GAC
Origin: preferably plant-based Activation: chemical Pore distribution: mesoporous-macroporous Contact time: 12–20 min Height: 1.6–5 m Filtration rate: 8–15 $m \cdot h^{-1}$ Particle size: 1.0–1.2 mm

Acceptable turbidity (NTU):
<0.5 NTU for a filtration rate of 15 $m \cdot h^{-1}$
0.5–1 NTU for a filtration rate <10 $m \cdot h^{-1}$

Table 7.21. *Main sizing parameters*

It is recommended to allow sufficient degassing time in the volume of water above the material to avoid gas embolism problems. For this purpose, a sufficient water clearance (minimum 1.8 m) should be provided above the GAC.

7.8. Treatment of ion exchange resins

Ion exchange is a physicochemical process characterized by a reversible exchange of ions between the solid and liquid phases, and there is no permanent change in the structure of the solid. The solid phase is the ion exchange resin, while the liquid phase is the water solution. A resin contains fixed ionic groups that define the positive or negative charge of the resin functional group. The electroneutrality of a resin is achieved by a balance between the charge of the resin functional group and the ions whose charge is opposite to that of the resin functional group.

In drinking water treatment, the most common applications are softening and removal of polyvalent cations (Fe, Mn, Ba, Sr, Cd, Hg, Pb), as well as nitrates, arsenates, perchlorates, chromates, selenites, fluorides, boron and other anionic contaminants.

7.8.1. *Use of resins for NOM removal: mechanism of NOM removal*

NOM is negatively charged in the pH range of natural waters and can therefore be removed by anionic exchange resins.

Ion exchange is a balanced reaction that can be written as:

$$\alpha i^{\beta}_{(r)} + \beta i'^{\alpha}_{(s)} \underset{2}{\overset{1}{\rightleftarrows}} \beta i'^{\alpha}_{(r)} + \alpha i^{\beta}_{(s)}$$

with:

– i: counterion that is initially fixed on the resin;

– i': ion to be exchanged;

– α and β: valence of the ions concerned (exponent) or the stoichiometric coefficient (multiplier);

– (r) and (s): phase in which the ions are located; resin (r) or solution (s).

In direction 1 of the reaction, the counter ions are exchanged with respect to the ion to be removed. The resin is thus saturated in i' ion. In direction 2, the resin is regenerated to regain its initial form.

NOM retention on anionic resins involves two mechanisms (Figure 7.88):

– ion exchange between the NOM and the counterions on the resin, involving electrostatic interactions between the carboxyl group of the NOM and the counterion;

– surface adsorption of part of the NOM in the pores of the resin, involving Van der Waals interactions between the non-ionic (hydrophobic) part of the NOM and the polymeric skeleton of the resin.

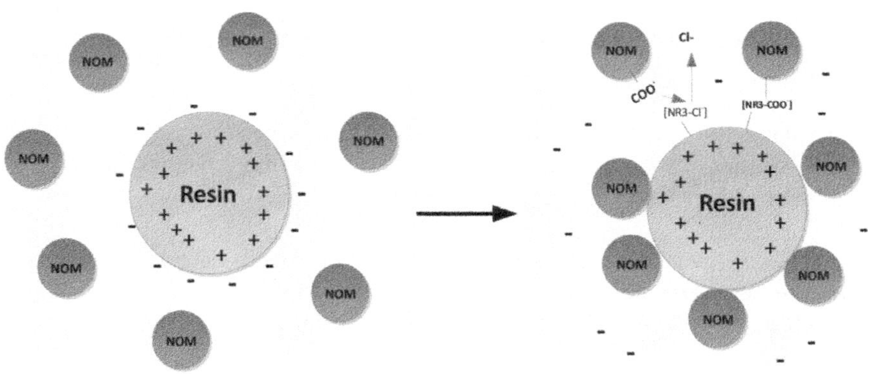

Figure 7.88. *Mechanism of NOM removal by the resin. For a color version of this figure, see www.iste.co.uk/gaid/watertreatment2.zip*

7.8.2. *Parameters affecting the performance of resins for the removal of NOM*

The operating parameters affecting the performance of resins with respect to NOM removal are contact time, pH, temperature and resin dosage (Table 7.22).

A contact time of less than 25 min is often insufficient to achieve good NOM removal. At high pH, ion exchange is the dominant mechanism for NOM removal by the resin, while at neutral pH, NOM removal is mostly achieved by adsorption.

However, there are resins on the market whose effectiveness is independent of pH and depends much more on the organic fractions characterizing the NOM present in water.

Operating parameters	Characteristics of the raw water	Characteristics of the resins
Contact time pH Temperature Dosage of resins	Different fractions constituting the NOM Presence of cations and anions	Porous structure Basicity

Table 7.22. *Operating parameters that affect the effectiveness of resins with respect to NOM removal*

Inorganic ions can enhance or inhibit NOM removal by ion exchange. The presence of competing ions, such as sulfates and nitrates, can negatively influence the effectiveness of the resin. Indeed, sulfates and nitrates are very well removed by resins. This leads to competition with the NOM and, consequently, to a lower removal rate.

Dissolved iron present in water can also have a negative impact on NOM removal by resins because it forms a clogging layer of iron oxide on the surface of the resin, blocking all access to the exchange and adsorption sites.

In contrast, the presence of calcium in water improves NOM removal because it forms calcium–NOM complexes that can, in turn, form bonds with the carboxyl groups of the NOM that is already retained on the resin.

The different organic fractions constituting the NOM do not react in the same way with the anion exchange resins. The nature, composition and quantity of NOM to be treated are factors that determine the efficiency of the resins. Resins preferentially remove high molecular weight humic molecules (HS and building blocks). The SUVA ratio of a water, defined as the ratio between the UV absorbance at 254 nm (m^{-1}) and the DOC concentration ($mg \cdot L^{-1}$), indicates the hydrophobic nature of the organic matter, that is, the presence of HS. The higher the SUVA value, the more hydrophobic the organic matter is, so it therefore contains more HS, leading to a better NOM removal performance by ion exchange resins. Strong-based resins remove more NOM than weak-based resins.

7.8.3. *Resin regeneration*

As previously mentioned, ion exchange consists of a succession of saturation and regeneration cycles of the resin.

Once the resin is saturated, in other words, the counter ions, i, have been exchanged with the ions to be removed, i', the resin is brought into contact with a regenerating solution that has a high concentration of counter ions (Cl^-, OH^-or HCO_3 ions for an anionic resin), which have a low affinity with the resin. The ions to be removed, i', are then exchanged with the counterions, i, allowing the resin to regain its initial shape and to be able to carry out a new exchange cycle.

Figure 7.89 shows the selectivity diagram of an anionic resin for NOM removal.

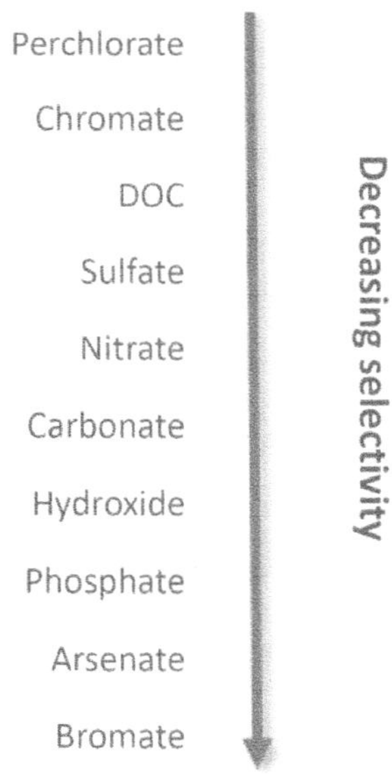

Figure 7.89. *Selectivity diagram of an anionic resin*

7.8.4. *The Opalix® process*

Veolia has developed the Opalix® process for the removal of NOM, color and various organic substances by ion exchange resins.

The resin used is A860S FM (Purolite), with the following characteristics:

– type: strong base anionic resin;

– polymeric structure: DVB cross-linked polyacrylic;

– functional group: quaternary ammonium;

– initial ionic form: chloride;

– median bead diameter: 368 μm;

– uniformity coefficient: 1.3;

– theoretical exchange capacity: 0.85 $eq \cdot L^{-1}$.

The Opalix® process is installed in the second stage of treatment, after a clarification stage (floatation or decantation). It is installed in the first stage if the turbidity is <5 NTU and does not contain algae.

7.8.4.1. *Operating principle of the Opalix® process*

The Opalix® process consists of two units: the treatment unit and the regeneration unit (Figure 7.91). These two units are complementary, and the performance of the treatment unit depends on the efficiency of the regeneration unit.

The treatment unit ensures:

– contact of the resin with the water to be treated in a contact tank;

– separation of the resin through a sieve (Hydrotech filter);

– recirculation of the resin from the wash water tank to the contact tank.

The regeneration unit ensures:

– continuous extraction of saturated resins from the wash water tank and their accumulation;

– regeneration of the saturated resin volumes at regular intervals with a brine solution;

– continuous injection of the regenerated resins in the contact tank.

To keep the amount of resin in the treatment unit constant, the saturated resin purge flow is equivalent to the supplementary regenerated resin flow.

The Opalix® process includes a reaction tank and a mechanical filter (Hydrotech filter) equipped with 100 μm filter panels and a wash water recovery tank.

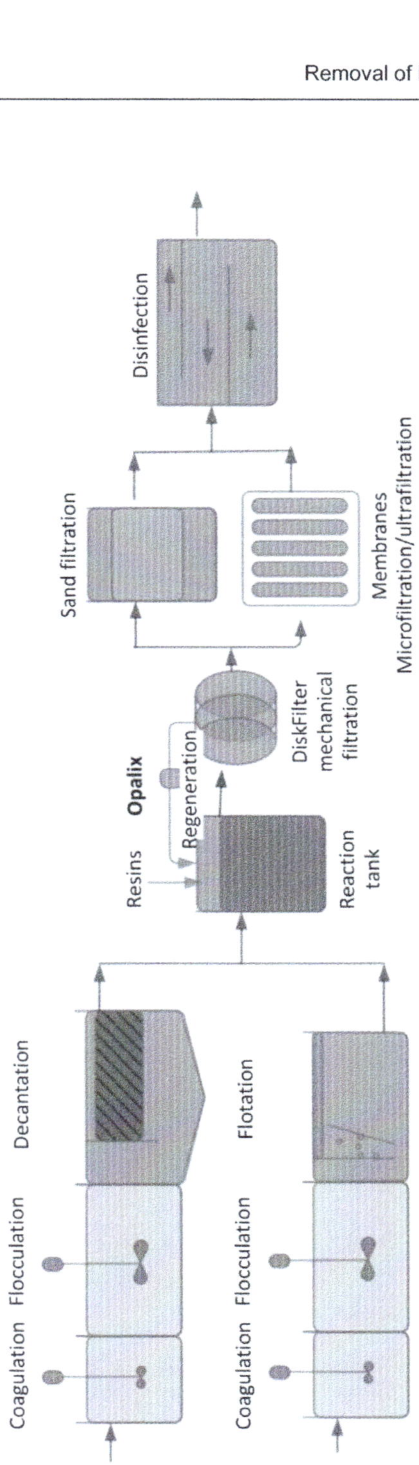

Figure 7.90. *Position of the Opalix® process in the treatment chain. For a color version of this figure, see www.iste.co.uk/gaid/watertreatment2.zip*

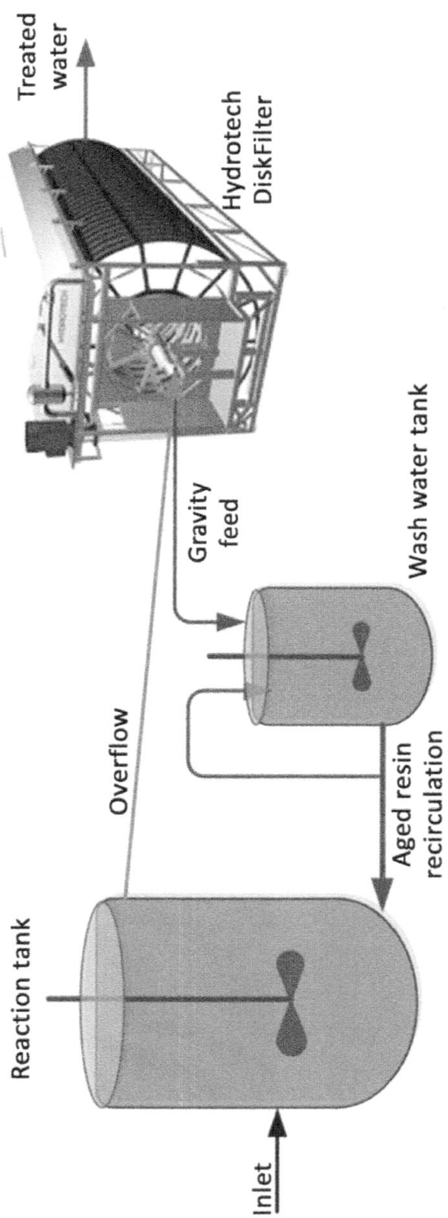

Figure 7.91. *Treatment unit equipment. For a color version of this figure, see www.iste.co.uk/gaid/watertreatment2.zip*

In the reaction tank, the water to be treated is brought into contact with the ion exchange resins (Figure 7.91). The contact time is between 30 and 45 min, depending on the DOC concentration of the water to be treated. The NOM is exchanged with chloride ions and fixed on resin beads. The Hydrotech filter is fed by overflow from the reaction tank and separates the treated water from the resins. The treated water is then sent to a sand filter or an MF or UF type of membrane.

The cleaning of the Hydrotech filter is combined with a drum and panel rotation. The rotation allows the resin that is accumulated at the bottom of the filter to be lifted up by gravity and evacuated with the wash water to the wash water tank.

The regeneration unit ensures the purging of saturated resin, the regeneration of this resin and the addition of regenerated resin. The treatment rate is the amount of regenerated resin used per volume of treated water.

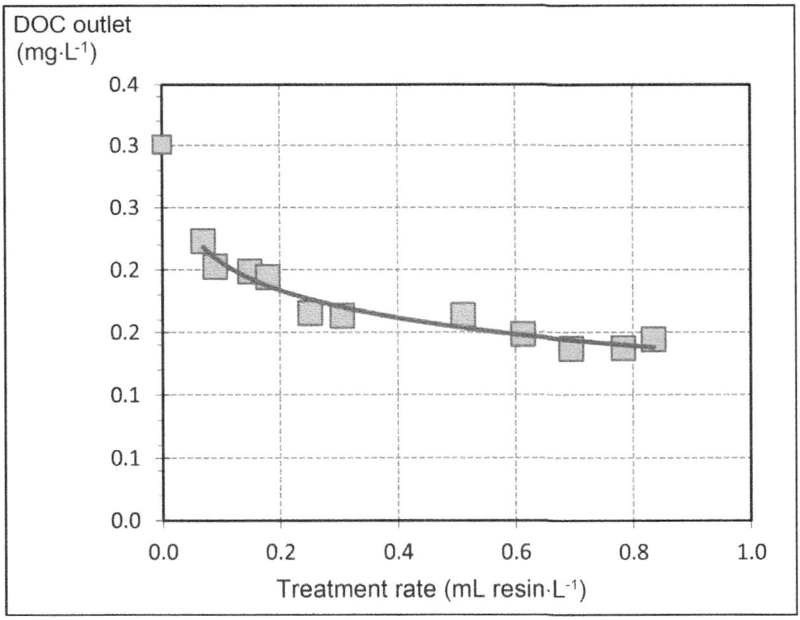

Figure 7.92. *Relationship between the treatment rate and the DOC concentration of treated water. For a color version of this figure, see www.iste.co.uk/gaid/watertreatment2.zip*

The Opalix® process has several advantages:

– the ability to absorb pollution peaks without the need to modify operating conditions;

– its positioning in the second treatment stage makes it possible to reduce the dosage of chemical products in the clarification stage, which then only has the objective of reaching a turbidity <5 NTU;

– a treatment of the eluates is possible.

7.8.4.2. *NOM removal performance*

Figure 7.92 shows the relation between the DOC concentration of treated water and the treatment rate for water with a DOC concentration of 3–4 mg·L^{-1} and a turbidity <5 NTU.

A DOC concentration in the treated water ≤2.0 mg·L^{-1} is guaranteed for the conditions given in Table 7.23.

DOC concentration of the water to be treated (mg·L^{-1})	Treatment rate (mL resin·L^{-1})	Resin concentration in the contact tank (mL resin·L^{-1})
≤ 4.0	≥2.0	40
4.0 < DOC ≤ 6.0	≥3.0	50
6.0 < DOC ≤ 10.0	≥4.0	60

Table 7.23. *Relationship between inlet DOC, treatment rate and resin concentration*

7.8.4.3. *Sizing parameters*

The operating parameters of the treatment and regeneration units are shown in Table 7.24.

Parameters	Threshold values
Resin–water contact time (min)	30–45
Filtration rate on Hydrotech filters (m·h^{-1})	7.5
R concentration in the wash water tank (mL resin·L^{-1})	400
Duration of the accumulation stage (min)	90
BV regeneration (L·L^{-1} resin)	2.5
BV rinse (L·L^{-1} resin)	1.5
Number of rinses	2
Volume of reused eluate (% of reuse)	80
Resin concentration of the slurry (mL resin·L^{-1})	400
Saturated brine concentration (g·L^{-1})	300
Brine concentration (g·L^{-1})	120

Table 7.24. *Operating parameters of a treatment unit*

7.9. NOM removal by high-pressure membranes

Two types of high-pressure membranes are currently used for conventional surface water treatment: nanofiltration (NF) and low-pressure reverse osmosis (LPRO).

The characteristics of these two types of membranes in relation to DOC are given in Table 7.25.

Parameters	Nanofiltration		Low-pressure reverse osmosis
	Nanofiltration	Tight nanofiltration	
Types	DOW: NF200, NF 270	NF70 and NF90	Dow XLE Toray TMH 20 A 400
Pore diameters (nm)	0.25–0.60	0.15–0.5	0.1–0.3
MWCO (Da)*	>200	>130	100
Removal of organic substances	88%	92%	95%
Transmembrane pressure (bar)	12–15	12–15	12–15

Table 7.25. *Characteristics of NF and OIBP membranes with respect to DOC (*for 90% removal)*

DOC removal by this type of membrane is very high, but, as NOM is considered responsible for membrane fouling, the DOC concentration at the inlet of the membrane must be examined very carefully. Indeed, the main limitation of membrane filtration is the fouling of the membranes, which generates a performance decrease, such as a decrease in the permeability and/or an increase in the longitudinal pressure.

It is mainly the hydrophilic neutral fraction, which comprises polysaccharides and proteins in macromolecular or colloidal form (biopolymers), that acts on fouling. It is assumed that once fouling is triggered by biopolymers, a decrease in electrostatic forces allows hydrophobic NOM to adsorb to the membranes, causing increased fouling.

The particle size distribution of NOM varies from source to source, but generally more than 50% of NOM molecules have a molecular weight <1 kDa, and 80% have a molecular weight <10 kDa. The optimal molecular weight threshold for NOM removal ranges from 0.2 to 0.3 kDa.

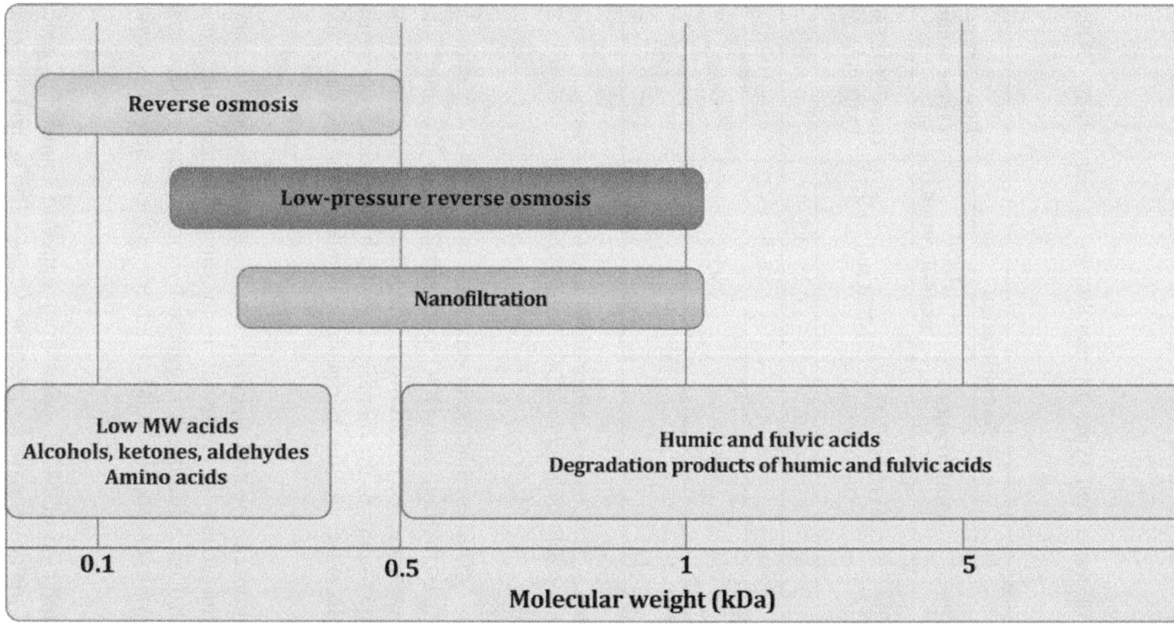

Figure 7.93. *Natural organic matter fractions removed by high-pressure membranes. For a color version of this figure, see www.iste.co.uk/gaid/watertreatment2.zip*

The hydrophobic fraction of NOM, as well as high molecular weight compounds, has a great impact on fouling and flow decline. Pretreatment with advanced coagulation is therefore necessary to prevent fouling-related consequences on high-pressure membranes.

Other factors affecting membrane fouling include membrane characteristics (membrane type, pore distribution, surface charge, hydrophobicity), operating conditions (flow, yield, pretreatment, backwash, chemical cleaning) and water quality (pH, ionic strength, particle concentration).

NOM can be efficiently removed by NF through a combination of diffusion, convection and electrostatic repulsion mechanisms. The transport mechanisms of NOM through high-pressure membranes (NF, LPRO, OI) depend on the operating conditions as well as on the solute and pore size.

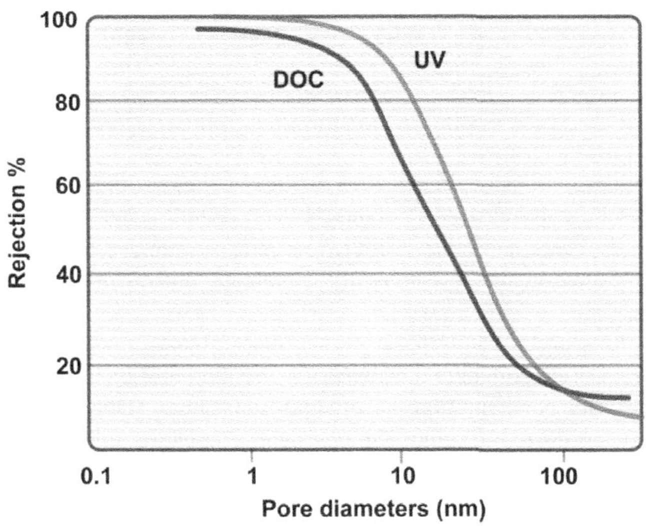

Figure 7.94. *Influence of pore diameter on DOC removal. For a color version of this figure, see www.iste.co.uk/gaid/watertreatment2.zip*

The TOC values shown in Figure 7.95 show that TOC removal by nanofiltration (DOW NF 200) is excellent, with TOC concentrations below 0.4 mg·L^{-1} for 95% of the values for a TOC water inlet varying between 1 and 3 mg·L^{-1}.

Due to this strong TOC removal, the chlorine injected into the drinking water decreased by 20–30%, and the chlorine concentration is also more stable in the system (less consumption).

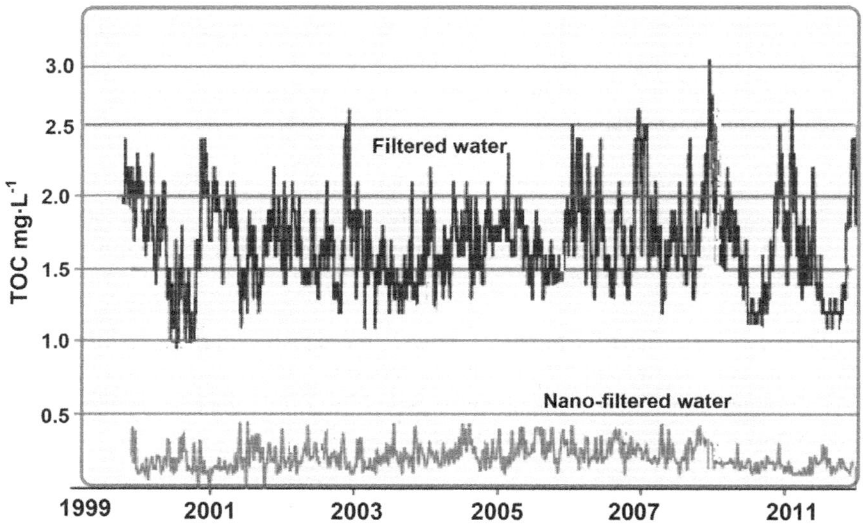

Figure 7.95. *TOC in water upstream and downstream of nanofiltration membranes. For a color version of this figure, see www.iste.co.uk/gaid/watertreatment2.zip*

Membranes must be cleaned periodically with chemical agents (such as biocides, surfactants, acids, bases and other chemical reagents) to remove organic and inorganic substances deposited on the surface of the membrane and to restore the membrane's characteristics.

The extent of pretreatment and its necessity are therefore dictated by the quality of the raw water. Its concentration in organic substances (DOC) requires that the pretreatment be optimized to remove the particles that cause turbidity and to partially reduce the DOC concentration.

In the majority of cases, the removal of organic substances expressed as UVA 254 absorbance is always higher than that of DOC, confirming the preferential release of high molecular weight aromatic/hydrophobic organic substances. Furthermore, the reduction in THM formation potential correlates better with DOC than with UVA, confirming that the main THM precursors are low molecular weight NOM with low UVA values.

The average values of the PTHM/DOC ratios are higher in the NF 270 membrane permeate than in the NF 90 membrane permeate. This means that small THM precursor molecules pass through the NF 270 membranes, whereas NF 90, which has a higher charge density, has the potential to reject small molecules better.

Nanofiltration and low-pressure reverse osmosis performance	
Ox; $KMnO_4$ mg $O_2 \cdot L^{-1}$	Approximately 90%
DOC ($mg \cdot L^{-1}$) (dissolved organic carbon)	>90%
BDOC ($mg \cdot L^{-1}$) (biodegradable dissolved organic carbon)	>90%
Chlorine demand for the system ($mgCl_2 \cdot L^{-1}$)	Reduction of 20–35%
PTHM ($\mu g \cdot L^{-1}$) Precursors of trihalomethanes formation	>90%

Table 7.26. *Nanofiltration and low-pressure reverse osmosis performances*

7.10. References

Archer, A.D. and Singer, P.D. (2006b). An evaluation of the relationship between SUVA and NOM coagulation using the ICR database. *J. Am. Water Works Assoc.*, 98(7), 110–123.

Amy, G.L., Minear, R.A., Cooper, W.J. (1987). Testing and validation of a multiple nonlinear regression model for predicting trihalomethane formation potential. *J. Water Research*, 21, 649–659.

Amy, G.L., Tan, L., Davis, M.K. (1991). The effects of ozonation and activated carbon adsorption on trihalomethane speciation. *J. Water Research*, 25(2), 191–202.

Baker, A. (2002). Spectrophotometric discrimination of river dissolved organic matter. *Hydrol. Process.*, 16, 3203–3213.

Bank, J. and Wilson, D. (2002). Low-cost solution for trihalomethanes compliance. *Journal of the Chartered Institution of Water and Environmental Management*, 16(4), 264–269.

Billica, J.A. and Gertig, K.R. (2000). Optimization of a coagulation process to treat high TOC, low alkalinity water and its impact on filtration performance. In *Proceedings of the AWWA Water Quality Technology Conference*. American Water Works Association, Denver/Salt Lake City.

Boere, J.A. (1992). Combined use of ozone and granular activated carbon (GAC) in potable water treatment: Effects on GAC quality after reactivation. *Ozone Sci. Eng.*, 14, 123–37.

Camel, V. and Bermond, A. (1998). The use of ozone and associated oxidation processes in drinking water treatment. *Water Research*, 32(11), 3208–3222.

Cancho, B. and Ventura, F. (2005). Optimization of methods for the determination of DBPs. *Global Nest: The International Journal*, 7(1), 16–36.

Carlson, K. and Amy, G. (2001). Ozone and biofiltration, optimization for multiple objectives. *J. Am. Water Works Assoc.*, 93(1), 88–98.

Carroll, T., King, S., Gray, R., Bolto, B.A., Booker, N.A. (2000). The fouling of microfiltration membranes by MON after coagulation treatment. *Water Research*, 34(11), 2861–2868.

Check, J.K. (2005). Characterization and removal of NOM from raw waters in coastal environments. PhD Thesis, Georgia Institute of Technology, Atlanta.

Cho, J., Amy, G., Pellegrino, J. (2000). Membrane filtration of natural organic matter: Factors and mechanisms affecting rejection and flux decline with charged ultrafiltration (UF) membrane. *J. Membr. Sci.*, 164(1–2), 89–110.

Chow, C.W.K., Fabris, R., Drikas, M. (2004). A rapid fractionation technique to characterise natural organic matter for the optimisation of water treatment processes. *J. Water Supply Res. Technol. Aqua*, 53(2), 85–92.

Chow, C., Fabris, R., Wilkinson, K., Fitzgerald, F., Drikas, M. (2006). Characterising NOM to assess treatability. *Water*, 33(2), 74–85.

Cipparone, L.A., Dielh, A.C., Speitel, G.E. (1997). Ozonation and BDOC removal: Effect on water quality. *J. Am. Water Works Assoc.*, 89(2), 84–97.

Crittenden, J.C., Trussell, R.R., Hand, D.W., Howe, K.J., Tchobanoglous, G. (2005). *Water Treatment: Principles and Design*. John Wiley & Sons, New York.

Croft, J. (2012). Natural organic matter characterization of different source and treated waters; implications for membrane fouling control. Master's Thesis, University of Waterloo, Waterloo.

Croué, J.P., Martin-Mousset, B., Lefebvre, E., Legube, B., Deguin, A. (1997). Impact des traitements de potabilisation sur le CODB et sur les substances humiques et non humiques. *Revue des sciences de l'eau/Journal of Water Sciences*, 10(1), 63–81.

Croué, J.P., Violleau, D., Bodaire, C., Legubre, B. (1999). Removal of hydrophobic and hydrophilic constituents by anion exchange resin. *J. Water Science Technology*, 40(9), 207–214.

Davis, C.C. and Edwards, M. (2014). Coagulation with hydrolyzing metal salts: Mechanisms and water quality impacts. *Crit. Rev. Environ. Sci. Technol.*, 44(4), 303–347.

De la Rubia, A., Rodriguez, M., León, V.M., Prats, D. (2008). Removal of natural organic matter and THM formation potential by ultra- and nanofiltration of surface water. *Water Research*, 42(3), 714–722.

Demeusy, A. (2007). Approche analytique et statistique de l'évolution de la qualité des eaux d'une retenue. Importance et origine des matières organiques dissoutes. PhD Thesis, University of Poitiers, Poitiers.

Dempsey, B.A. (1984). Removal of naturally occurring compounds by coagulation and sedimentation. *CRC Critical Reviews in Environmental Control*, 14(4), 311–331.

Déry, A. (2012). Comparaison à l'échelle réelle des performances d'un filtre conventionnel et d'un filtre au charbon activé en grains précédés d'une interozonation. Master's Thesis, Laval University, Quebec.

Dubowski, Y., Greenberg-Eitan, R., Rebhun, M. (2018). Removal of trihalomethane precursors by nanofiltration in low-SUVA drinking water. *Water*, 10, 1370–1381.

Edwards, M. (1997). Predicted DOC removal during enhanced coagulation. *J. Am. Water Works Assoc.*, 89(5), 78–89.

Edzwald, J.K. (1993). Coagulation in drinking water treatment: Particles, organics and coagulants. *Water Sci. Technol.*, 27(11), 21–35.

Edzwald, J.K. and Kaminski, G.K. (2009). A practical method for water plants to select coagulant dosing. *J. N. Engl. Water Works Assoc.*, 123(1), 15–31.

Edzwald, J.K. and Tobiason, J. (1999). Enhanced coagulation: US requirements and a broader view. *J. Water Science and Technology*, 40(9), 63–70.

Edzwald, J.K. and Van Benschoten, J.E. (1990). *Chemical Water and Waste Water Treatment*. Springer-Verlag, Berlin.

Eikebrokk, B. and Saltnes, T. (2001). Removal of natural organic matter (NOM) using different coagulants and lightweight expanded clay aggregate filters. *Water Sci. Technol. Water Supply*, 1(2), 131–140.

Eikebrokk, B., Juhna, T., Østerhus, S.W. (2006). Water treatment by enhanced coagulation – Operational status and optimization issues. Techneau. D 5.3.1, December.

Eikebrokk, B., Juhna, T., Melin, E., Østerhus, S.W. (2007). Water treatment by enhanced coagulation and oxonation-biofiltration: Intermediate report on operation optimization procedure and trials, Techneau. D 5.3.2A, December.

Espinoza, L.A. and Frimmel, F.H. (2009). A simple simulation of the degradation of natural organic matter in homogeneous and heterogeneous advanced oxidation processes. *Water Res.*, 43, 3902–3909.

Fabris, R., Chow, C.W.K., Drikas, M., Eikebrokk, B. (2008). Comparison of MON character in selected Australian and Norwegian drinking waters. *Water Research*, 42(15), 4188–4196.

Garcia, I. and Moreno, L. (2003). Drinking water treatment plant in Nicaragua: A short review. In *Proceeding of the IWA Asia-Pacific Regional Conference*. Bangkok.

Gibert, O., Lefèvre, B., Fernández, M., Bernat, X., Paraira, M., Pons, M. (2013). Fractionation and removal of dissolved organic carbon in a full-scale granular activated carbon filter used for drinking water production. *Water Research*, 47(8), 2821–2829.

Gilmore, P.L. and Summers, R.S. (2015). Organic matter removal via biological drinking water filters: Removal efficacy based on quantifiable system factors. In *Proceedings of the AWWA Water Quality Technology Conference*. American Water Works Association, Denver/Salt Lake City.

Gimbel, N.J.D. and Collins, M.R. (eds). *Recent Progress in Slow Sand and Alternative Biofiltration Progress*. IWA Publishing, London.

Gregory, R. and Edzwald, J.K. (2011). Sedimentation and flotation. In *Water Quality and Treatment*, Edzwald, J.K. (ed.). McGraw Hill, New York.

Hozalski, R.M., Bouwer, E.J., Goel, S. (1999). Removal of natural organic matter (NOM) from drinking water supplies by ozone-biofiltration. *Water Sci. Tech.*, 40(9), 157–163.

Huber, S.A., Balz, A., Abert, M., Pronk, W. (2011). Characterisation of aquatic humic and non-humic matter with size-exclusion chromatography – organic carbon detection – organic nitrogen detection (LC-OCD-OND). *Water Research*, 45(2), 879–885.

Humbert, H., Gallard, H., Suty, H., Croué, J-P. (2008). Natural organic matter (NOM) and pesticides removal using a combination of ion exchange resin and powdered activated carbon (PAC). *Water Research*, 42(6–7), 1635–1643.

Jacangelo, J.G., DeMarco, J., Owen, D.M., Randtke, S.J. (1995). Selected processes for removing NOM: An overview. *J. Am. Water Works Assoc.*, 87, 64–77.

Juhna, T. and Melin, E. (2006). Ozonisation and biofiltration in water treatment – Operational status and optimization issues. Techneau. D.5.3.1 B:80, December.

Kalbitz, K., Geyer, S., Geyer, W. (2000). A comparative characterization of dissolved organic matter by means of original aqueous samples and isolated humic substances. *Chemosphere*, 40, 1305–1312.

Kastl, G., Sathasivan, A., Fisher, I., Van Leeuwen, J. (2004). Modeling DOC removal by enhanced coagulation. *J. Am. Water Works Assoc.*, 96(2), 79–89.

Leenheer, J.A. and Crouè, J.-P. (2003). Characterizing aquatic dissolved organic matter. *Environ. Sci. Tech.*, 37(1), 18A–26A.

Marconnet, C., Djafer, M., Fournier, L., Metteil, X., Heim, V., Chagneau, G. (2011). Nanofiltration in Méry-sur-Oise drinking water treatment plant: Lessons learnt after 10 years. In *American Water Works Association Membrane Technology Conference Proceedings*.

Marhaba, T. and Pipada, N. (2000). Coagulation: Effectiveness in removing dissolved organic matter fractions. *J. Environmental Engineering Science*, 17(2), 107–115.

Martínez-Lladó, I. (2013a). Characterising biofilm development on granular activated carbon used for drinking water production. *Water Research*, 47(3), 1101–1110.

Matilainen, A., Lindqvist, N., Korhonen, S., Tuhkanen, T. (2002). Removal of MON in the different stages of the water treatment process. *Environ. Int.*, 28, 457–65.

Matilainen, A., Vieno, N., Tuhkanen, T. (2006). Efficiency of the activated carbon filtration in the natural organic matter removal. *Environ. Int.*, 32, 324–331.

Matilainen, A., Vepsäläinen, M., Sillanpää, M. (2010). Natural organic matter removal by coagulation during drinking water treatment: A review. *Advances in Colloid and Interface Science*, 159, 189–197.

McCreary, J.J. and Snoeyink, V.L. (1980). Characterization and activated carbon adsorption of several humic substances. *Water Research*, 14, 151–60.

Melin, E. and Odegaard, H. (2000). The effect of biofilter loading rate on the removal of organic ozonation by-products. *Water Research*, 34,18,4464.

Merlet, N., Prévost, M., Merlet, Y., Coallier, J. (1992). Enlèvement de la matière organique dans les filtres CAB removal of organic matter in BAC filters. *Revue des sciences de l'eau*, 5, 143–164.

Newcombe, G., Morrison, J., Hepplewhite, C. (2002). Simultaneous adsorption of MIB and MON onto activated carbon I. Characterization of system and MON adsorption. *Carbon*, 40, 2135–2146.

Nishijimaa, W. and Speitel, G.E. (2004). Fate of biodegradable dissolved organic carbon produced by ozonation on biological activated carbon. *Chemosphere*, 56(2), 113–119.

Nkambule, T.I., Krause, R.W.M., Haarhoff, J., Mamba, B.B. (2011). Treatability and characterization of natural organic matter (NOM) in South African waters using newly developed methods. *Phy. Chem. Earth*, 36, 1159–1166.

Ødegaard, H., Thorsen, T., Melin, E. (2000). Practical experiences from membrane filtration plants for humic substance removal. *Water Sci. Tech.*, 41(10/11), 33–41.

Ødegaard, H., Melin, E., Leiknes, T. (2006). Ozonation/biofiltration for treatment of humic surface water. In *Recent Progress in Slow Sand and Alternative Biofiltration Processes*, Gimbel, R., Graham, N.J.D., and Collins, M.R. (eds). IWA Publishing, London.

Ødegaard, H., Østerhus, S., Melin, E., Eikebrokk, B. (2010). NOM removal technologies Norwegian experiences. *Drink. Water Eng. Sci.*, 3, 1–9.

Omelia, C.R., Becker, W.C., Au, K.K. (1999). Removal of humic substances by coagulation. *J. Water Science and Technology*, 40(9), 47–54.

Owen, D.M., Amy, G.L., Chowdhury, Z.K. (1993). Characterisation of natural organic matter and its relationship to treatability. *J. Am. Water Works Assoc.*, 85, 72–78.

Owen, D.M., Amy, G.L., Chowdhury, Z.K., Paode, R., McCoy, G., Viscosil, K. (1995). NOM characterization and treatability. *J. Am. Water Works Assoc.*, 87, 46–53.

Pernitsky, D. and Edzwald, J.K. (2003). Solubility of polyaluminium coagulants. *Journal of Water Supply: Research and Technology. Aqua*, 52, 395–406.

Plourde-Lescelleur, F., Papineau, I., Carrière, A., Gadbois, A., Barbeau, B. (2015). NOM removal: Evaluating five process alternatives to alum coagulation. *J. Water Supply Res. Technol. Aqua*, 64(3), 278–289.

Servais, P., Anzil, A., Ventresque, C. (1989). Simple method for determination of biodegradable dissolved organic carbon in water. *App. Env. Microbio.*, 55(10), 2732–2734.

Servais, P., Prévost, M., Laurent, P., Joret, J.C., Summers, S., Hamsch, B., Ventresque, C. (2005). Biodegradable organic matter in drinking water treatment. In *Biodegradable Organic Matter in Drinking Water Treatment and Distribution*, Prévost, M. (ed.). AWWA, Denver, CO.

Sharpe, E.L., Jarvis, P., Parsons, S.A., Jefferson, B. (2006). Impact of fractional character on the coagulation of NOM. *Colloids and Surfaces A: Physicochem. Eng. Aspects*, 286, 104–111.

Siddiqui, M.S., Amy, G.L., Murphy, B.D. (1997). Ozone enhanced removal of natural organic matter from drinking water sources. *Water Research*, 31(12), 3098–3106.

Singer, P.C. (2006). Effect of SUVA and enhanced coagulation on removal of TOX precursors. *J. Am. Water Works Assoc.*, 98(8), 97–107.

Van der Kooij, D., Visser, A., Hijnen, W.A.M. (1982). Determining the concentration of easily assimilable organic carbon in drinking water. *J. Am. Water Works Assoc.*, 74, 540–545.

Ventresque, C., Gisclon, V., Bablon, G., Chagneau, G. (2000). An outstanding feat of modern technology: The Méry-sur-Oise nanofiltration treatment plant (340,000 m3/d). *Desalination*, 131, 1–16.

Volk, C., Renner, C., Roche, P., Paillard, H., Joret, J.C. (1993). Effects of ozone on the production of biodegradable dissolved organic carbon (BDOC) during water treatment. *Ozone: Science & Eng.*, 15, 389–404.

Volk, C., Bell, K., Ibrahim, E., Verges, D., Amy, G., Lechevallier, M. (2000). Impact of enhanced and optimized coagulation on removal of organic matter and its biodegradable fraction in drinking water. *Water Research*, 34(12), 3247–3257.

Wang, G.-S., Kang, S.-F., Yang, H.-J., Pai, S.-Y., Chen, H.-W. (2002). Removal of dissolved natural organic matter from source water with alum coagulation. *Env. Technol.*, 23, 1415–1423.

Yavich, A.A., Lee, K.-H., Chen, K.-C., Pape, L., Masten, S.J. (2004). Evaluation of biodegradability of NOM after ozonation. *Water Research*, 38, 2839–2846.

8

Filtration

Filtration is a technology used to remove suspended particles that are present in the water to be treated. These particles either come from the raw water or are generated in the water treatment chain: this is, for example, the case of metal hydroxide flocs produced by a coagulation–flocculation stage, insoluble carbonate precipitates resulting from chemical softening, or manganese and/or iron oxide precipitates formed during preoxidation.

Granular media filtration can be done by gravity (rapid gravity filtration) or under pressure (pressure filtration).

The flow through the rapid filters is due to gravity or to an imposed pressure (2–5 bar). The basic mechanisms of particle removal are fundamentally the same in both gravity and pressure modes.

The filtration device is implemented:

– after coagulation, flocculation and decantation or flotation;

– in direct filtration with coagulation and sometimes flocculation.

Filters are either made up of a single media of material (single-media filter) or two to three materials (dual media or multimedia filter).

The quality of the decanted water varies on average between 0.5 and 3 NTU, with some peaks between 5 and 10 NTU. Filtration on filter material is then used as a refining step and makes it possible to deliver water <0.5 NTU. Filter systems should be operated in such a way as to reduce turbidity levels as much as possible. Filtration is thus an important barrier for the removal of particles causing turbidity, as well as some microorganisms. To achieve the most effective inactivation of pathogens during disinfection, the filtered water must be at a low turbidity. The most important point with respect to turbidity is to reduce it to as low a level as reasonably achievable and to minimize its fluctuation. Filtration systems should be designed to be reliable enough to provide continuous log removal credits for pathogens.

8.1. Rapid filters and very high-rate filters (TGV)

Sand filtration is the most widely used process in drinking water plants. When installed behind a clarification process, the load of suspended solids (SS) that is applied to the filters is reduced and allows for longer filtration cycles and high-quality effluents at higher flow rates.

The main filtration mechanism of rapid filters is surface retention, in other words, in the first third of the material's height. Typically, sand filters have a cutoff of 10 μm, that is, they can cohesively remove 90–95% of particles ≥10 μm in size. This is what allows turbidity to be less than 0.4 NTU for an inlet turbidity of 3–5 NTU.

The shape of the filter material also affects the retention capacity, in that angular particles are preferable to rounded particles. Regardless of the choice of supporting material, the size of the material is limited to between 0.6 and 1.2 mm for most common cases. Figure 8.1 shows the design of a sand filter with the ground supporting the filtering material, material height, water production circuit and backwash circuit.

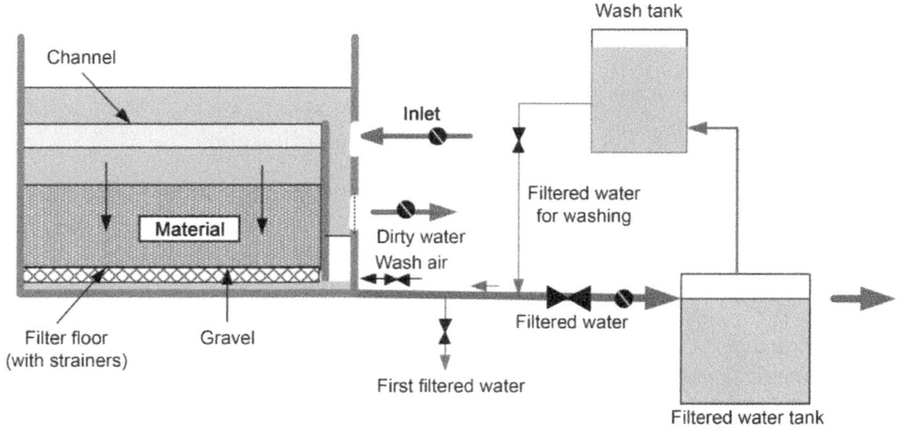

Figure 8.1. *Operating principle of a filter. For a color version of this figure, see www.iste.co.uk/gaid/watertreatment2.zip*

Over the last two decades, Veolia has developed very high-rate filters (TGV filters), the advantage of which is that they operate at high rates (>15 $m \cdot h^{-1}$) and have a high material height that allows for the deep diffusion of particles, ensuring greater retention capacity and longer cycle times (Figure 8.2).

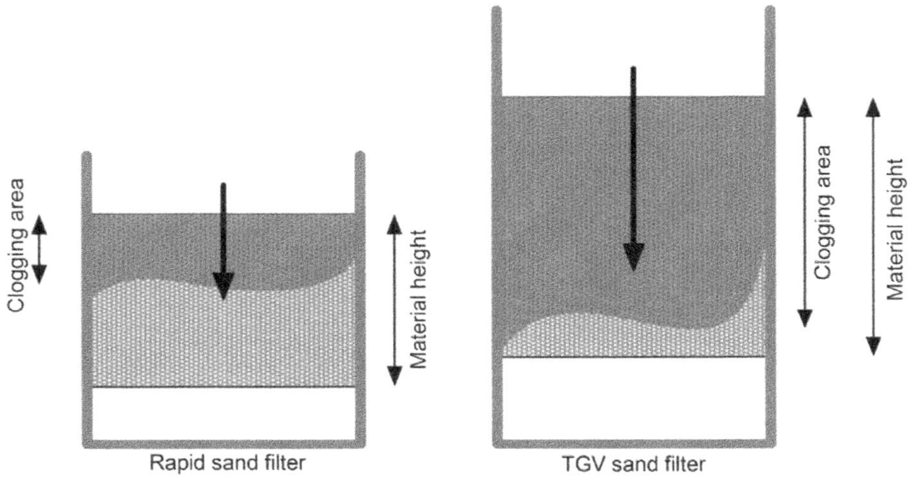

Figure 8.2. *Clogging depth between a rapid sand filter and a TGV sand filter. For a color version of this figure, see www.iste.co.uk/gaid/watertreatment2.zip*

8.2. Multimedia filters

Multimedia filters consist of two or three media of different densities and grain sizes, which filter deeper and operate at higher rates than conventional rapid sand filters. These are known as dual media or three-media filters (Figure 8.3). The gravel layer is not considered because it does not act as a filtering barrier due to its too high particle size (5–7 mm).

For example, in dual media anthracite-sand or pumice-sand filters, the lower density of the anthracite or pumice allows for a larger grain size to be used so that after backwashing, the larger size anthracite is on top of the smaller effective-sized sand. In this way, the water is first filtered through the media with the larger effective size (ES) and then reaches the media with the smaller ES, thus making better use of the total filtration height.

Multimedia filters have longer cycle times because their retention capacity is greater due to a large retention area.

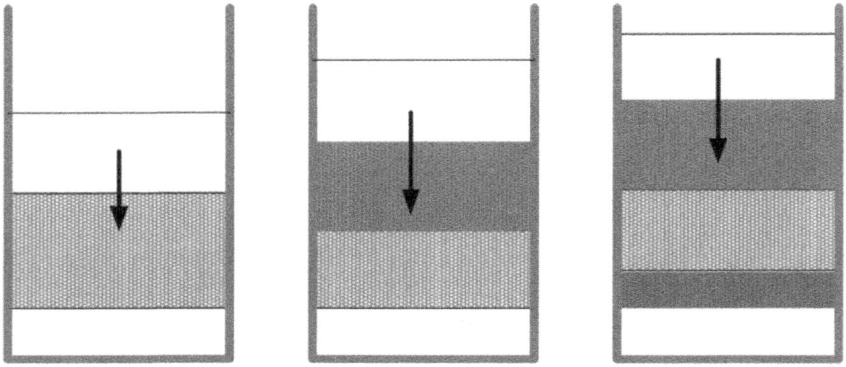

Figure 8.3. *Diagram of the different types of filters: single-media, dual media and three-media. For a color version of this figure, see www.iste.co.uk/gaid/watertreatment2.zip*

When the filter is clean, the particles are retained by the transport mechanisms. During the filter fouling period, the particles that are already retained serve as a collector for other particles, which increases the removal rate. Subsequently, this rate stabilizes for a certain period of time but ultimately degrades, inducing a breakthrough of the filter through an increase in turbidity at the outlet.

These filters operate at filtration rates of up to 25 $m \cdot h^{-1}$, above which pressure drops develop too quickly. This results in cycle times that are too short.

Dual media anthracite-sand or pumice-sand filters have a cutoff of 5 μm; in other words, they can consistently remove 95–98% of particles that have a size ≥5 μm. This is due to the maturation of the upper filter layer, which is able to retain small particles containing turbidity, as well as due to an effective sand size (lower layer), which is often smaller than that of a conventional sand filter.

In addition to materials such as sand and anthracite, which are often used, the filters can be filled with pumice, activated carbon, manganese dioxide or similar, garnet and limestone.

In contrast to the rapid sand filter, for which clogging occurs in a layer of approximately 30–40 cm (resulting in an exponential increase in pressure loss), the dual media filter allows for optimal use of the filtering layer.

Due to its particle size, the top layer of anthracite works according to the principle of volume filtration, that is, it retains particles over its entire depth. The height of the larger material (top layer) acts as a reservoir and storage for the retained particles. The sand layer (bottom layer) ensures the completion of the filtration by retaining the small particles. It also helps to ensure protection against the risk of breakthroughs.

As a result of the deep particle retention, the pressure drop of a dual media filter increases more slowly than that of a sand filter.

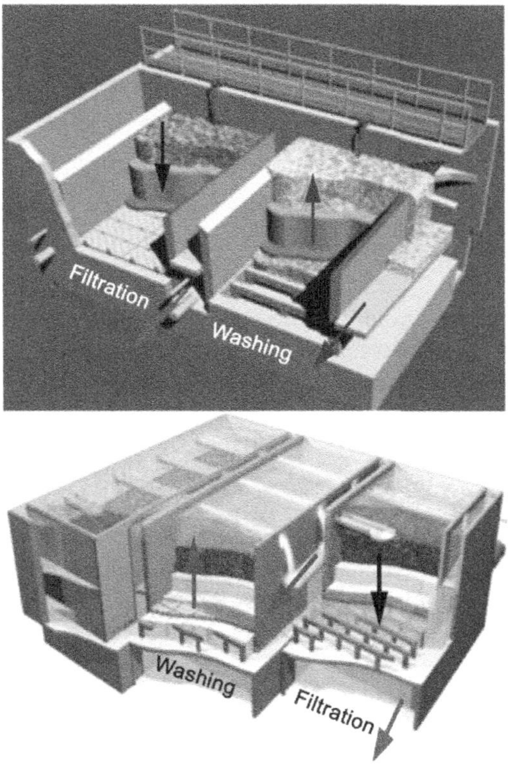

Figure 8.4. *Single-media filter and dual media filter, in both filtration mode and wash mode. For a color version of this figure, see www.iste.co.uk/gaid/watertreatment2.zip*

In some cases, dual media filtration allows sand filter rehabilitation. Thus, from an initial sand layer of 1 m, only 40 cm of sand can be kept, and the remaining 60 cm is replaced by anthracite. As a result, the filtration capacity of the plant is increased without building new filters.

Table 8.1 shows the maximum acceptable turbidities for different filter types, as well as the average cycle time. The type of gravity filter is selected based on the particle load and treatment objectives.

Types of filters	Maximum acceptable turbidity NTU	Average cycle duration H
Rapid sand filter	5	24
TGV sand filter	12–15	24–48
Rapid dual media filter	15–20	24
Dual media TGV filter	20–25	24–48

Table 8.1. *Maximum acceptable turbidities for different types of filters (TGV = very high-rate filters; H = hours)*

Knowing that water turbidity varies regularly throughout the year in a range between 1 and 5 NTU or between 5 and 25 NTU, these cycle times are shown for a turbidity output ≤0.4 NTU.

8.3. Direct filtration

In the direct filtration of raw water feeding a sand filter, the turbidity removal performance is not very efficient if no reagent (coagulant, flocculant) is added beforehand, and the colloidal particles pass through the filter without being retained in the filter mass.

Direct filtration implies that there is no decantation stage in the treatment chain. However, the water must be of fairly good quality because the only physical treatment barrier is filtration. Direct filtration is a relatively simple and economically attractive filtration process. It consists of in-line or tank coagulation, followed by in-line or tank flocculation. The flocculated water goes directly to filtration (Figure 8.5). The quality of the source water can be very variable in a system that includes decantation and filtration, whereas in a direct filtration system, the turbidity of the raw water must be less than 5 NTU for a single-media filter and 15 NTU for a dual media filter.

Flocculation improves the quality of the filtered water, increases the cycle time and reduces the elevation rate of the pressure drop. The flocculation tank is usually designed for contact times of 10 min, with a velocity gradient (G) of up to 100 s^{-1}. Low dosages of coagulant are often injected, and flocculation is either carried out in the water volume above the medium (with or without a flocculation adjuvant) or by means of a reaction tank with the addition of a flocculation adjuvant. The contact times in the tank are 10–20 min long, depending on the water temperature.

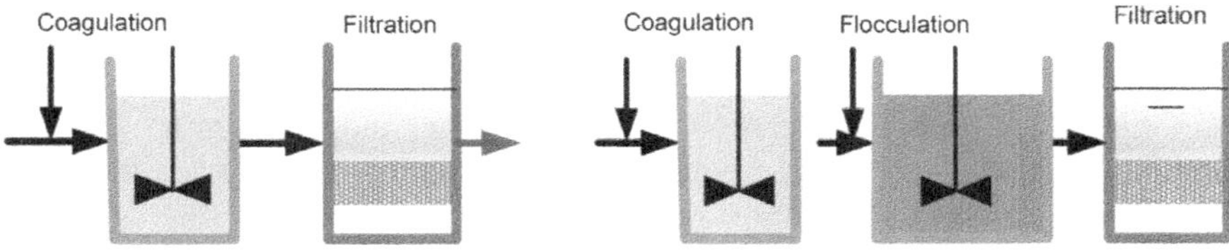

Figure 8.5. *Direct filtration with coagulation, with or without flocculation. For a color version of this figure, see www.iste.co.uk/gaid/watertreatment2.zip*

The presence of algae (>2,000 cells per mL) compromises the filtration cycles of sand filters and dual media filters.

The main disadvantage is that degradation in the feed water quality no longer allows the same production rates to be met, as the chemical reagent dosages increase, and the contact times and cycle times can be too short.

In direct filtration, instrumentation is very important for the automatic control of the treatment. Monitoring water resources is also imperative to warn the operators of variations and to allow them to adjust the reagents.

The operating parameters need to be controlled with respect to pH, temperature and alkalinity, as they affect the mechanisms of coagulation and flocculation. pH affects coagulant speciation and solubility. The temperature also has an impact on flocculation because, on the one hand, it modifies the viscosity of water, and, on the other hand, it is involved in the kinetics of hydrolysis and precipitation. This is why, at low temperatures, coagulants such as polyaluminum chloride are preferred over aluminum and iron salts because $AlCl_3$ is already hydrolyzed, and temperature has less effect on the coagulation–flocculation process.

Direct filtration may require the use of a flocculation adjuvant (polymer), which is dosed in very small quantities (<0.1 mg·L^{-1}) and which allows a good distribution of particles throughout the depth of the filter. This type of reagent promotes particle agglomeration by facilitating the formation of a chemical or ionic bond with the filtering material. For example, a filter is capable of removing 98% of particles ≥2 µm with the addition of reagents.

Direct filtration can be preceded by chemical preoxidation (for iron and manganese) or combined with granular activated carbon (GAC) adsorption for the removal of organic substances and micropollutants.

The parameters used for the sizing of direct filtration are identical to those of conventional filtration: filtration rate, height and type of material (x), pressure drop and retention capacity.

Filtration rates vary from 6 to 8 $m \cdot h^{-1}$ for an F sand filter, 14–16 $m \cdot h^{-1}$ for an F-TGV filter, 8–12 $m \cdot h^{-1}$ for a fast DC dual media filter and 14–18 $m \cdot h^{-1}$ for a DC dual media TGV filter. For each case, the height of the material must be considered to avoid overly high-rate filtration quickly breaking through the filter due to a material height that is too low.

Since there is no intermediate decantation or flotation in direct filtration, all particles present in the raw water end up on the filter. This affects the operation of these filters if the recommended feed water qualities for this process are changed. Raw water turbidity exceeding its acceptable limit for direct filtration results in a reduction in operating time and a rapid increase in pressure loss. For algae such as diatoms beyond the tolerated concentration threshold, wash water volumes are induced that may require up to 20–25% of the filtered water.

When the turbidity of the water is low throughout the year, direct filtration is the most beneficial solution compared to a decanter or a flotation device, both in terms of investment cost and operating cost.

Direct dual media filtration can achieve *Giardia* cyst and *Cryptosporidium* oocyst removal credits of 2 log (99%), provided that the filtered water turbidity is consistently less than 0.2 NTU, and at 0.1 NTU 80% of the time.

8.4. Pressurized filters

Pressure filtration is carried out in closed cylindrical tanks, generally working at an absolute pressure of approximately 3–5 bar. The filtration speeds used are 8–15 $m \cdot h^{-1}$ and can be increased to 20 or 25 $m \cdot h^{-1}$ in certain applications, such as iron biological treatment.

Whether it is a single-media, dual media or three-media filter, the same materials used in gravity filtration are used in pressure filters.

In the case of groundwater applications, pressure filters often save the pumping of treated water to reservoirs. In fact, the pressure provided by the dewatering pump is maintained in the treatment chain and allows the water to be pumped back to the distribution system. The mechanisms of particle retention and pressure loss development, as well as the pretreatment, are similar to those of gravity filtration in open filters.

The filter is backwashed after being put under atmospheric pressure. The washing sequences are similar to those of a gravity flow sand filter.

The main advantage of a pressurized filter is that it has a very high maximum available load, which allows for longer cycle times.

The large surface pressure filters are horizontal, and the distribution of the input water is carried out in a formless manner over the entire cross-section of the filter to ensure that there are no preferential channels for the water. The size of the sand grains is between 0.85 and 1.5 mm, and the height of the grains varies between 0.9 and 2 mm.

Pressure filters are backwashed in three phases: air, air/water and rinse. They are similar to those of a gravity flow sand filter.

During backwash, there is a 20–30% volume expansion through a backwash flow for a single-layer sand filter of 25 $m^3 \cdot m^{-2} \cdot h^{-1}$. The air/water backwash times must be well controlled because the water clearance above the medium is lower than that in gravity filters.

The backwash process often results in sand loss, although it is not significantly noticeable, thus requiring a periodic top-up of sand.

8.5. Filtration mechanisms

8.5.1. *General principle*

Filtration can be defined as the interception of a particle by a grain of material known as a *collector*. The transport of suspended particles in the immediate vicinity of the filter grains depends on physico-hydraulic factors, including flow rate, size and shape of the filter medium, filter configuration, filter porosity and finally, the physical characteristics of the suspended particles (shape, size, density).

The filtration process has two stages. The first is called the "initial stage", which is the introduction of particles into a clean filter and where retention occurs on clean material grains. The second stage is a transitional stage, in which particle deposition occurs on the surfaces of material grains that are beginning to be covered by previously deposited particles. During this period, particle retention can improve or degrade if the chemical characteristics of the particles have been altered in terms of electrical charges. During this transitory stage, the retention improvement is explained by a maturation of the filter, which corresponds to the filter being filled by particles, reducing its cutoff threshold, and ensuring a refining of the filtered water.

At the end of the cycle, the saturation of the filtering material and the hydrodynamic conditions tend to degrade the retention and, on the contrary, induce filter breakthrough by particle leakage.

Particles and material grains have surface charges that may be different. If they have opposite charges, filter retention conditions are favorable due to their mutual attraction.

If their charges have the same sign, filter retention conditions are unfavorable due to repulsive interactions. Generally, materials such as sand have negative charges, while particles (flocs) have positive charges after their destabilization by the coagulant. This explains why flocs are well retained during filtration.

On the other hand, if no chemical destabilization of the particles is achieved and these particles have negative signs, retention in the filter is difficult.

8.5.2. *Mechanisms involved in filtration*

Particle retention therefore depends on surface forces (Van der Waals forces) between the particles and the material grains when the distance between them is in the nanometer range.

The particles are generally smaller than the interstices. As a result, they follow the water circulation currents; many of them should not touch the surface of the material grain and should end up in the filtered water.

However, due to the various transport mechanisms of the particles, they move through the circulation currents and end up next to the surface of a grain. At this point, adhesion forces must be present for the particles to be retained on the material grains or on the particles that have already been deposited. If the deposited particles are carried away by the circulating water, a detachment mechanism is at work.

8.5.2.1. *The transport mechanism*

The transport mechanism implies that particles must be transported close to the filter material grains, and various transport mechanisms are involved in repelling particles from the filter grains. Particle retention by rapid filtration involves complex mechanisms that include decantation, interception, diffusion and attachment. Detachment is an additional phenomenon to hydrodynamic effects.

Four transport mechanisms are involved, depending on the size of the particles.

If particles large enough to be filtered reach the surface of the filter, they form a cake that tends to clog the filter quickly. This surface clogging also occurs if the particle concentration is too high. This phenomenon is known as early clogging.

When a large particle (5–25 µm) approaches a collector (material grain), it tends to decant since its density is greater than that of water (particle B in Figure 8.6). The efficiency of the mechanical decantation mechanism depends on the type and size of the particles. This is the process by which a particle is deflected from the aerodynamic path through a gravitational effect, resulting from the difference between its weight and buoyancy.

This effect is characterized by N_G, which is the ratio of the Stokes velocity of the particle/filtration rate:

$$\mathrm{NG} = \frac{(\rho p - \rho w)dp^2 g}{18\mu \mathrm{v}}$$

with:

– ρp: particle density ($kg \cdot m^{-3}$);

– ρw: water density ($kg \cdot m^{-3}$);

– dp: particle diameter (m);

– g: acceleration of gravity ($m \cdot s^{-2}$);

– v: filtration rate ($m \cdot s^{-1}$);

– µ: dynamic viscosity of the water ($kg \cdot m^{-1} \cdot s^{-1}$).

A granular filter has an action that can be considered a shallow decantation unit. For a medium-sized particle (1 and 5 µm), if its flow line passes at a distance equal to or less than its radius near the collector, it is intercepted by the collector (particle A in Figure 8.6). This is the interception mechanism that occurs when a particle carried by the pore stream approaches an existing filter grain or deposits cling to it. This is characterized by the particle diameter to grain diameter ratio (dp/dg).

The flow lines of water approaching a grain of filter material must diverge as the flow passes around it. If the particles have sufficient inertia, they maintain their trajectory, which allows them to collide with the material grain. This is the mechanism of inertia. This action is characterized by N_I:

$$\mathrm{NI} = \frac{\rho p dp^2 v}{18\mu dg}$$

Small particles (<1 μm) are influenced by Brownian motion; in other words, they follow a random trajectory that deviates from the flow lines due to diffusion. The smaller the particle and the higher the temperature, the greater the diffusion will be (particle C in Figure 8.6). This is the mechanism of diffusion. The diffusion resulting from these movements increases the probability that the particles will encounter the filtering mass and increases the probability of contact between the particles. It is therefore a transport mechanism in filtration. This mechanism is expressed in terms of the Peclet number, which is the ratio of transfer by convection to transfer by diffusion:

$$\text{NPe} = \frac{dpv}{D}$$

D is the Stokes–Einstein diffusion coefficient.

As there are often particles of a mineral or biological nature that are very small in size and are not well stopped by the two previous mechanisms, it is the attachment mechanism that completes the overall efficiency of the filter.

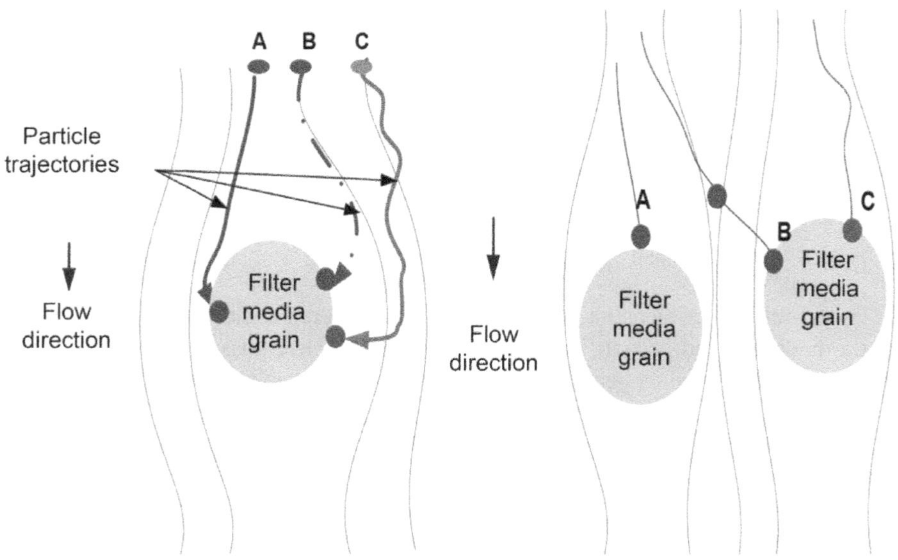

Figure 8.6. *Particle retention mechanisms on a filtering material. For a color version of this figure, see www.iste.co.uk/gaid/watertreatment2.zip*

The particle-to-grain attachment mechanism depends on physicochemical mechanisms. These mechanisms may involve electrostatic interactions, chemical bridging or accelerated adsorption, all of which are influenced by pH, ionic composition of the water, floc volume, nature and dosage of coagulants and composition of the feed water. The mechanisms of attachment can be considered analogous to those of colloidal particle destabilization. It is therefore the process by which particles stick to the surface of the filter media grains or to particles that are already present. The method of particle attachment is similar to the flocculation process. A small particle in the vicinity of a solid surface is subject to electrical attraction or repulsion (depending on the charges of the particle) and to the attraction caused by Van der Waals forces. It is also subject to hydraulic forces resulting from water movement. Since the coagulation–flocculation stages allowed an optimal destabilization of the colloids, the particles are adequately retained by filtration.

As the particles are retained, the porosity of the filter medium decreases. As a result, the interstitial velocities increase and cause shear forces to be exerted on the deposits. When these shear forces reach the order of magnitude of the adhesion forces, the particles detach and migrate to deeper layers of the filter (in the case of downward filtration). This is the detachment mechanism.

Under the action of hydrodynamic forces caused by the flow of water through the media with increasing pressure drops, the structure constituting the particle deposit is partially degraded. Some of the previously adhered particles, which are less strongly bound to the others, detach from the grains. The space left by the detached particles then restarts the cycle of attachment mechanisms for new particles brought by the filter feed water.

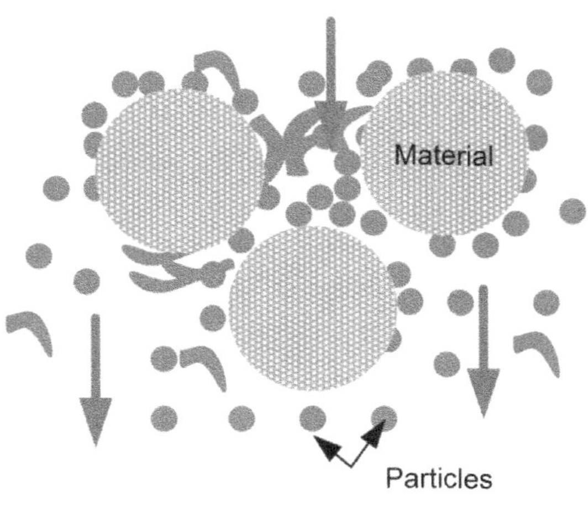

Figure 8.7. *Detachment mechanism under the action of shear forces. For a color version of this figure, see www.iste.co.uk/gaid/watertreatment2.zip*

This detachment phenomenon can also be caused by an increase in the filtration speed or by a sudden and large addition of particles in suspension. Detachment can be the cause of particle leakage into the filtered water.

8.6. Implementation parameters

8.6.1. *Materials*

The mechanical filtration materials used must be insoluble, non-brittle and must not release any substance that could alter the qualities of the water.

The materials depend on the type of filter used (single-layer, dual media or multimedia layer) and the purpose of the treatment. The various materials used in drinking water are quartz sand, gravel, garnet, anthracite, pumice, expanded shale and pozzolan (Biodagene volcanic stones). Other materials, such as Filtralite (expanded clay), are available on the market.

For each filter material, the physical characteristics determined in the laboratory or indicated by the supplier are:

– the ES d_{10} (opening allowing the passage of 10% (w/w) of particles) in mm;

– the uniformity coefficient (UC) (d_{60}/d_{10});

– the maximum grain size d_{90} in mm;

– the solid density (absolute), dry density (apparent in the air), wet density (saturated in water) and porosity ($g \cdot cm^{-3}$);

– the hardness (mhos);

– geometric shape (spherical, ellipsoidal, granular, cylindrical, irregular, angular);

– composition, which is determined, depending on the case, by the level of purity (silica content, etc.), solubility in acid or water, ash content (activated carbon), etc.

The equivalent unit is the mesh (US or BS with the following conversion: mesh size = 15/size in mm).

Table 8.2 summarizes the characteristics of various materials used in drinking water treatment.

Materials	Physical characteristics						
	ES d_{10} mm	UC (d_{60}/d_{10})	d_{90} mm	Mhos Hardness	Porosity ε	Absolute density $g{\cdot}cm^{-3}$	Apparent density $g{\cdot}cm^{-3}$
Gravel	5	<2	9.00	–	–	2.65	1.65
Fine sand 1	0.55–0.65	≤1.3	0.85	–	0.45	2.65	1.50
Fine sand 2	0.85	≤1.3	1.25	–	0.45	2.65	1.50
Medium sand	1.0	≤1.3	1.55	–	0.45	2.65	1.50
Large sand	1.35	≤1.45	2.10	–	0.45	2.65	1.50
Anthracite	1.4	<1.50	2.50	>4	0.55	1.45–1.60	0.85–0.95
Pumice	1.2–1.3	<1.45	2.00	>4	0.50–0.52	0.8–1.2	1.30–1.55
Schist or (Biodagene)	2.5	<1.7	5.00	>4.5	0.40	1.60	0.96
Fine GAC	0.7	<1.4	0.80	>3	0.5–0.7	1.20–1.35	0.80
Medium GAC	1.0	<1.4	1.80	>3	0.5–0.7	1.20–1.35	0.80
Large GAC	1.2	<1.4	1.25	>3	0.5–0.7	1.20–1.35	0.80
Limestone	3.0	≤1.4	4.00	–	0.56	2.70	1.19
Manganese dioxide	0.8	<1.5	3.00	>4–4.5	0.50	3.80	1.90
Garnet	0.4	<1.4	0.60	≥7	0.45–0.50	3.80–4.10	1.90

Table 8.2. *Characteristics of various filtration materials*

The choice of material depends on the water quality and upstream processes, the quality objectives of the filtered water, the cleaning method, the filtration rate and the duration of the filtration cycles. The sand used in filtration is a natural material based on silica from rivers, natural deposits, dunes, etc. It is available in two forms:

– so-called rolled sand (alluvial or loose rock), whose rounded shape has been acquired by erosion, is a natural sand sifted after washing and drying. They are found in rivers or near or on shallow seabeds;

– crushed sand is obtained by cutting and crushing massive rocks, which gives it angular shapes favorable to the retention of particles during filtration. This sand has a low sphericity but a better porosity.

Anthracite is a very hard mineral coal characterized by its high carbon content of up to 95%. It comes in the form of hard, angular grains. It is positioned in a top layer with a grain size large enough to hold particles of different sizes.

Light materials such as pumice (light gray volcanic lava made very light by the numerous gas bubbles trapped during cooling), pozzuolana (volcanic rock with a brown–black to brick-red honeycomb structure) and expanded clay or shale are materials used in filtration and as a support for biological development.

Garnet is a very dense and chemically inert material. It is used as the bottom layer of (more than 85%) dual media or multimedia filters. It occupies approximately 15% of the total filter height, and the ES can be as low as 0.25–0.35 mm.

Activated carbon is a material with a high carbon content obtained by calcination and activation of wood, coal, peat or coconut. Its high internal specific surface allows it to be a very good adsorbent of organic substances and other micropollutants. It can also be used as a biological support (see Chapters 7 and 9).

Manganese dioxide is mainly composed of manganese dioxide and other mineral constituents. It is characterized by high catalytic activity. The concentration of manganese dioxide varies from 75 to 85%. It is positioned in the lower layer due to its very high density and is used for the fixation of heavy metals.

The support gravel is not a filter material. It is used to retain the media of the filter above it, to improve water distribution during washing and to protect the nozzles during washing. Its characteristics depend on those of the filtering floor, as well as the material.

For all these materials, an ES d_{10} is defined, which represents the size of the opening through which 10% of the material weight passes.

The diameter d_{50} is obtained by the relation: $d_{50} = ES \cdot UC^{0.834}$.

The diameter d_{90} is given by: $d_{90} = d_{10} \cdot UC^{1.67}$.

The UC is the ratio between the diameter that allows 60% of the particles to pass and the diameter that allows 10% to pass. Therefore,

$$UC = \frac{d60}{d10}$$

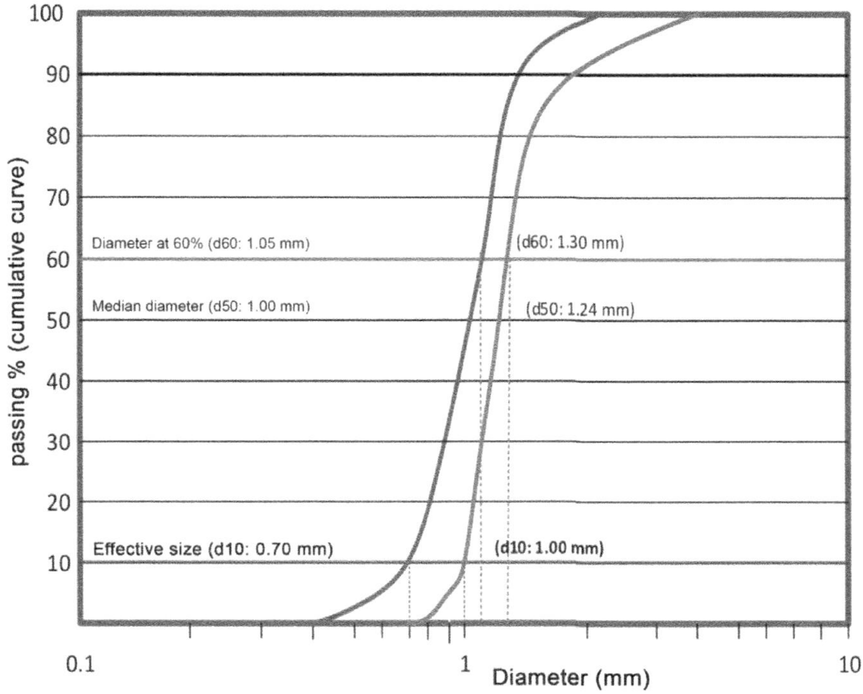

Figure 8.8. *Cumulative distribution curve of the diameters of sand grains (d_{10}: 0.70 mm and d_{10}: 1.00 mm). For a color version of this figure, see www.iste.co.uk/gaid/watertreatment2.zip*

Here, d_{60} is the size of the opening through which 60% of the sand passes. This coefficient gives an indication of the particle size homogeneity of the filtering mass. UC values should be less than 1.6 and are generally between 1.3 and 1.4. Lower UC values make the material expensive because a high proportion of fine material is lost during washing, and higher values would reduce the porosity of the material.

The porosity ε is defined as the fraction of void volume to total material volume. It therefore represents the proportion of void volume existing between stacked material grains. Hence, the volume occupied by the material is $(1 - \varepsilon)$. Depending on the nature of the porous medium, ε varies between values close to 0 and 1. Thus, the higher ε is, the lower the resistance to water flow. The porosity of a material does not depend on its size but on its shape and geometric characteristics alone. The porosity of sand, for example, is between 42 and 45%, while that of anthracite is 50 and 55%.

Weight loss by fire at 450°C should be $<2\%$, and weight loss in hydrochloric acid washing (20% v/v for 24 h at 20°C) should be $<2\%$.

Density (often called apparent density) is the ratio of the mass of the material divided by the total volume occupied. It is therefore the mass of dry bulk sand occupying the unit volume. It is between 1,400 and 1,600 $kg \cdot m^{-3}$ for the rolled silica-limestone sands.

In reality, the apparent density is a dimensionless parameter because it is the ratio of the density to the mass of the same volume of water (at a given T°C). It is equal to one thousandth of the density. The absolute density of the material, in other words, the volume of excluded voids, is much higher. It is, for example, equal to 2,500–2,600 $kg \cdot m^{-3}$ for the same rolled silica-limestone sands.

The size of the grains determines the specific surface of a material, which is the sum of all the external surfaces of the particles contained in a unit of mass. The specific surface of a material As (m^{-1}) is the surface developed by a grain reduced to the volume of this grain.

For a sphere, it is equal to:

$$As = \frac{6}{d}$$

The sphericity factor of a grain measures the similarity of its shape to that of a sphere:

$$\psi = \frac{Grain\ volume}{Volume\ of\ the\ circumscribed\ sphere}$$

If the grain is nearly spherical, it almost fills the entire circumscribed sphere. If, instead, the grain is more like a thin disk, it occupies a small piece inside the sphere. Sphericity increases as the grain occupies more space. The sphericity factor is 1 when the two volumes coincide. If a grain is not spherical, it has more surface area for the same volume and, consequently, offers more resistance to transport.

The hardness of the material is defined by its resistance to attrition for its application in filtration. The measurement of hardness according to the Mohs scale depends on the ability of a material to scratch a reference material or to be scratched by a reference material. The measurement is based on a scale of 0–10, the latter value corresponding to the diamond that can only be scratched by another diamond. The hardness should preferably be >3 to ensure that washing operations do not produce fine activated carbon particles.

8.6.2. *Material height/d_{10} ratio*

The objective of this ratio is to maintain a constant quality of filtered water for a low particle input concentration and filtration rate.

This H/d_{10} ratio allows for consistent particle removal performance as the filtration height and material grain diameter increase.

A ratio between the height of the filter layer (H) and the diameter of the grains (d_{10}) must be respected.

For a sand filter, the H/d_{10} ratio must be greater than 1,000. For dual media and three media filters, this ratio must be greater than 1,200.

For multimedia filters, the L/d ratio is calculated for each material and then added to obtain the total filter ratio. Gravel is not counted. We thus have the following ratios for dual media and three-media filters:

$$\frac{(H1)}{(d10)1} + \frac{(H2)}{(d10)2} > 1{,}200; \frac{(H1)}{(d10)1} + \frac{(H2)}{(d10)2} + \frac{(H3)}{(d10)3} > 1{,}200$$

For depth filtration that is applicable to very high-rate filters with an ES greater than 1.35 mm, the H/d_{10} ratio should be greater than 1,400.

Table 8.3 shows some typical cases used in the treatment plants.

Filter type	Materials	Height (mm)	d_{10} (mm)	UC	H/d_{10}	Recommended value
Rapid sand filter	Sand	900	0.85	1.3	1,058	>1,000
TGV sand filter	Sand	2,000	1.35	<1.4	1,481	>1,400
Rapid dual media filter	Anthracite Sand	600 600	1.20 0.85	<1.4	1,205	>1,200
Dual media TGV filter	Anthracite Sand	1,500 500	1.40 0.65	<1.4	1,840	>1,400
Rapid three-media filter	Anthracite Sand MnO_2	500 500 300	1.20 0.85 1.00	<1.4	1,304	>1,200

Table 8.3. *Examples of H/d_{10} ratio for different types of filters*

An ideal filter would give a turbidity–diameter relationship of the material, such that the turbidity reduction would evolve directly with the particle size along a linear line. The large particles would be in the upper part, and the fine particles that would allow refining would be in the lower part. However, due to the reclassification of particles during a wash, this configuration is not possible in the presence of a single material (Figure 8.10).

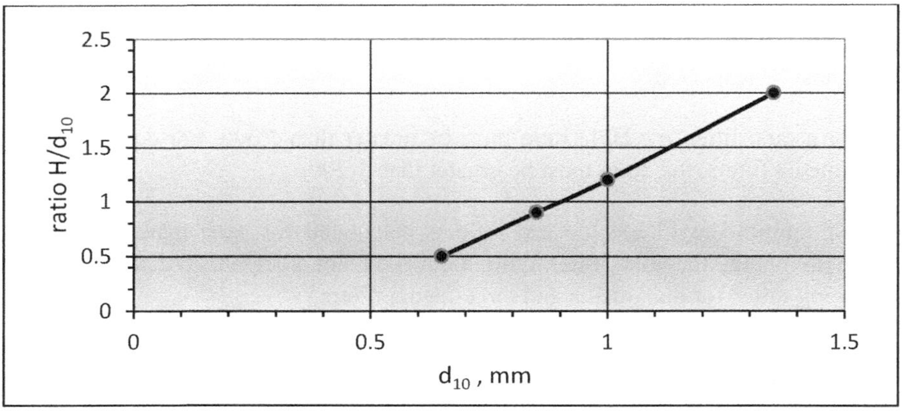

Figure 8.9. *H/d_{10} ratio for sand filters*

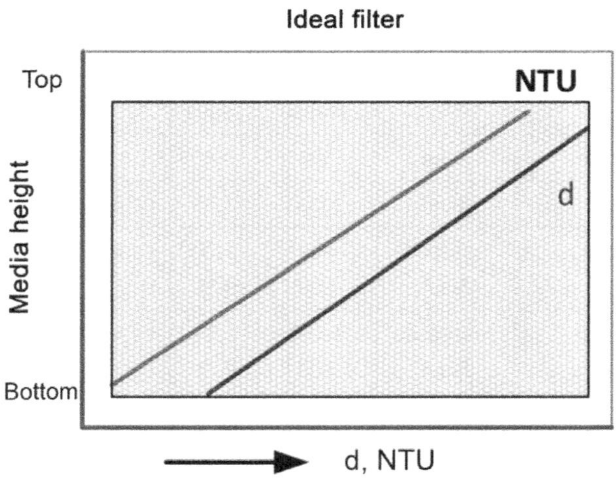

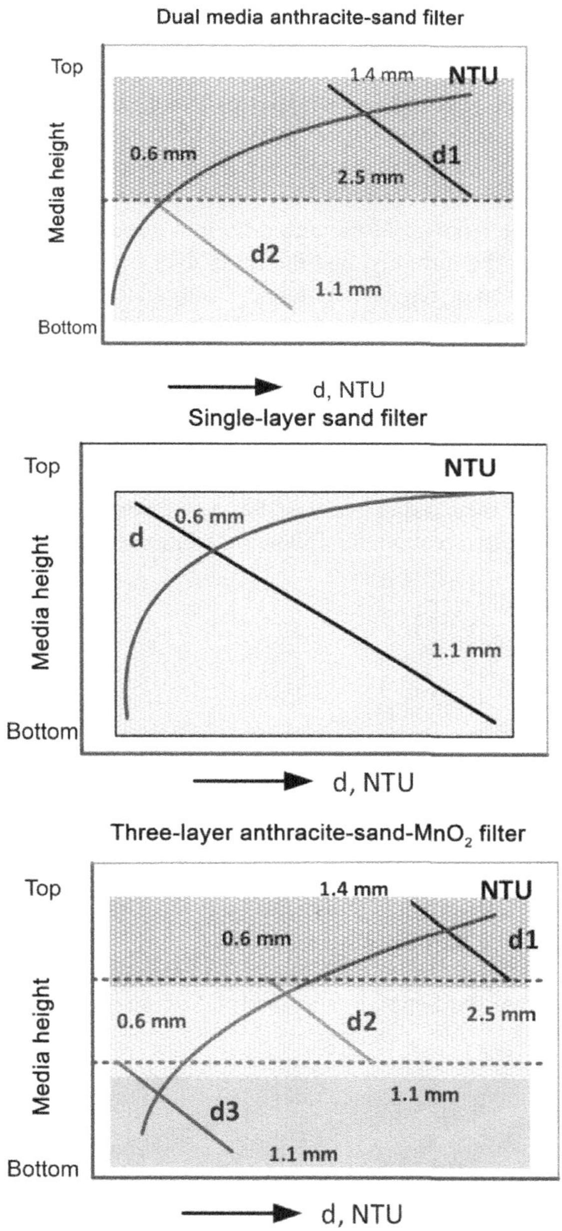

Figure 8.10. *Turbidity evolution profiles according to material diameter. For a color version of this figure, see www.iste.co.uk/gaid/watertreatment2.zip*

In fact, in a single-media filter in which the distribution of the material grains is based on a classification where the fine particles are in the upper part and the coarser ones are in the lower part of the filter, the reduction of turbidity follows a curve that shows that it is better removed in the upper part (with the finer grains) than in the lower part. However, this curve can be corrected by using materials with a UC < 1.3. For two- and three-media filters, two parameters come into play: the diameter of the grains and their respective densities. In a dual media filter, the sand plays the polishing role for the removal of the turbidity. In a three-media filter, the sand acts as a final barrier for the turbidity, and the third layer provides additional polishing. Manganese dioxide can be replaced by garnet. Anthracite can be replaced by pumice or activated carbon.

8.6.3. *ES ratio*

In dual media and three-media filters, the classification of the materials after washing depends on their respective densities and dimensions. To avoid mixing of materials or poor reclassification, the following relationship must be observed:

$$\frac{d1}{d2} = \left(\frac{\rho 2 - \rho water}{\rho 1 - \rho water}\right)^{0.667}$$

with:

– d_1 and d_2: d_{10} of each material;

– ρ_1 and ρ_2: densities of the two materials;

– ρ_{water}: density of water.

For example, in a dual media anthracite-sand filter, the ES of the anthracite can be deduced by knowing the size of the sand.

If $d_2 = d_{10}$ of sand: 0.85 mm, $\rho_2 = \rho$ sand = 2.65 g·cm^{-3} and $\rho_1 = \rho$ anthracite = 1.55 g·cm^{-3}.

For ρwater = 1 g·cm^{-3}, we deduce that:

$$d1 = d2\left(\frac{\rho 2 - \rho water}{\rho 1 - \rho water}\right)^{0.667} = 0.85\left(\frac{2.65 - 1}{1.55 - 1}\right)^{0.667} = 1.76 \text{ mm}$$

To select the ESs of the materials in the multimedia filters, the ratio between the ESs and the densities of the media and water is used.

Index 1 refers to the finest material, and index 2 refers to the coarsest material. For example, in a dual media sand-anthracite filter, the d_1/d_2 ratio corresponds to the ratio between the ES of the sand and that of the anthracite.

This equation is used to avoid mixing of the layers during washing. Indeed, if the ratio between the ES of the two materials is insufficient, the lower layer material could mix with the upper layer material when the filter bed is expanding.

A ratio of 1.75–3.0 is recommended for dual media anthracite/sand filters.

For a three-media filter, the same provisions are made by considering the two upper layers and then the two lower layers. The material layer is used twice.

8.7. Sizing parameters: filtration rate and material height

On all Veolia filters (Filtraflo®), filtration takes place in a downward direction at a speed equal to:

– $v = Q/S$

– v in $m \cdot h^{-1}$, Q in $m^3 \cdot h^{-1}$, S in m^2.

The application of a high filtration rate implies a higher material height, a larger effective grain size and a significant increase in the washing speed.

The use of a larger grain size and higher filtration rate results in an increase in the depth of SS penetration, without a decrease in cycle time or filtered water quality.

For rapid single-media filters, the filtration rates are between 6 and 8 $m \cdot h^{-1}$. It depends on the quality of the water to be treated, the sand grain size and the height of the material.

For a conventional filtration carried out after flocculation-decantation, the nominal rate generally adopted for a 1 m layer of sand (ES = 0.8–1 mm and UC < 1.3) is approximately 6 $m \cdot h^{-1}$ on N filters and 8 $m \cdot h^{-1}$ on N-1 filters.

For TGV single-media filters (very high-rate filters), filtration rates are between 14 and 16 $m \cdot h^{-1}$ on N filters. It depends on the quality of the water to be treated, the sand grain size and the height of the material.

For filtration carried out after flocculation–decantation, the nominal speed generally adopted for a 2 m layer of sand (ES = 1.35 mm and UC < 1.4) is approximately 15 $m \cdot h^{-1}$ on N filters and 20 $m \cdot h^{-1}$ on N-1 filters.

The empirical relation between the filtration rate and material height (m) is as follows:

$$v \cdot (m \cdot h^{-1}) = 6.8169\ H - 0.0057$$

Higher filtration rates present two contradictory effects: first, the clean bed pressure drop is increased. Thus, for a hydraulic profile with available pressure loss, a lower clogging pressure loss is available to maintain the filtration cycle. This shortens the cycle time as the maximum available pressure drop is reached. In addition, for a given material, the pressure drop increases when the filtration rate is greatly increased because the particle load is higher, and these particles are therefore retained more quickly.

Figure 8.11 shows the effect of the filtration rate on water production for equivalent cycle times. Water losses are not taken into account.

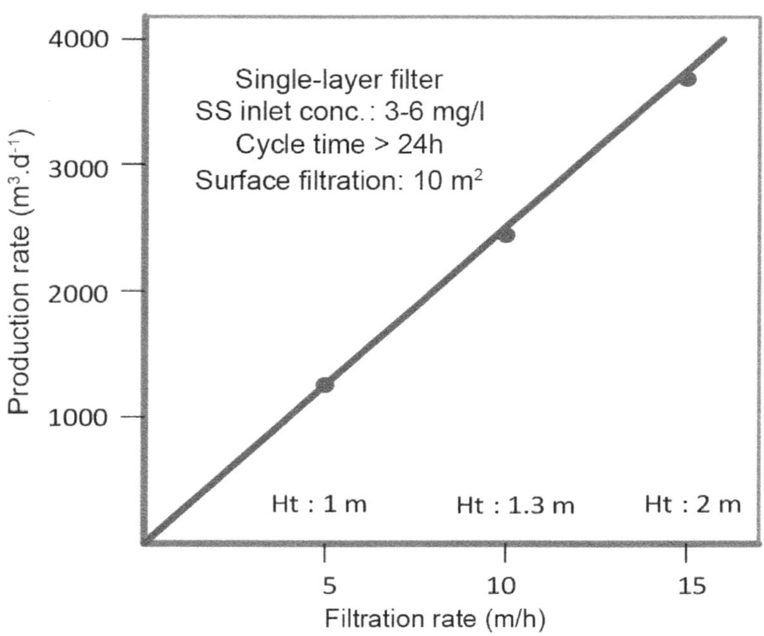

Figure 8.11. *Filtered water production rate depending on the filtration rate. For a color version of this figure, see www.iste.co.uk/gaid/watertreatment2.zip*

A classic dual media filter is composed of a lower layer of sand, varying from 40 to 80 cm, and an upper layer of anthracite that is 40–80 cm thick. The filtration speeds used are between 8 and 12 $m \cdot h^{-1}$, depending on the nature of the pretreatment and the quality of the water to be filtered.

A dual media TGV filter is composed of a lower layer of sand, varying from 40 to 150 cm, and an upper layer of anthracite that is 150–160 cm in height. The filtration rates used are between 15 and 18 $m \cdot h^{-1}$, depending on the nature of the pretreatment and the quality of the water to be filtered.

Filtraflo® filters have a constant operating water level due to the downstream regulation device. It is specific to each type of filter and must satisfy different conditions:

– avoid scouring during filtration (depends on the type of intake and material);

– ensure sufficient time for flocculation (if reagents are added) and degassing (after ozonation);

– ensures sufficient hydraulic load for optimal cycle time.

The form factor Φ is defined as the ratio of the grain surface (spherical equivalent) to the actual grain surface.

For a spherical particle, the specific area is:

$$\frac{Ap}{Vp} = \left(\frac{\pi dg^2}{\frac{\pi dg^3}{6}}\right) = \frac{6}{dg}$$

with:

– Ap: grain surface (m^2);

– Vp: volume (m^3);

– dg: diameter (m).

For non-spherical grains, we introduce the shape factor Φ. Therefore,

$$\frac{Ap}{Vp} = \frac{6}{\Phi dg}$$

8.8. Operating parameters

8.8.1. *Pressure loss in a clean filter*

The acceptable pressure loss is equal to the difference between the liquid level in the filter during operation and the restitution level (retention) in the tank.

It includes:

– pressure loss across the clean filter;

– pressure loss due to clogging and control equipment;

– pressure loss due to the filtering floor strainers.

The calculation of the pressure drop in a clean filter involves considering several parameters, such as the filtration speed, filtration height and characteristics of the filter material.

The equation developed by Carman–Kozeny for a single layer filter is:

$$\Delta P = \frac{Kv\,(1-\varepsilon)^2}{\rho w\; g\; d^2 g\; \varepsilon^3}\,\mu\, H\, \mathrm{v} + \frac{Ki\,(1-\varepsilon)\, H\, \mathrm{v}^2}{g\; dg\; \varepsilon^3}$$

with:

– ΔP: pressure loss (m water);

– H: filtration height (m);

– ε: porosity of the clean filter;

– dg: d_{10} of the filtering material (m);

– μ: dynamic viscosity of water ($kg \cdot m^{-1} \cdot s^{-1}$);

– v: filtration velocity ($m \cdot s^{-1}$);

– ρw: density of water at T°C ($kg \cdot m^{-3}$);

– g: acceleration of gravity ($m \cdot s^{-2}$);

– Kv: coefficient due to viscous forces;

– Ki: coefficient due to inertial forces.

Table 8.4 shows the values of Kv, Ki and ε for sand and anthracite.

Filtering media	Kv	Ki	ε
Sand	150	1.75	0.45
Anthracite	228	4.4	0.55

Table 8.4. *Parameters for pressure loss calculations*

Thus, for a filter with a sand height of 1 m operating at a filtration rate of 7 m·h^{-1} (0.00192 m·s^{-1}), the clean water pressure loss is equal to:

$$\Delta P = \frac{150\,(1-\;0.45)^2}{(1000)(9.81)(0.001^2)(0.45^3)}\,(0.00114)(1)(0.00192) + \frac{(1.75)(1-0.45)(1)\,(0.00192^2)}{(9.81)\,(0.001)(\,(0{,}45^3)}$$

$\Delta P = 0.12$ m

with $\varepsilon = 0.45$; dg = 0.001 m; $\mu = 0.00114$ kg·m^{-1}·s^{-1}; $\rho w = 1{,}000$ kg·m^{-3}; g = 9.81 m·s^{-2}; Kv = 150 and Ki = 1.75, T = 20°C.

Figure 8.12 shows the evolution of ΔP/H according to the filtration rate and for different effective sand sizes.

For a dual media filter, the sum of the respective pressure losses contributed by the two media must be calculated.

We therefore have:

$$\Delta P1\;(media\;1) = \frac{Kv1\,(1-\;\varepsilon1)^2}{\rho w\;g\;d^2g_1\;\varepsilon1^3}\,\mu\,H\,v + \frac{Ki1\,(1-\varepsilon1)\,H\,v^2}{g\;dg_1\;\varepsilon1^3}$$

$$\Delta P2\;(media\;2) = \frac{Kv2\,(1-\;\varepsilon2)^2}{\rho w\;g\;d^2g_2\;\varepsilon2^3}\,\mu\,H\,v + \frac{Ki2\,(1-\varepsilon2)\,H\,v^2}{g\;dg_2\;\varepsilon2^3}$$

The total clean filter pressure loss is the sum of $\Delta P_1 + \Delta P_2$, using the respective Kv and Ki coefficients for each medium.

Thus, for a filter with a total height of 1.6 m, including 0.8 m of anthracite and 0.8 m of sand, operating at a speed of 12 m·h^{-1} (0.00333 m·s^{-1}), the pressure loss in the clean filter is equal to:

$$\Delta P1 = \frac{228\,(1 - 0.55)^2}{(1000)(9.81)(0.0014^2)(0.55^3)}\,(0.00114)(0.8)(0.00333) + \frac{4.4\,(1 - 0.55)(0.8)\,(0.00333^2)}{(9.81)\,(0.0014)\,(0.55^3)}$$

$$\Delta P2 = \frac{150\,(1 - 0.45)^2}{(1000)(9.81)(0.00085^2)(0.45^3)}\,(0.00114)\,(0.8)(0.00333) + \frac{1.75(1 - 0.45)(0.8)\,(0.00333^2)}{(9.81)(0.00085)\,(0.45^3)}$$

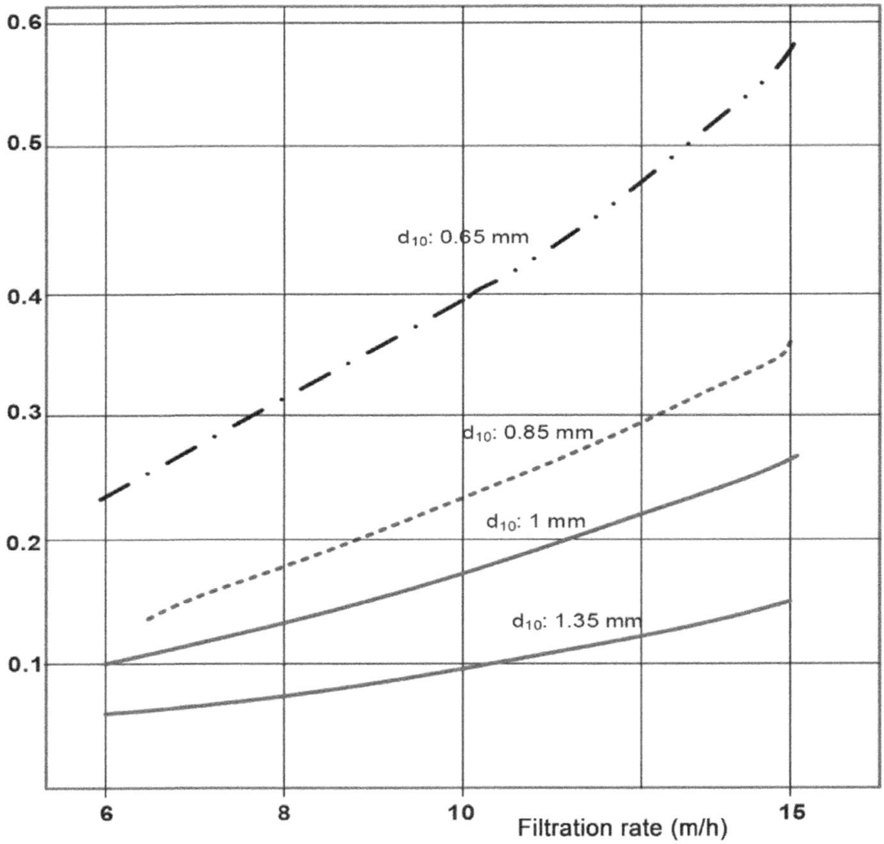

Figure 8.12. *ΔP/H obtained for different effective sand sizes and at different filtration rates. For a color version of this figure, see www.iste.co.uk/gaid/watertreatment2.zip*

Therefore:

$$\Delta P = 0.05 + 0.22 = 0.27 \text{ m } (T°C = 20°C)$$

with:

– $\varepsilon = 0.55$; dg = 0.0014 m; $\mu = 0.00114$ $kg{\cdot}m^{-1}s^{-1}$; $\rho w = 1{,}000$ $kg{\cdot}m^{-3}$; $g = 9.81$ $m{\cdot}s^{-2}$; Kv = 228 and Ki = 4.4 for anthracite;

– $\varepsilon = 0.45$; dg = 0.00085 m; $\mu = 0.00114$ $kg{\cdot}m^{-1}s^{-1}$; $\rho w = 1{,}000$ $kg{\cdot}m^{-3}$; $g = 9.81$ $m{\cdot}s^{-2}$; Kv = 150 and Ki = 1.75 for sand.

The calculation is similar for a three-media filter, as long as the Kv and Ki values for the media are integrated.

8.8.2. *Pressure loss during clogging*

The development of pressure loss in a clean filter depends on the physical characteristics of the filter media. For an operating filter, the pressure loss also depends on the quality of the water at the filter inlet and the filtration rate. Water loaded with SS involves an accumulation of the latter on the material according to the previously described mechanisms and increases the pressure loss. This results in an increase in the water height above the media because clogging reduces the water flow velocity. The higher the filtration rate, the more this phenomenon is accentuated.

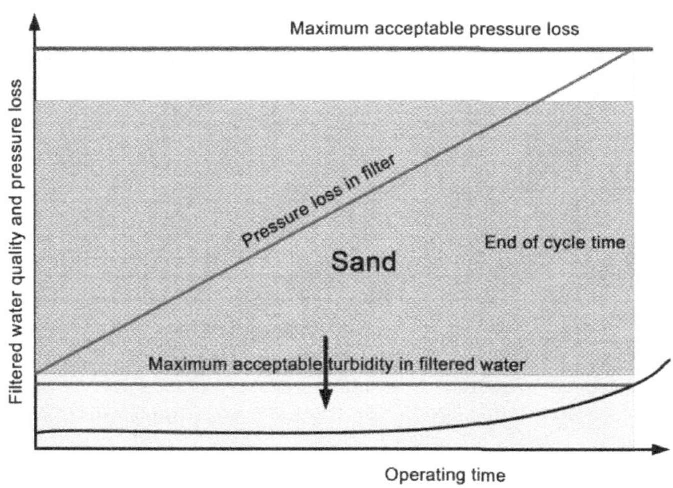

Figure 8.13. *Changes in filtered water quality and pressure loss. For a color version of this figure, see www.iste.co.uk/gaid/watertreatment2.zip*

The maximum available pressure loss is pressure loss due to the height of the water above the material, pressure loss due to the height of the material and pressure loss due to the filter equipment such as the filter floor.

When this limit is reached, the filter should be washed. If the filter continues to operate under these conditions, it will enter a negative pressure condition. Negative pressure occurs when the pressure in the filter falls below atmospheric pressure. This can occur near the end of the filtration cycle. The negative pressure creates air suction, which tends to form air bubbles that attach to the material and clog the pores of the filter media. Uneven flow occurs, which causes the particles to fall out and increases the turbidity of the filtered water. Preferential channels created by air bubbles in the filter mass can also be observed, which promote the flow of unfiltered water, thus leading to an increase in turbidity.

Figures 8.14 and 8.15 illustrate the principle of pressure loss development in a filter until negative pressure is reached. The maximum acceptable pressure loss should not be exceeded, and a wash should be initiated just before it reaches this limit.

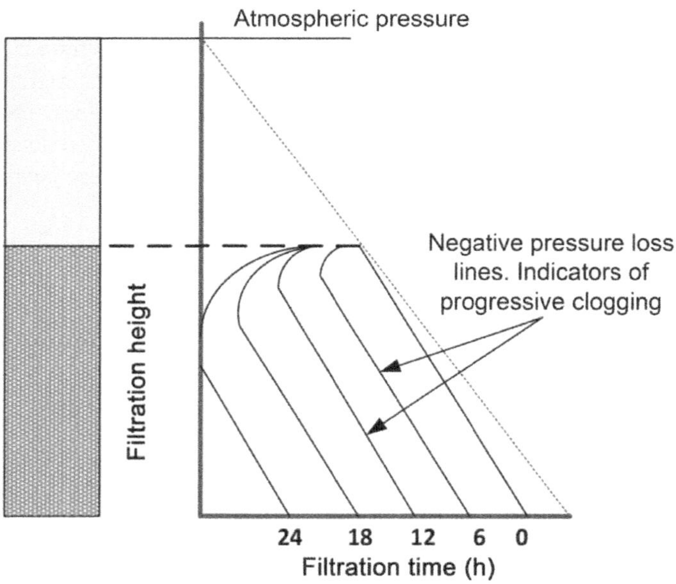

Figure 8.14. *Development of negative pressure losses in a filtration cycle. For a color version of this figure, see www.iste.co.uk/gaid/watertreatment2.zip*

The maximum pressure loss to clogging is therefore calculated by considering the height of the media (m), the height of water above the media (m) and the pressure loss due to the filtering floor.

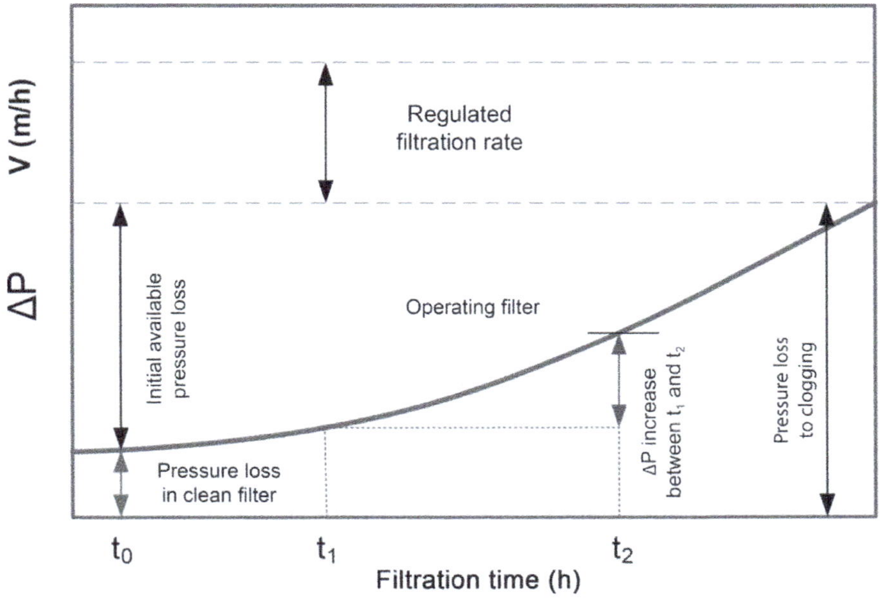

Figure 8.15. *Evolution of pressure losses in a filtration cycle. For a color version of this figure, see www.iste.co.uk/gaid/watertreatment2.zip*

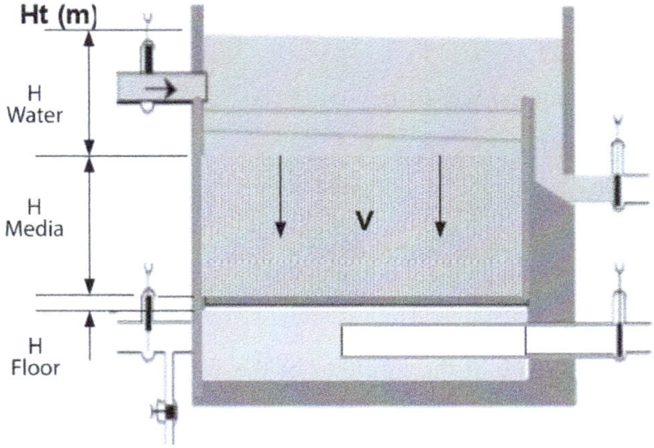

Figure 8.16. *Characteristics of different heights in a filter. For a color version of this figure, see www.iste.co.uk/gaid/watertreatment2.zip*

Thus, for a filter with a height of 1.00 m and a water height of 1.2 m above the media, the maximum available pressure loss (ΔP max) is 2.20 m. The pressure loss to clogging (ΔP clogging) is:

$$\Delta P \text{ max} - \Delta P \text{ clean filter} = 2.20 - 0.12 = 2.08 \text{ m}$$

Sometimes, operators only take 90% of the ΔP max, which, in this case, is a ΔP clogging of 1.98 m.

8.8.3. *Estimation of the turbidity of filtered water in single-layer filters*

The turbidity of filtered water at time *to* can be estimated using the following relation:

$$Log\,\frac{(Turb)out}{(Turb)in} = -\frac{3(1-\varepsilon)\,\alpha\,(\rho s-\rho water)gdp^2H}{82.8\,\mu\,vd_{10}}$$

with:

– (Turb)out and (Turb)in: turbidities (or SS) in filtered water and filter feed water;

– ε: material porosity;

– α: parameter related to particle collisions in the material collision factor (0.065 for a $d_{10} < 1$ mm).

For TGV filters, the value of α increases to 0.1 for a d_{10} of 1.35 mm, with:

– H: material height (m);

– d_{10}: effective material size (m);

– g: acceleration of gravity ($m \cdot s^{-2}$);

– μ: dynamic viscosity ($kg \cdot m^{-1} \cdot s^{-1}$);

– dp: particle diameter (m), estimated at 10 μm;

– v: filtration rate ($m \cdot s^{-1}$).

This gives an estimate of the turbidity of the filtered water during the first few hours of filtration (Figure 8.17).

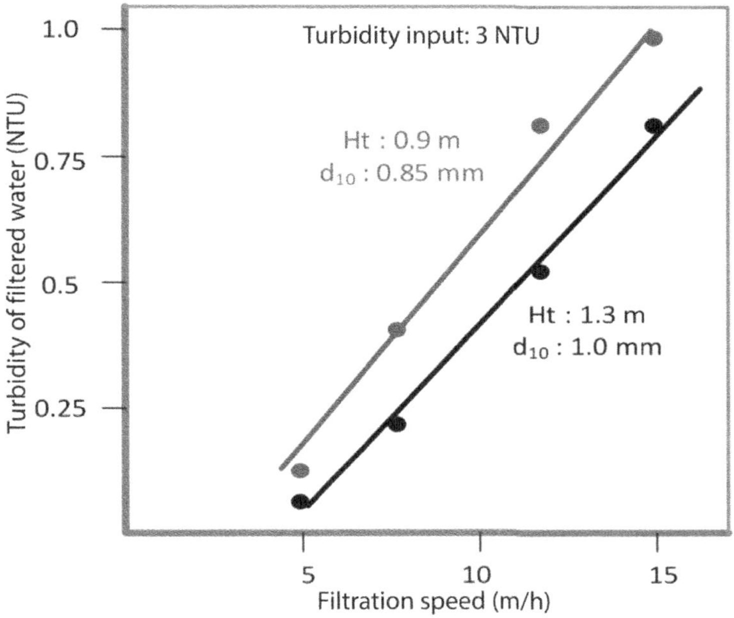

Figure 8.17. *Estimation of the turbidity of filtered water in relation to the speed and height of filtration. For a color version of this figure, see www.iste.co.uk/gaid/watertreatment2.zip*

8.8.4. Retention capacity

It represents the quantity of material that the filter is likely to retain during a filtration cycle. It is usually expressed in kg of material per m^2 of material for rapid filters and in $kg{\cdot}m^{-3}$ for TGV filters.

The retention capacity of a filter depends on the porosity of the material. In a clogging situation, it can be assumed that the interstices form an inner zone, which is itself inscribed in a circle.

Each arc of this area is calculated according to:

$$\text{arc A} = 0.0174 \times 60\ (d_{10}/2), \text{ with an angle of } 60°.$$

This area $A_1A_2A_3$ (Figure 8.18) has an even surface:

$$S = 0.25\ \sqrt{[(3 \times \text{arc}) \times (\text{arc } A_1) \times (\text{arc } A_2) \times (\text{arc } A_3)]}$$

knowing that arcs A_1, A_2 and A_3 are equal.

The perimeter of this area is given by:

$$P = 0.5 \times (3 \times \text{arc } A)$$

The radius of the inscribed circle in this area is Rz = S/P.

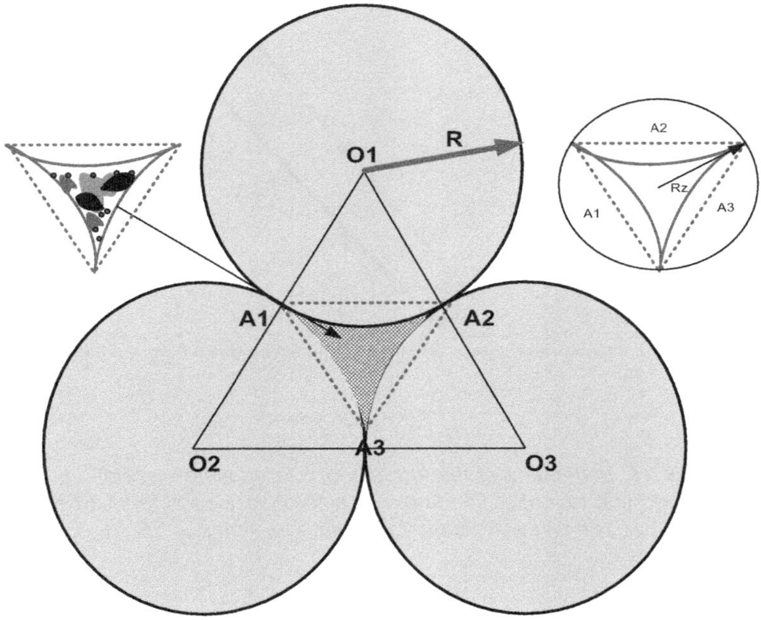

Figure 8.18. *Interstices of sand grains saturated with retained particles. For a color version of this figure, see www.iste.co.uk/gaid/watertreatment2.zip*

The area of the inscribed circle is $\pi\ Rz^2$. Therefore, the diameter is equal to 2 Rz. This shows (see example in Table 8.5) that the diameter of the inscribed circle is equal to 30% of the d_{10} of the material.

The pore volume available in the filter is given by: $Vp = S \cdot H \cdot \varepsilon$ for a surface S and a height H.

The available volume is therefore 30% of the pore volume, in other words, Vd = 0.30 Vp. For a concentration of particles in the interstices varying between 15 and 20 $g \cdot L^{-1}$, we deduce the theoretical retention capacity ($kg \cdot m^2$) depending on the type of flocs (hydroxide flocs, silt, coarse particles, etc.).

Parameters	Calculations $d_{10} = 0.85$ mm	Calculations $d_{10} = 1.35$ mm
Arc A	$0.0174 \times 60 \times (0.85/2) = 0.443$	0.704
S (mm²)	$0.25\ \sqrt{[(3 \times 0.443) \times (0.443) \times (0.443) \times (0.443)]} = 0.0852$	0.215
Perimeter (mm)	$0.5 \times (3 \times 0.443) = 0.665$	1.057
Inscribed circle radius (mm)	$0.0852/0.665 = 0.128$	0.203
Diameter (mm)	$2 \times 0.128 = 0.256$	0.406
Inscribed circle area (mm²)	$3.14 \times 0.128^2 = 0.0515$	0.129
Inscribed diameter/d_{10} ratio	$0.256/0.85 = 0.30$	$0.406/1.35 = 0.30$

Table 8.5. *Example of the calculation of the interstice diameter/d10 ratio of the material*

The actual capacity of gravity sand filters operating at rates ≤ 8 m·h^{-1} for a height < 1.20 m is in the range of 40% of the theoretical capacity because clogging is established in the first 40 cm of material height. This accumulation results in a rapid increase in pressure loss. However, there may still be a slow diffusion of particles to the base areas of the filter. It is a theoretical approach because particles are also coating the grains of the filter. This calculation can just give an idea of the capacity of a filter depending of the size of the grain.

The retention capacity of sand filters is estimated to be between 800 and 1,100 g·m^2. For dual media filters (anthracite/sand or pumice/sand), an identical calculation can be made by adding the capacities calculated for the two materials. However, since the porosity of anthracite is 0.55, the total pore volume is used because it is assumed that particle diffusion occurs over the entire height of the material. For sand, the calculation is performed as in the case of the single-media filter.

Thus, a dual media filter (anthracite/sand) has a retention capacity of 2,000–2,200 g·m^{-2}. Pressure loss on anthracite will increase faster than on sand due to the greater accumulation of particles.

Parameters	Calculations
Material height (m)	1
Surface (m^2)	1
ε	0.45
Pore volume (interstices) (mm^3)	$1 \times 1 \times 0.45 = 0.45$
Available volume (m^3) (0.30: ratio d interstice/d_{10})	$0.45 \times 0.30 = 0.135$
Floc concentration in the interstices ($kg \cdot m^{-3}$)	15–20
Theoretical retention capacity over the entire height ($kg \cdot m^{-2}$)	$0.135 \times 15 = 2.025$
Real retention capacity or 40% of the height ($kg \cdot m^{-2}$)	$2.025 \times 0.4 = 0.81$

Table 8.6. *Calculation of the retention capacity of sand filters*

For single-media or dual media filters receiving fine $CaCO_3$ particles, their retention capacity is between 3,000 and 4,500 $g \cdot m^{-2}$ because the concentration in the interstices is higher (60–80 $g \cdot L^{-1}$) due to the thickening of carbonates during their deposit.

For other types of filters (biological iron removal, activated carbon, etc.), it is advisable to refer to the chapters related to this application.

8.8.5. *Filter washing conditions: theory and calculations*

The basic purpose of backwashing the filter is to remove the maximum number of particles accumulated during a filter cycle and to flush them out of the filter. The filter then starts a new cycle under favorable pressure loss conditions. The key purpose in backwashing is to achieve the highest possible particle removal rate. Proper backwash is a very important step in the operation of a filter. If the filter is not properly backwashed, operational problems occur, such as overly short cycles, high water losses, lower production, high turbidity output, etc.

The filter should be washed when one of the following conditions is met:

– the pressure loss is so high that the filter no longer produces water at the desired flow rate;

– flocs begin to break through the filter, and turbidity in the filtered water increases;

– the filter cycle reaches the expected operating time.

If the water supplied to a filter is of very good quality, the filtration cycles can be very long. Some filters can run for more than a week before needing to be washed. However, this is not recommended because long filter cycles can cause the filter material to compact, making it difficult to expand the media height during washing.

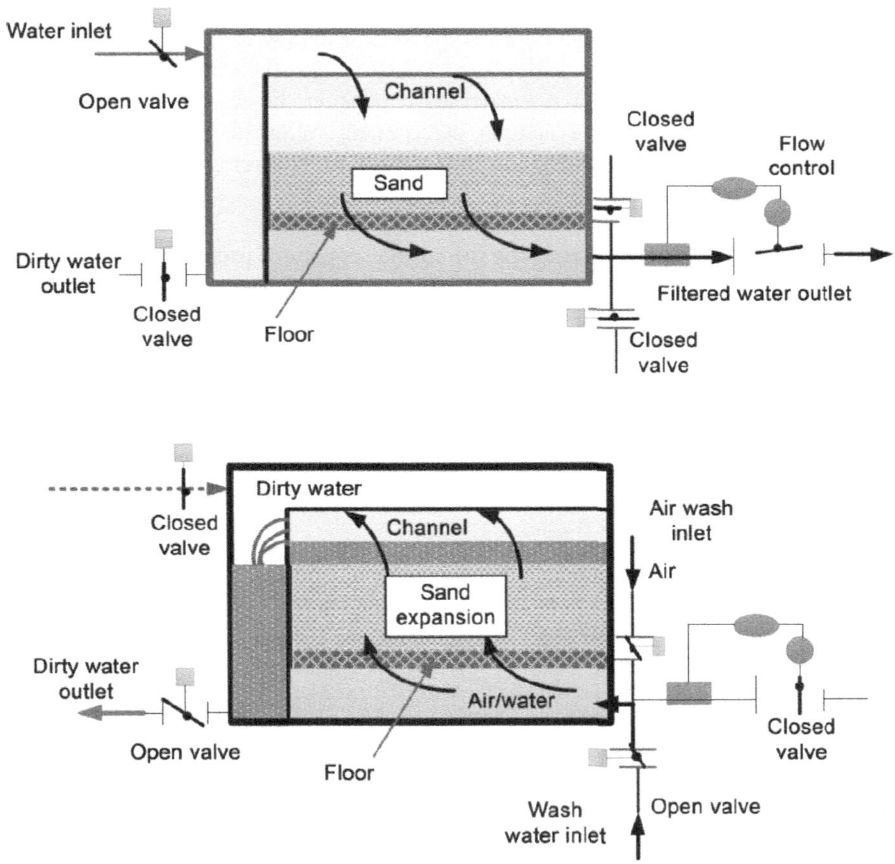

Figure 8.19. *General diagram of how the filters operate. For a color version of this figure, see www.iste.co.uk/gaid/watertreatment2.zip*

The release of particles requires the application of an energetic thrust that is greater than the forces that bind them to the material. The release rate depends on both the attachment forces and the forces that allow the release. All of these forces depend on the structure of the particles and their composition. The profile of a clogged filter shows a disparity and a heterogeneity in the distribution of particle deposits.

The shear forces required to break up the deposit and allow it to detach are greater at the top of the filter, where there is a large amount of accumulation, compared to the bottom of the filter. In the upper section of the filter, the particle load implies that the optimal porosity is not far from being reached.

The rate of the wash water must be high enough to remove all particles without being excessive to avoid media loss. It is therefore higher than the minimum fluidization rate. A material expansion of 25–30% is sufficient. A minimum fluidization rate is then determined, followed by a washing speed that respects this level of expansion and allows for the removal of the detached particles.

This washing speed is obtained by the various equations presented in Table 8.7.

	Minimum fluidization rate (V_{mf}) ($m \cdot h^{-1}$)	Backwash rate(v_{bw}) ($m \cdot h^{-1}$)
Leva	$Vmf = \frac{22.12(\rho w/(\rho s - \rho w))^{0.94}}{\mu^{0.88}} d10^{1.82}$	$v_{bw} = Vmf\left(\frac{\varepsilon exp}{\varepsilon}\right)^n$
Ergun (ref a)	$Vmf = -b + \frac{\sqrt{\Delta}}{2a} 3{,}600$	$v_{bw} = Vmf\left(\frac{1-\varepsilon}{\varepsilon^3}\right)\left(\frac{\varepsilon\, exp^3}{1-\varepsilon exp}\right)$
Kozeny	$Vmf = \left(\frac{1{,}000\,(\rho s - \rho w).\,g.\,\varepsilon^3}{(hK.\,\mu.\,0.001)(1-\varepsilon)\left(\frac{6}{0.001.\,\psi.\,d10}\right)^2}\right) 3{,}600$	$v_{bw} = Vmf\left(\frac{1-\varepsilon}{\varepsilon^3}\right)\left(\frac{\varepsilon\, exp^3}{1-\varepsilon exp}\right)$
When and Yu	$Vmf = \frac{\mu}{\rho w d10}(33.7^2 + 0.0408\, Ga)^{0.5} - \frac{33.7\mu}{\rho w d10}$ $Ga = \frac{d^3 \rho w(\rho s - \rho w) g}{\mu^2}$	$v_{bw} = Vmf\left(\frac{1-\varepsilon}{\varepsilon^3}\right)\left(\frac{\varepsilon\, exp^3}{1-\varepsilon exp}\right)$

Table 8.7. *Equations to calculate the minimum fluidization rate and the washing rate*

with:

– ρw: water density at T °C (g·cm^{-3});

– ρs: media density (g·cm^{-3});

– μ: dynamic viscosity of water at T °C (cPa);

– Ψ: form factor of the media;

– d_{10}: effective media size (mm);

– hK: Kozeny constant;

– ε: media porosity;

– v_{dec}: particle settling rate (m·s^{-1});

– v_{bw}: backwash rate (m·s^{-1}).

Given that the volume of filter material per unit area is constant, we have:

$$\frac{Hexp}{H} = \frac{1 - \mathcal{E}}{1 - \mathcal{E}exp} = \frac{1 - \mathcal{E}}{1 - \left(\frac{v_{bw}}{vexp}\right)^{0.22}}$$

with:

– ε_{exp}: porosity of the expanded media:

$$\mathcal{E}exp = \left(\frac{v_{bw}}{vdec}\right)^{0.22}$$

$$v_{bw} = V_{dec}\, \mathcal{E}_{exp}{}^{4.5};$$

– H: media height (m);

– H_{exp}: height of expanding media (m)$Hexp = H\ (1 + \frac{expansion}{100})$;

– Re: Reynolds number, Re = √ [33.72 + (0.0408)Ga] – 33.7;

– Ga: Galileo number;

– Parameters a, b and c:

$$a = \frac{36{,}000\ hB\ \rho w}{\Psi\ d10/1{,}000}$$

$$b = \frac{36\ kH(0.001\mu)(1 - \mathcal{E})}{(\Psi\ \frac{d10}{1{,}000})^2}$$

$$c = -1{,}000(\rho s - \rho w)9.81 \cdot \mathcal{E}^3$$

Pressure loss required to expand the filter:

$$h = Hexp\,(1 - \mathcal{E}exp)\frac{(\rho s - \rho w)}{\rho w}$$

H_{exp} is the expansion height, $\mathcal{E}_{exp}$ is the porosity of the expanded filter, and ρ_s and ρ_w are the densities of the filter material and water, respectively.

Figures 8.20 and 8.21 show the backwash rates versus the ES (d_{10}) for different media, as well as the backwashing rates for anthracite and pumice versus their expansion.

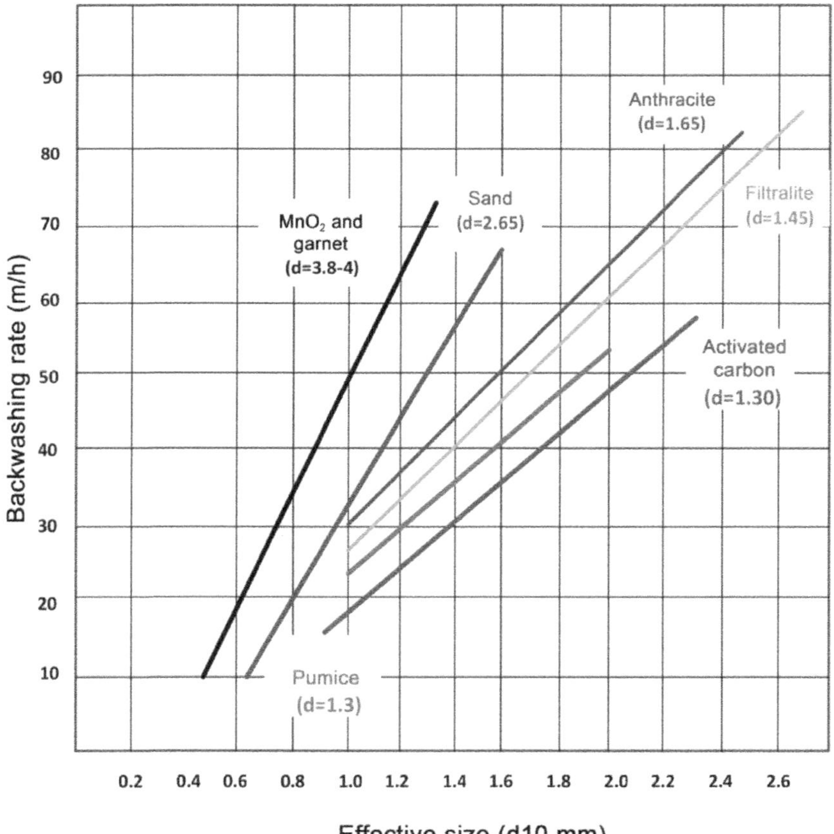

Figure 8.20. *Backwashing rates versus ES for different media at 15°C. For a color version of this figure, see www.iste.co.uk/gaid/watertreatment2.zip*

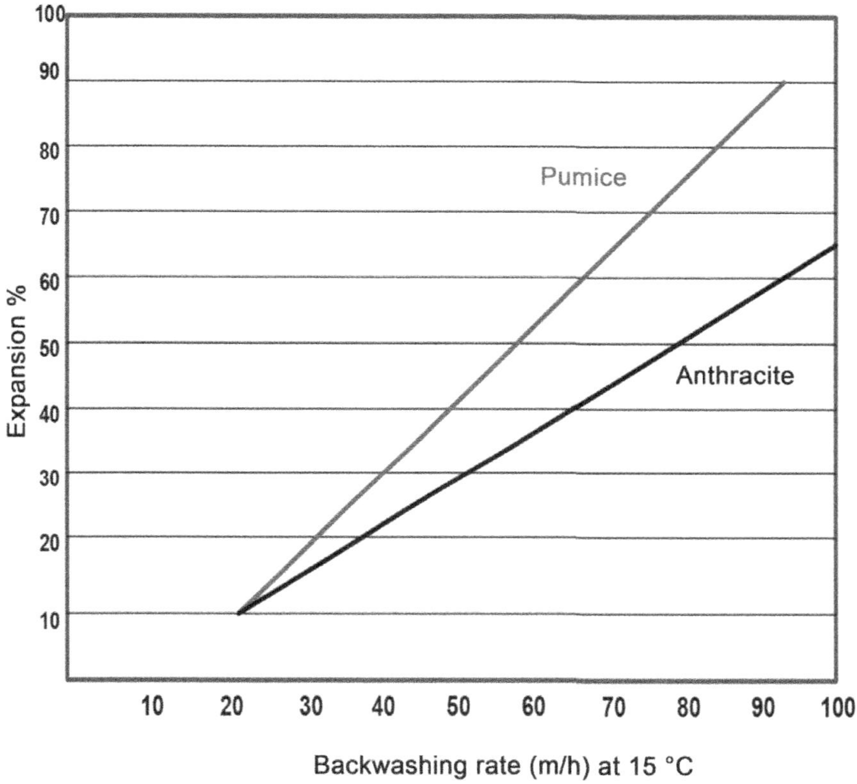

Figure 8.21. *Expansion percentage versus different backwashing rates. For a color version of this figure, see www.iste.co.uk/gaid/watertreatment2.zip*

The minimum fluidization rate is the rate required for the start of material fluidization. The calculation of the minimum fluidization rate is important for determining the wash flow conditions. For an expansion of 25–30%, the diameter d_{10} is sufficient. This is because a diameter d_{90} involves the largest materials and leads to a complete fluidization of the filter.

8.8.5.1. *Implementation of washing conditions*

The implementation of backwashing conditions includes three phases for filters arranged in the first filtration stage or in direct filtration, that is, air, air + water, and water alone. For filters in the second stage or treating water without turbidity, the number of phases can be reduced to two, in other words, air and water alone.

8.8.5.1.1. Gravity filters in the first stage or direct filtration

First, the water level is lowered to the height of the media. This makes it possible to acquire all the water above the filtering media as filtered water.

First phase: air scouring

During this phase, the injection of air allows the filter to be loosened, as it has become progressively compacted during the filtration cycle. It also destructures the particles retained between the interstices of the material. In addition, air allows the crust on the surface, where there are the most particle deposits, to be broken. The air injection system can be a completely separate pipe from the backwash water system, or it can be made through nozzles that distribute both air and water. If the air is introduced directly under the gravel, provisions must be made to avoid movement of the fine gravel by the air, especially by the air and water during the air/water phase. In this regard, the nozzle system is more efficient and less problematic.

Air is never used during overflow (backwashing with water alone), as this results in media suspension and media loss with the dirty water.

Cleaning by air scouring at the base of the bed is carried out for 1–2 min at 50 $Nm^3 \cdot h^{-1} \cdot m^{-2}$. It is advisable to visually check the good distribution of the air on the whole filter to detect possible anomalies (broken nozzles, faulty seal).

Second phase: air scouring and water

Backwashing with water alone, without air scouring, is not very effective because the shear forces are not sufficient to break the particle–media bonds, causing limited collisions and detachments of the particles. During this air/water phase, a slow expansion of the media grains occurs, caused by the simultaneous flow of air and water, leading to abrasion between the grains. This abrasion and the high interstitial water speeds result in an effective backwash.

Water is flushed at a low rate simultaneously with air for approximately 8 min, dislodging the particles from between the interstices of the media and suspending them in the area above the filter. The forces of adhesion between the deposited particles and the filter grains can become so strong that the backwashing conditions must be adapted so that the particles can be dislodged. If this does not happen, the particles and media grains can agglomerate to form sludge balls that are very resistant to washing.

During this phase, the shear forces are high, and the rising water allows the particles to be carried out of the interstices. Under these conditions, the porosity of the expanded media increases. The consequent grain movement promotes collisions and attrition between grains. This results in particles falling out of the media and being carried by the rising water toward the surface.

Air/water backwashing does not produce sludge aggregates because there is no bed expansion and because the surface crusts are broken up with air.

During air scouring, the flow rate of the wash water can vary in a very wide range between 8 and 20 $m^3 \cdot h^{-1} \cdot m^{-2}$, depending on the media height. It is lower than the minimum fluidization rate. The air flow rate is maintained between 50–60 $Nm^3 \cdot h^{-1} \cdot m^{-2}$.

The time frame for this second phase is calculated so that the dirty water reaches the filter base level without overflowing (Figure 8.22). If this time is exceeded, there is a risk of losing media because some of it is suspended.

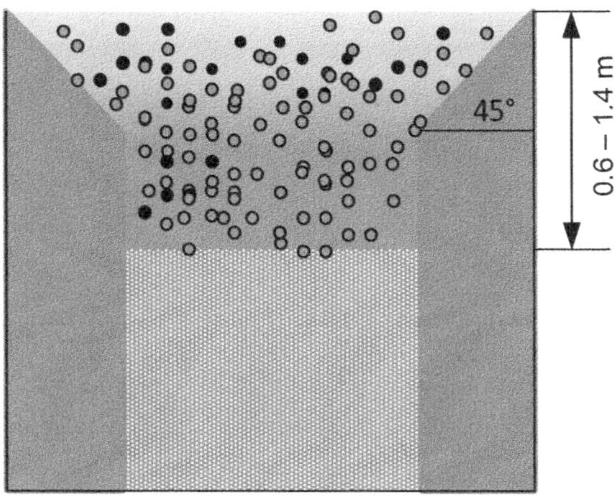

Figure 8.22. *Layout of the filter walls against media losses during the air/water phase. For a color version of this figure, see www.iste.co.uk/gaid/watertreatment2.zip*

Several precautions are recommended to avoid sand or anthracite loss during the air/water phase. Water and air flow rates vary depending on the size of the filter media. In general, the dirty water reaches the top end of the filter walls or at the edge of a channel (or channels), and the vertical distance between the sand surface and the wash water overflow varies between 0.6 and 1.4 m, depending on the filter.

The upper edge of the overflow wall is sloped downward at 45° toward the filter bed so that sand grains that fall onto the sloped wall during backwash return to the filter (Figure 8.24).

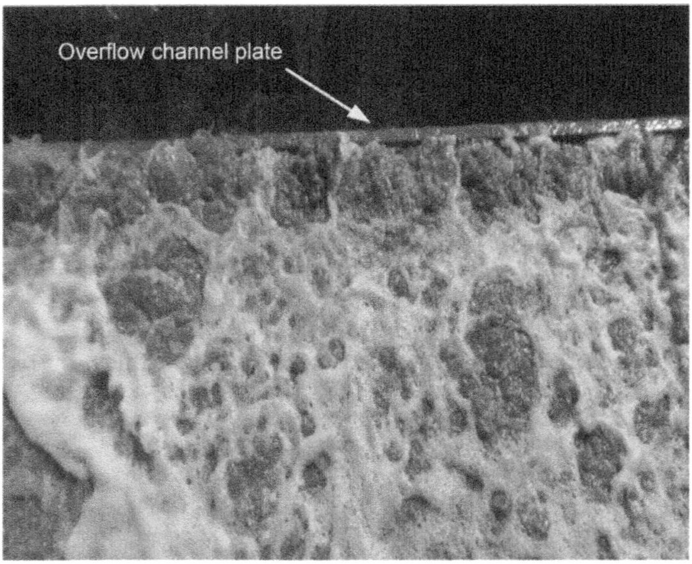

Figure 8.23. *Air/water washing. For a color version of this figure, see www.iste.co.uk/gaid/watertreatment2.zip*

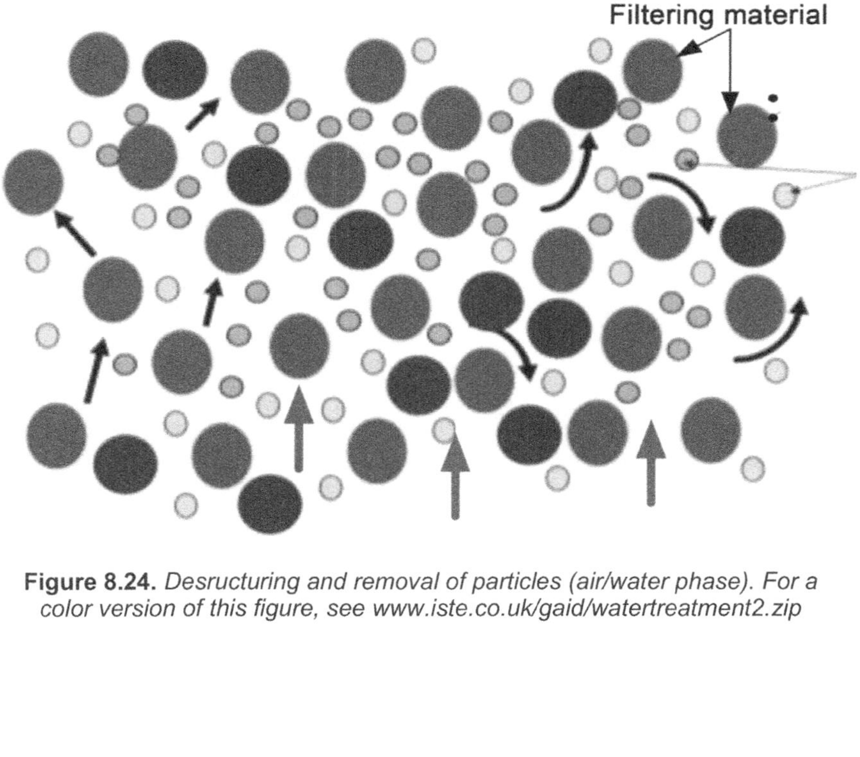

Figure 8.24. *Desructuring and removal of particles (air/water phase). For a color version of this figure, see www.iste.co.uk/gaid/watertreatment2.zip*

An expansion of 25–30% is recommended, as an expansion that is too high results in an increase in the distance between the grains and reduces the frequency of collisions. The particle separation efficiency is very low.

Third phase: rinsing with water

The third phase involves rinsing the filter with water alone. A sufficient flow of water is used to achieve an expansion of 25–30%. The sludge is removed from the surface of the sand particles as they move through the colliding water flow.

The water flow rate varies between 20 and 60 $m \cdot h^{-1}$, depending on the particle size of the material, the temperature and the type of filter (single-layer, dual media or multimedia). This operation takes 8–15 min. Figure 8.25 shows the evolution of the SS concentration in the dirty water during the first few minutes of removal. The wash water volume according to wash time is also shown. Rinsing continues until mottling appears on the filter surface or clear rinse water is obtained. A 25–30% expansion is recommended, as an expansion that is too high results in an increase in the distance between the grains and reduces the frequency of collisions. The particle separation efficiency is very low if the percentage of expansion is too low.

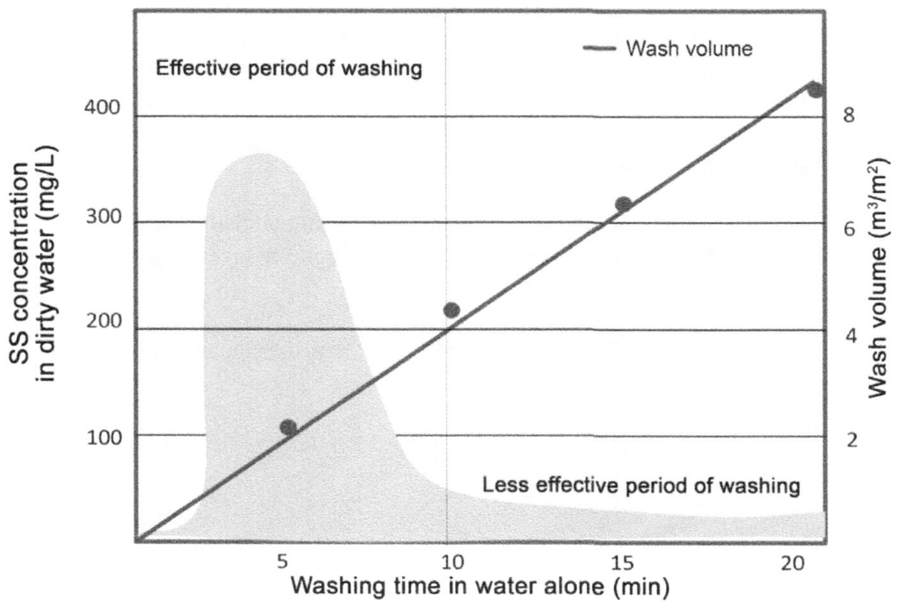

Figure 8.25. *Example of the evolution of SS concentration in dirty water during backwashing with water alone (dual media filter). For a color version of this figure, see www.iste.co.uk/gaid/watertreatment2.zip*

This type of washing is effective if the backwashing rate is higher than the minimum fluidization rate of the media. If this is not the case, mud balls may be formed, and pressure losses may develop more rapidly.

It should be noted that the backwashing rates are different between summer and winter, as it depends on the viscosity of the water. To maintain the same expansion, a lower flow rate is normally used during cold temperatures than during warm periods.

Depending on the design of the filter, the washing can sometimes be coupled with a surface wash (surface sweep), which consists of horizontal additional water injection above the upper layer media and is involved in the removal of flocs. This is particularly important when the water level above the filter is high (>1.4 m). This avoids a reflocculation of the sludge, which becomes larger and heavier and cannot be removed with conventional washing speeds. The volume of rinsing water varies from 2 to 5% of the treated flow.

In dual media and multimedia filters, the backwash sequences are identical in terms of operating protocol. During the air/water phase, the air flow remained injected for 1–2 min. The associated water flow is mostly between 12 and 20 $m{\cdot}h^{-1}$ for 8 min, depending on the type of material, its ES and the material height.

During the rinsing phase (water only), the washing rates are strongly increased to 50–60 $m{\cdot}h^{-1}$ to partially fluidize the different materials, making it possible to restore the stratification by gravity separation.

In dual media and multimedia filters, it is not recommended to have a mixture of material layers because the finer material in the bottom layer acts as a turbidity refiner.

The equation to calculate the difference between the apparent densities of the two materials is given as:

$$\rho filter = (1 - \varepsilon 1)(\rho s1 - \rho water) - (1 - \varepsilon 2)(\rho s2 - \rho water)$$

with:

- ρfilter: apparent density of the filter;
- ρs_1: grain density of media 1;
- ρs_2: grain density of media 2.

Here, ε1 is associated with the coarser material, and ε2 is associated with the finer material. When material 2 has a higher density than material 1, material 1 tends to be below material 2 at low backwash water flow rates. For a sufficiently high flow rate, coarser material 1 is above finer material 2. This is the case between anthracite or pumice with sand.

When the filter is put back into operation after the third stage (rinsing), the turbidity of the filtered water may rise to 5 NTU, or even 8 NTU, for a short period of time, lasting approximately 10–30 min. This is due to the displacement of the residual wash water, containing particles that have been removed from the filter.

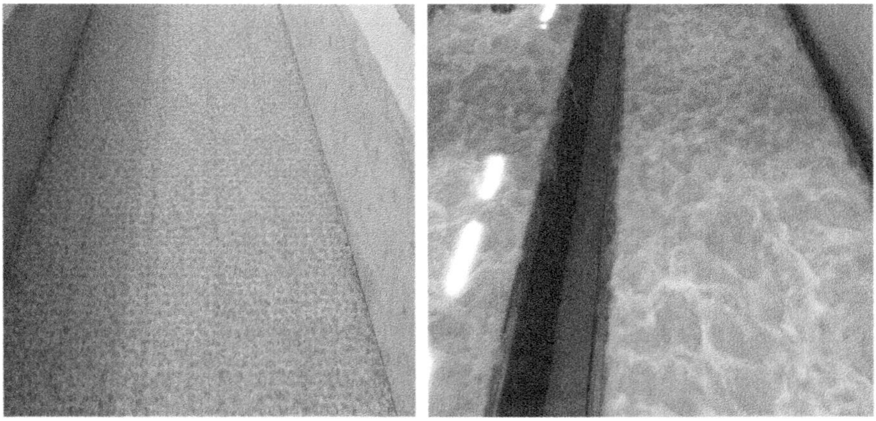

a) b)

Figure 8.26. *First phase of air scouring (a) and second phase of washing (b) (air/water). For a color version of this figure, see www.iste.co.uk/gaid/watertreatment2.zip*

Fourth phase: putting the filter back to work

It is recommended that the initial filtered water during the first 10–30 min of filter production be drained. When draining the initial filtered water is not possible, it is recommended that the filter be restarted slowly to limit the release of particles into the filtered water.

Care must be taken not to overspeed the filtration when restarting the filter that has just been cleaned and, generally speaking, not to create hydraulic jerks on the entire filtering bank when restarting the cleaned filter.

Table 8.8 summarizes the different washing conditions for some single-media and dual media filters.

Parameters	**Sand filter** **H:1 m** **ES: 0.85 mm**	**TGV sand filter** **H: 2 m** **ES: 1.35 mm**	**Dual media anthracite-sand filter** **H: (0.8 + 0.8)m** **ES: 1.4 mm anthracite** **ES: 0.85-1 mm sand**	**Dual media pumice-sand filter** **H: (0.8 + 0.8)m** **ES: 1.4 mm pumice** **ES: 0.85-1 mm sand**
Air	50 $Nm^3 \cdot m^{-2} \cdot h^{-1}$ (1 min)	55–60 $Nm^3 \cdot m^{-2} \cdot h^{-1}$ (1 min)	50 $Nm^3 \cdot m^{-2} \cdot h^{-1}$ (1 min)	50 $Nm^3 \cdot m^{-2} \cdot h^{-1}$ (1 min)
Air + water	50 N $m^3 \cdot m^{-2} \cdot h^{-1}$ for air 8 $m^3 \cdot m^{-2} \cdot h^{-1}$ for water (8 min)	55–60 $Nm^3 \cdot m^{-2} \cdot h^{-1}$ for air 15 $m^3 \cdot m^{-2} \cdot h^{-1}$ for water (8 min)	50 $Nm^3 \cdot m^{-2} \cdot h^{-1}$ for air 12 $m^3 \cdot m^{-2} \cdot h^{-1}$ for water (8 min)	50 $Nm^3 \cdot m^{-2} \cdot h^{-1}$ for air 12 $m^3 \cdot m^{-2} \cdot h^{-1}$ for water (8 min)
Rinse	25 $m^3 \cdot m^{-2} \cdot h^{-1}$ for water (7–10 min)	60 $m^3 \cdot m^{-2} \cdot h^{-1}$ for water (10–15 min)	50–55 $m^3 \cdot m^{-2} \cdot h^{-1}$ for water (7–10 min)	35 $m^3 \cdot m^{-2} \cdot h^{-1}$ for water (7–10 min)

Table 8.8. *Cleaning conditions for single-media and dual media filters*

8.8.5.1.2. Expansion of the material during washing

When backwashing with a fluidization of approximately 20–30%, this expansion depends on many material and water parameters. The material parameters are diameter, shape and density. The water parameters are viscosity and density.

There is an importance in estimating the expansion, such as knowing how high the material will rise and prevent it from reaching the filter channels.

For this reason, a modified Reynolds number is calculated:

$$Re\ mod = \frac{\mathrm{v}\ \varepsilon\ exp\ \rho w}{\varepsilon\ exp\ \mathrm{Sv}(1-\varepsilon exp)\mu} = \frac{\mathrm{v}\rho w}{\mathrm{Sv}(1-\varepsilon exp)\mu}$$

where Sv is the specific surface of the material (6/d for spheres and 6/ψd for nonspherical grains). ε exp is the porosity of the expanded media.

With A as a dimensionless number related to the porosity and the specific surface:

$$A = \frac{\varepsilon exp^3 \rho w(\ \rho s - \ \rho w)g}{(1-\varepsilon exp)^2 Sv^3 \mu^2}$$

log A is correlated with log Re mod to obtain:

$$\log A = 0.56543 + 1.09348\ (\log Re\ mod) + 0.17971\ (\log Re\ mod)^2 - 0.00392\ (\log Re\ mod)^4 - 1.5(\log \psi)^2$$

By replacing the equation of A and Re mod including the porosity, the value of expanded ε is deduced.

Therefore:

$$\log\left(\frac{\varepsilon exp^3 \rho w(\ \rho s- \ \rho w)g}{(1-\varepsilon exp)^2 Sv^3 \mu^2}\right) = 0.56543 + 1.09348\ \left(\log\left(\frac{\mathrm{v}\rho w}{Sv(1-\varepsilon exp)\mu}\right)\right) + 0.17971\ \left(\log\left(\frac{\mathrm{v}\rho w}{Sv(1-\varepsilon exp)\mu}\right)\right)^2 - 0.00392\ \left(\log\left(\frac{\mathrm{v}\rho w}{Sv(1-\varepsilon exp)\mu}\right)\right)4 - 1.5(\log \psi)^2$$

As a result, the expansion height (Hexp) is deduced from the following equation:

$$\frac{Hexp}{H} = \frac{1-\ \varepsilon}{1-\ \varepsilon exp}$$

8.8.6. *Cycle time*

This is the time between the start of filtration (clean filter) and the moment when the washing is triggered.

The cycle time is usually governed by one of the two following parameters: the pressure loss in the filter or the quality of the filtered water. The most limiting parameter triggers the backwash sequence.

Generally, washing becomes necessary when:

– the planned cycle time has been exhausted (washing on a timer);

– the volume of filtered water before washing (VFW) has been reached;

– the quality of the filtered water has deteriorated, the filter has “burst” (burst criteria: turbidity, particle count, oxidizable matter, aluminum, iron);

– the maximum pressure loss has been reached.

The cycle time is influenced by the filtration rate, as well as the content of SS, organic matter, algae, etc.

For SS retention filters, the duration of the filtration cycle depends on the retention capacity, the material volume and the SS flow to be separated:

$$\text{Cycle time [hours]} = \frac{\text{Surface } [m^2] x \text{ Retention capacity } [\text{kg}\cdot\text{m}^{-2}]}{\text{SS flow}[\text{kg}\cdot\text{h}^{-1}]}$$

The retention capacity is calculated in $\text{kg}\cdot\text{m}^{-2}$ for rapid filters and in $\text{kg}\cdot\text{m}^{-3}$ for TGV filters.

8.8.7. *Main parameters involved in filter sizing*

Filtration parameters	Comments
Filter water quality	Set by local regulations
Characteristics of the filter feed water	
SS concentration	Affects operating conditions and sizing
Turbidity	
Type of flocs	
Floc dimensions	

Filtration parameters	Comments
Characteristics of the filtering material(s) Effective size d_{10} Uniformity coefficient Form factor, density, type **Characteristics of the filter**	Affects particle removal and pressure loss
Filtration height (s)	Affect(s) the initial pressure loss and cycle time
Porosity (s)	Affects the amount of material that can be retained
Stratification	Affects filter performance and cycle time
Filtration rate Maximum acceptable pressure loss	Affects cycle time
Addition of reagents	Improves the turbidity of the filtered water
Washing conditions	Affects equipment sizing and material classification after washing

Table 8.9. *Main parameters involved in filter design*

8.9. Veolia filtration technologies: general information

Veolia has a family of filters used for the production of drinking water (Figure 8.27):

– filters used in the final clarification of water that was previously flocculated and decanted or floated or in direct filtration of raw water with a simple addition of reagent (**Filtraflo®F** or **F-TGV**);

– filters used for the filtration of water after decantation or flotation with a surface sweep (**Filtraflo® SV**);

– filters used on water that is loaded with dissolved oxygen (after flotation, aeration or inter ozonation (**Filtraflo® MD**)). They are used whenever a high water level is required to allow degassing of ozone or air in the case of aeration to oxidize iron and manganese upstream. They allow a contact time in the volume of water above the media that can be useful for chemical oxidation;

– filters used for direct filtration of preconditioned raw water with a turbidity of less than 20 NTU, for filtration after simple aeration of groundwater with an iron content of more than 3 mg/L (Fe removal), and for filtration after decantation or flotation: these are the dual media filters known as **Filtraflo® DC** (rapid or TGV).

These filters have the following common characteristics:

– a downward-direction filtration and an upward-direction washing (backwash);

– a false bottom made up of a filtering floor composed of slabs and nozzles, or Leopold-type filter bottoms or equivalent;

– a 100 mm layer of 4–7 mm gravel, covering the heads of the nozzles for the filtering floors;

– one or more layers of filtering materials;

– a constant water height during the filtration phase, maintained by downstream regulation;

– a 2-level channel (located between two filters), the lower part of which is designed to collect the filtered water and to distribute the dirty water, and the upper part is essentially used to collect the dirty washing water;

– washing equipment adapted to the characteristics of the filter.

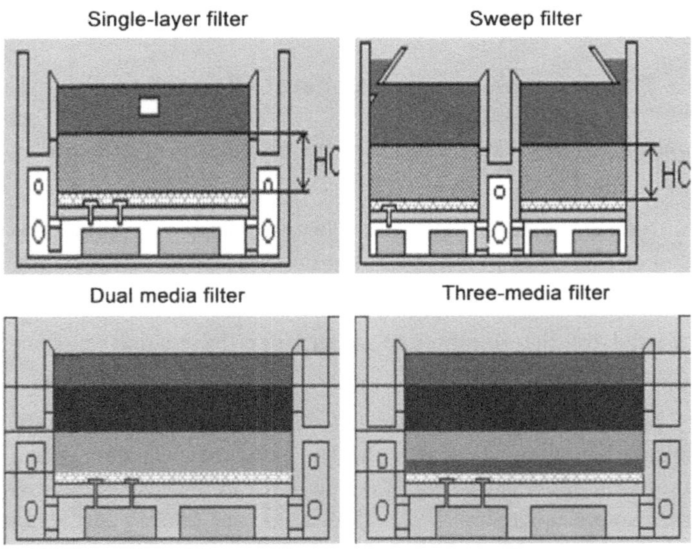

Figure 8.27. *Different types of Veolia filters. For a color version of this figure, see www.iste.co.uk/gaid/watertreatment2.zip*

According to the operation technology, we can distinguish:

– open filters, which are usually the most used, where water arrives by gravity at atmospheric pressure;

– pressure filters, where the water is subjected to pressure through the filter material.

Most filters are classified according to filtration rate, characteristics of the filter material(s) and operating conditions.

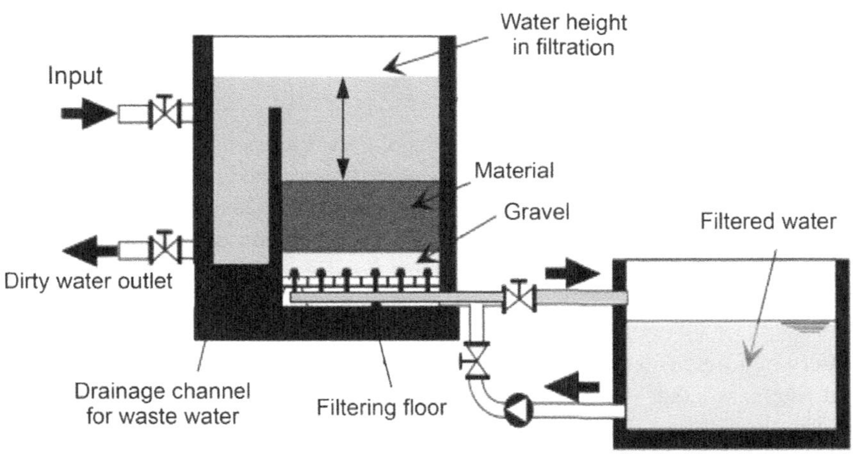

a)

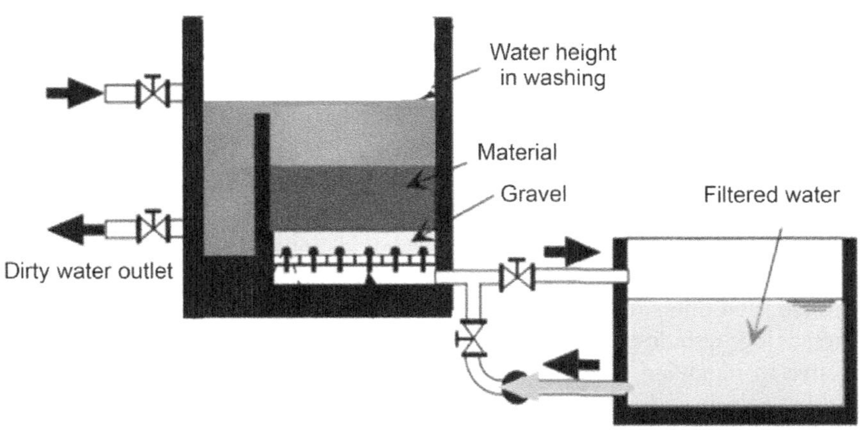

b)

Figure 8.28. *Single-media filter in filtration mode (a) and backwash mode (b). For a color version of this figure, see www.iste.co.uk/gaid/watertreatment2.zip*

We distinguish:

– rapid (Figure 8.28) or very high-rate single-media filters, multimedia filters based on rapid filtration or very high-rate filtration;

– rapid or high-rate sand filters are differentiated by the characteristics of the sand, the material height, the filtration rate, the retention capacity and the backwashing conditions.

The media is placed on an underdrain floor or on Leopold floors (or equivalent).

The design of the drained floor consists of a surface made of reinforced concrete slabs that are equipped with nozzles. The size of the nozzles depends on the sand grain size and the slab thickness.

The Leopold system consists of a series of PVC blocks that make up the surface of the floor and are equipped with meshes of sufficient size to retain the whole while letting the filtered water pass through. When the pressure loss reaches its acceptable limit due to the clogging of the filter by the accumulation of particles, the filter is backwashed with filtered water. The washing conditions depend on the particle size of the filter material.

Very high-rate filters (TGV filters) operate at rates three to four times faster than rapid filters, whether they are single or multimedia filters. Due to the diffusion of particles through the filter mass at greater depths, the retention capacities of these filters are much higher than those of rapid filters.

The distinction between rapid filters and TGV filters also comes from the particle removal mechanism: rapid filters retain particles on the surface of the filter material, whereas TGV filters retain them over a large part of the height of the filter mass.

Whether they are of the rapid or TGV type filters, they are preceded by a coagulation–flocculation–settling or flotation stage. This stage can itself be combined with other stages, such as chemical oxidation and softening.

Rapid sand filters remove relatively large suspended particles. They are effective in removing turbidity, iron and manganese. In addition, they are moderately effective in removing odors, bacteria and organics but have no effect on viruses, fluorides, arsenic and salts.

Table 8.10 shows the combinations of various Filtraflo® filter types with materials used in drinking water treatment.

	F	DC	TC	MD	SV
Sand	X	X	X		X
Anthracite		X	X		X
Pumice		X	X		X
Limestone	X				
MnO_2	X	X	X		
Granular activated carbon	X		X	X	X
Schist				X	X

Table 8.10. *Choice of the Veolia filter type according to its application*

COMMENTS ON TABLE 8.10.– *F: single-media filter; DC: dual-media filter; TC: three-media filter; MD: single-media filter with degassing; SV: single-media, dual-media or three-media filter with surface sweeping.*

Table 8.10 is used to select the most suitable filter and material(s) for the desired treatment and the position of the filter in the system.

8.9.1. *Characteristics of open gravity filters*

8.9.1.1. *Single-media filters: Filtraflo®F*

These filters are the most widely used due to their simplicity and low investment costs. They are characterized by an operating water height of 0.6–1.3 m, the advantage of which is a rapid removal of the wash water and low water loss.

Filtraflo®F are open gravity flow filters with a rectangular casing usually made of reinforced concrete. They are characterized by:

– a single filtering layer of homogeneous particle size resting on an underdrain floor (or on a Leopold type floor, or equivalent) that supports the filter media, collects filtered water, and distributes backwash water and air scouring.

– a variable media height between 0.9 and 1.3 m;

– a three-phase backwash (air, air/water and water alone for rinsing).

They are installed after coagulation–flocculation–settling, regardless of the type of settler. They are also proposed in direct filtration after coagulation (in line or in a specific tank) and flocculation (in a specific tank or in line).

Type and height of material applicable to rapid filters and TGV filters:

– sand: 0.9 to 2.0 m (d_{10} = 0.8–1.35 mm and UC $\leq$ 1.3);

– limestone: 0.9–3.0 m;

– MnO_2: 0.9–2.0 m (d_{10} = 1 mm and UC < 1.5);

– activated carbon: the contact time is the sizing parameter. The height can vary between 0.8 and 3.5 m (d_{10} = 1 mm and UC < 1.5) (see Chapter 9).

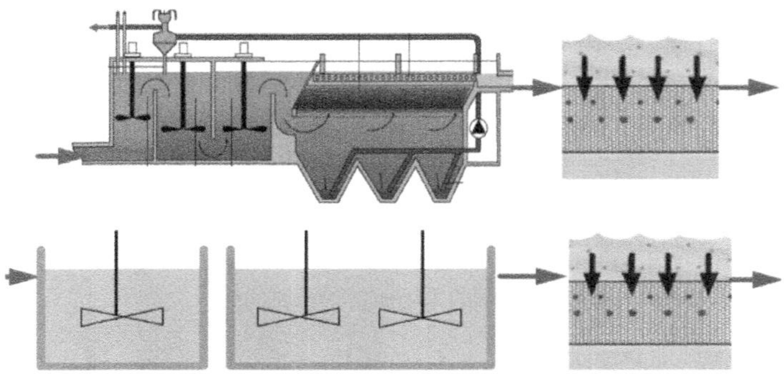

Figure 8.29. *Some layouts of the Filtraflo®F after an Actiflo® or in direct filtration. For a color version of this figure, see www.iste.co.uk/gaid/watertreatment2.zip*

The filtration layer of Filtraflo® sand is between 0.9 and 2 m.

The following correlation can be applied between the sand height and the maximum filtration rate based on N-1 filters:

$$H_{mat} = 0.077 \times V_{filtlim} + 0.47$$

H_{mat} is given in m and $V_{filtlim}$ in $m \cdot h^{-1}$. The filtration rates are therefore between 6 and 20 $m \cdot h^{-1}$. Figure 8.30 shows a description of the Filtraflo®F.

The grain size of the sand depends on the field of application (settled water filtration, direct filtration, etc.) considering the filtration rate and height:

– ES of 0.85 mm: filtration of clarified water on rapid filters. Height of 0.9 m and rate in N filters of 6 $m \cdot h^{-1}$;

– ES of 1.00–1.2 mm: filtration after coagulation/flocculation/settling or direct filtration after coagulation and flocculation without settling (turbidity < 5 NTU) for rapid filters with a height of 1.0–1.3 m and a rate in N filters of 8–10 $m \cdot h^{-1}$;

– ES of 1.35 mm: filtration after coagulation/flocculation/settling or direct filtration after coagulation and flocculation without settling (turbidity < 5 NTU), for very high-rate filters, for a height of 2 m and a rate in N filters of 15 $m \cdot h^{-1}$.

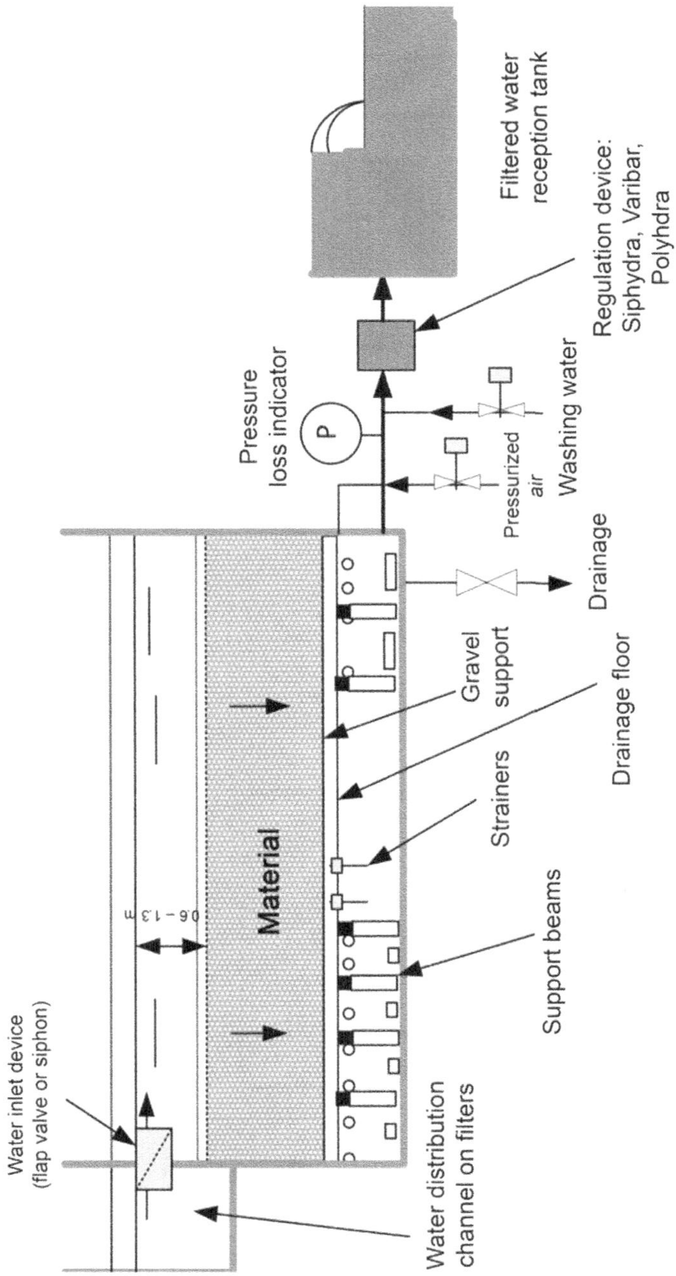

Figure 8.30. *Filraflo®F. For a color version of this figure, see www.iste.co.uk/gaid/watertreatment2.zip*

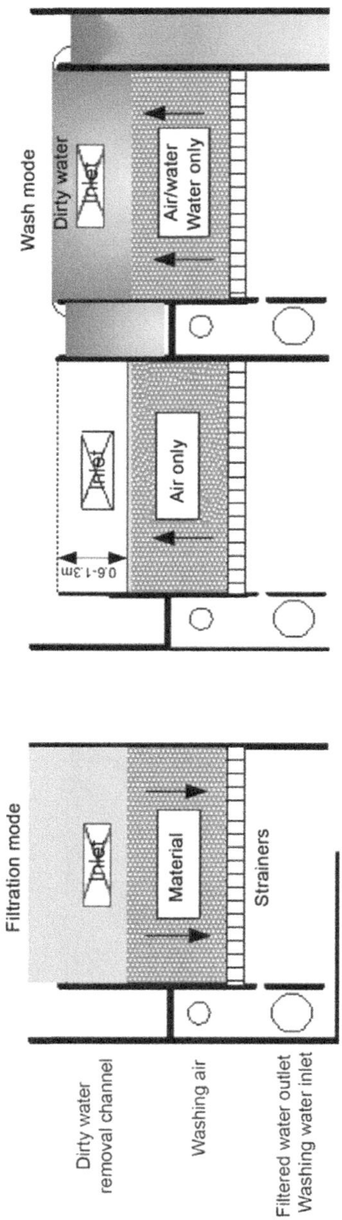

Figure 8.31. *Filtraflo®F washing procedure. For a color version of this figure, see www.iste.co.uk/gaid/watertreatment2.zip*

The floor of the filter of reinforced concrete slabs is equipped with nozzles at a density of 45–50 nozzles/m^2 of floor slab. These nozzles are embedded in a layer of gravel. The slabs are supported by beams. Holes are provided in the beams (by means of rigid PVC tubes embedded in the concrete) to facilitate the redistribution of the air washing along the length of the filter.

The backwash cycle includes:

– a lowering of the water level to the level of the material, followed by the injection of air (50 $Nm^3 \cdot m^2 \cdot h^{-1}$) for 1–2 min;

– the air/water phase is then implemented with filtered water;

– the air/water phase is carried out with an injection of air (50 $Nm^3 \cdot m^2 \cdot h^{-1}$) and a backwash of filtered water at a flow rate of 8–15 $m^3 \cdot m^2 \cdot h^{-1}$. The time allotted to this phase is 8 min but can be extended to 12 min if necessary. This phase must not allow the water to overflow above the filter bases, as this could result in loss of material;

– the rinse phase is carried out with filtered water alone (25–60 $m^3 \cdot m^2 \cdot h^{-1}$), depending on the type of material and the operating conditions. The time allotted to this phase is 8 min but can be extended to 15 min if necessary. Typically, it is recommended to keep the backwash water flowing until the drained water is clear;

– the washing water is removed on 1 or 2 sides, depending on the width of the cell, as shown in Figure 8.31.

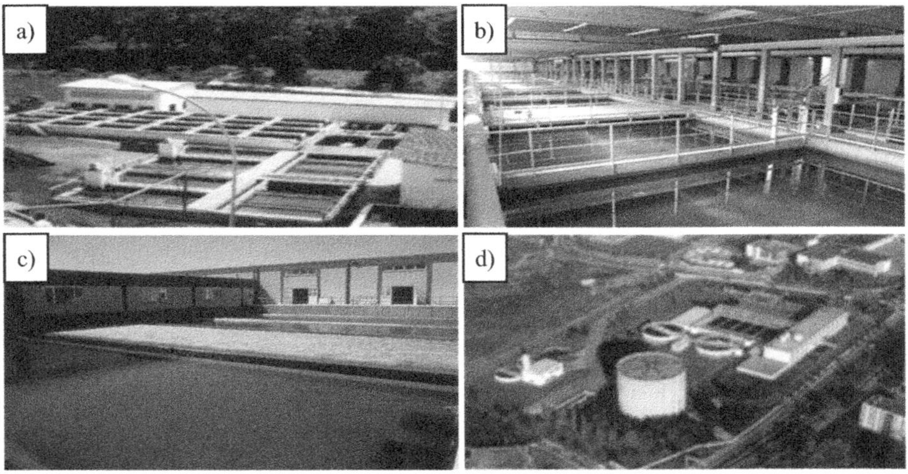

Figure 8.32. *Filtraflo®F installed in (a) Yaoudé (Cameroon), (b) Annet-sur-Marne (France), (c) Omerli (Turkey) and (d) Rennes (France). For a color version of this figure, see www.iste.co.uk/gaid/watertreatment2.zip*

The volume of the dirty water reception tank is calculated on the basis of 2–3 volumes of material.

The initial filtered water after backwashing is usually directed toward the sewer or sometimes recirculated (depending on local regulations). A lyre with a vent is installed on the wash water collector after the initial water outlet valve (one valve for all filters), with the same diameter as the filtered water outlet valve of each filter.

However, this initial filtered water can also be sent directly to the disinfection stage without any treatment. This protocol depends on the regulations applied in each country.

8.9.1.2. *Very high-rate filters: Filtraflo® F-TGV*

A variation of the Filtraflo® F, Filtraflo® F-TGV are open gravity-flow filters with a rectangular tank that is usually made of reinforced concrete. They are characterized by:

– a single filtering layer of homogeneous particle size resting on an underdrain floor (or on a Leopold type floor, or equivalent) that supports the filter media, collects filtered water, and distributes backwash water and air scouring;

– a variable material height up to 2 m;

– an ES of 1.35 mm and a UC < 1.4;

– a three-phase backwash (air, air/water and water alone for rinsing, see Table 8.8).

They are installed after coagulation–flocculation–decantation, regardless of the type of settler. They are also proposed in direct filtration after coagulation (in line or in a specific tank) and flocculation, in a specific tank or in line.

The principle of very high-rate filtration is to increase the filtration height and the particle size of the filter material to increase the storage capacity of the SS in the filters. This is because the operating principle is deep floc diffusion into the filter mass, aided by a high filtration rate. The pressure loss observed in the filter is linear.

Increasing the height of the material means that a higher pore volume is available for particle retention, which in turn improves the overall performance of the filter. As such, the recommended H/d_{10} ratio should be $>1{,}300$. However, the increase in sand height is not significant if not accompanied by a high hydraulic load; otherwise, most of the SS will simply accumulate in the upper part of the sand bed.

Increasing the water load improves the filtration cycles because a larger pressure drop is available in this case.

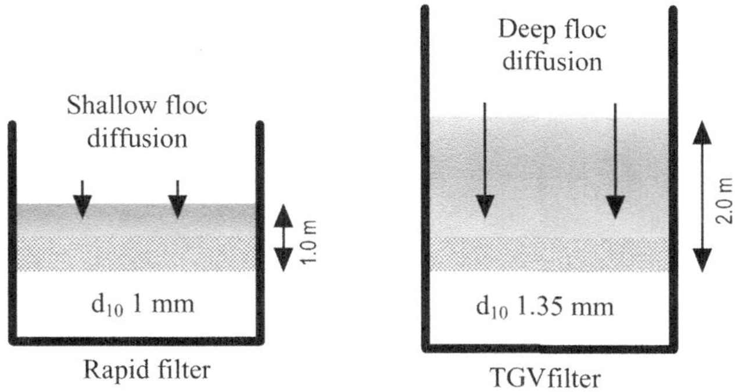

Figure 8.33. *Floc diffusion diagram in a rapid filter and an F-TGV filter. For a color version of this figure, see www.iste.co.uk/gaid/watertreatment2.zip*

The F-TGV filters operate at speeds of 15–17 m in N and 20 $m \cdot h^{-1}$ in N-1. As a result, the floor surface and the number of filters are reduced.

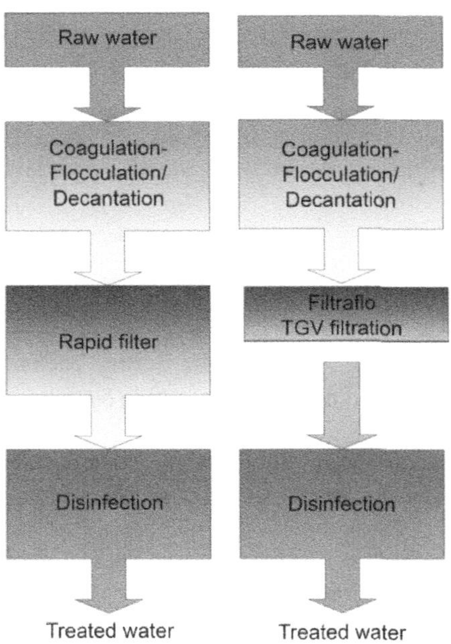

Figure 8.34. *Treatment process with filters on material for the F-TGV filter. For a color version of this figure, see www.iste.co.uk/gaid/watertreatment2.zip*

The retention capacity is expressed in kg·m^{-3} because the high filtration rate makes the particles diffuse in depth on at least 1.5 m. This value is approximately 3–4 kg SS·m^{-3} of material volume.

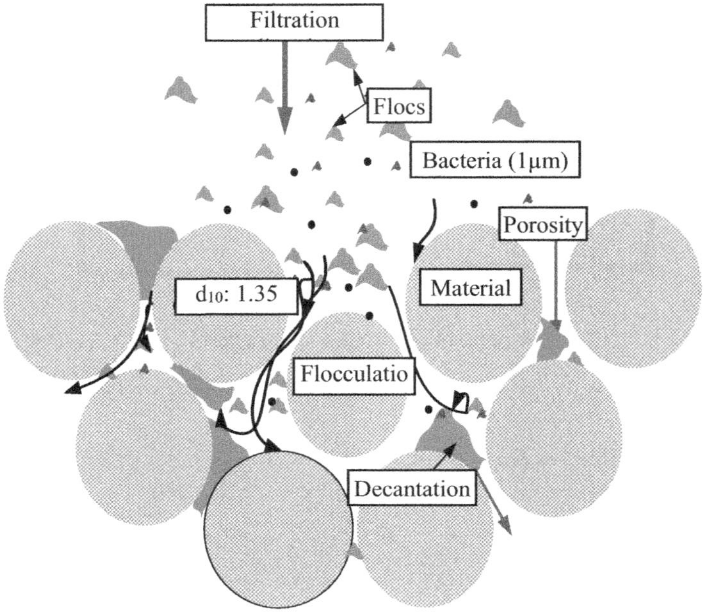

Figure 8.35. *Rapid particle attachment mechanisms and F-TGV filters. For a color version of this figure, see www.iste.co.uk/gaid/watertreatment2.zip*

The height of water above the filter mass is calculated by considering:

– the minimum water height to avoid excessive depression in the filter mass;

– the minimum acceptable water height during backwashing to ensure good removal of the SS accumulated during an operating cycle.

The minimum water height in filtration must not be less than 0.8 m above the level of the media. The acceptable wash water height for good SS removal is 1.4 m.

The available pressure loss on the filtering mass (clean filter pressure loss + pressure loss during the filtration cycle) is then calculated at the maximum flow rate on N-1 filters while also taking into account the pressure loss contributed by the nozzles of the floor, as well as that of the valves and filtered water pipes.

In the case of an individual restitution tank, the maximum available pressure loss on the filtering mass is in the range of:

– 2.6 m water column for a media height of 2 m;

– 2 m water column for a media height of 1.5 m.

In the case of a significant flow variation, during a low flow, the F-TGV filter can behave like a conventional filter at 10 $m \cdot h^{-1}$ and therefore clog the surface, which leads to faster clogging of the filter and makes it necessary to wash more often. Therefore, the degree of clogging must be taken into account according to the following formula:

$$d° \, (\%) = 100 \frac{\Delta P(t)}{\Delta Pmaxi} \left(\frac{vmaxi}{v(t)} \right)^{\alpha}$$

with:

– vmaxi: maximum filtration rate (in $m \cdot h^{-1}$);

– v(t): filtration speed at time t (in $m \cdot h^{-1}$);

– ΔPmaxi: maximum acceptable pressure loss of the material (in m water column);

– ΔP(t): pressure loss of the material at time t (in m water column).

This degree of clogging makes it possible to integrate the physical clogging of the filters with a few series of washes when returning to high flow rates. The coefficient α can be set at start-up with a value between 0.5 and 1.0.

The cleaning cycle includes:

– a lowering of the water level to the level of the material, followed by the injection of air (50 $Nm^3/m^2 \cdot h$) for 1–2 min;

– the air/water phase is then implemented with filtered water;

– the air/water phase is carried out with an injection of air (50 $Nm^3 \cdot m^2 \cdot h^{-1}$) and backwash filtered water with a flow rate of 12–15 $m^3 \cdot m^2 \cdot h^{-1}$. The time allotted to this phase is 8 min but can be extended to 12 min if necessary. This phase must not allow the water to overflow above the filter bases, as this could result in loss of media;

– the rinse phase is carried out with filtered water alone (55–60 $m^3 \cdot m^2 \cdot h^{-1}$), depending on the type of media and the operating conditions. The time allotted to this phase is 8 min but can be extended to 15 min if necessary. Typically, it is recommended to keep the backwash water flowing until the drained water is clear.

The filter inlet consists of pipes equipped with isolating butterfly valves and a device for optimizing flow distribution and anti-scouring. Due to the high speeds, scouring phenomena can occur.

In addition, good diffusion of the flow by means of diffusion cones must be implemented to avoid excessive turbulence.

The floor of the reinforced concrete slab filter layer is equipped with long-tail nozzles at a density of 50–65 nozzles/m² of floor slab. These nozzles are embedded in a layer of gravel. The slabs are supported by beams. Holes are provided in the beams (by means of rigid PVC tubes embedded in the beams) to facilitate the redistribution of the washing air along the length of the filter. The volume of the dirty water reception tank is calculated on the basis of 2–3 volumes of material.

The initial filtered water after washing is usually directed to the sewer or sometimes recirculated (depending on local regulations). A lyre with a vent is installed on the wash water collector after the initial water outlet valve (one valve for all filters), with the same diameter as the filtered water outlet valve of each filter.

However, this initial filtered water can also be left untreated and sent to the disinfection stage. This protocol depends on the regulations applied in each country.

F-TGV filters have longer cycle times than rapid filters due to their higher retention capacity.

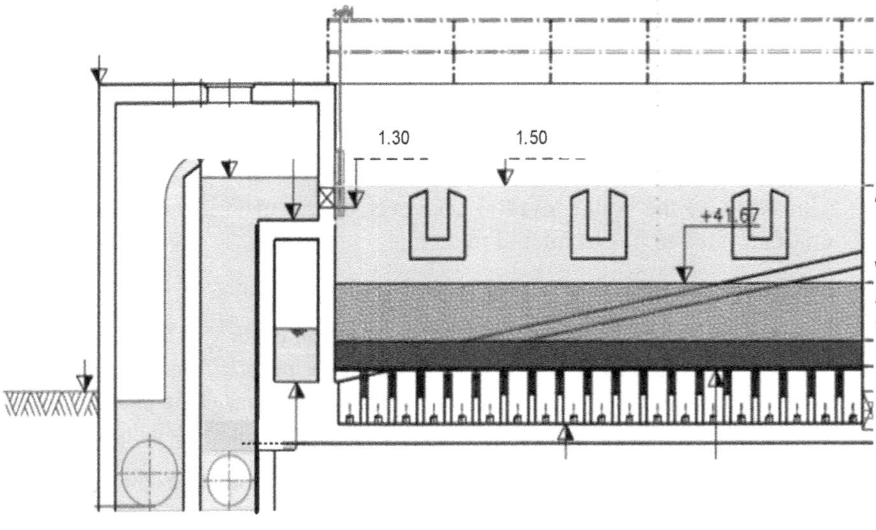

Figure 8.36. *Filtraflo® F-TGV. For a color version of this figure, see www.iste.co.uk/gaid/watertreatment2.zip*

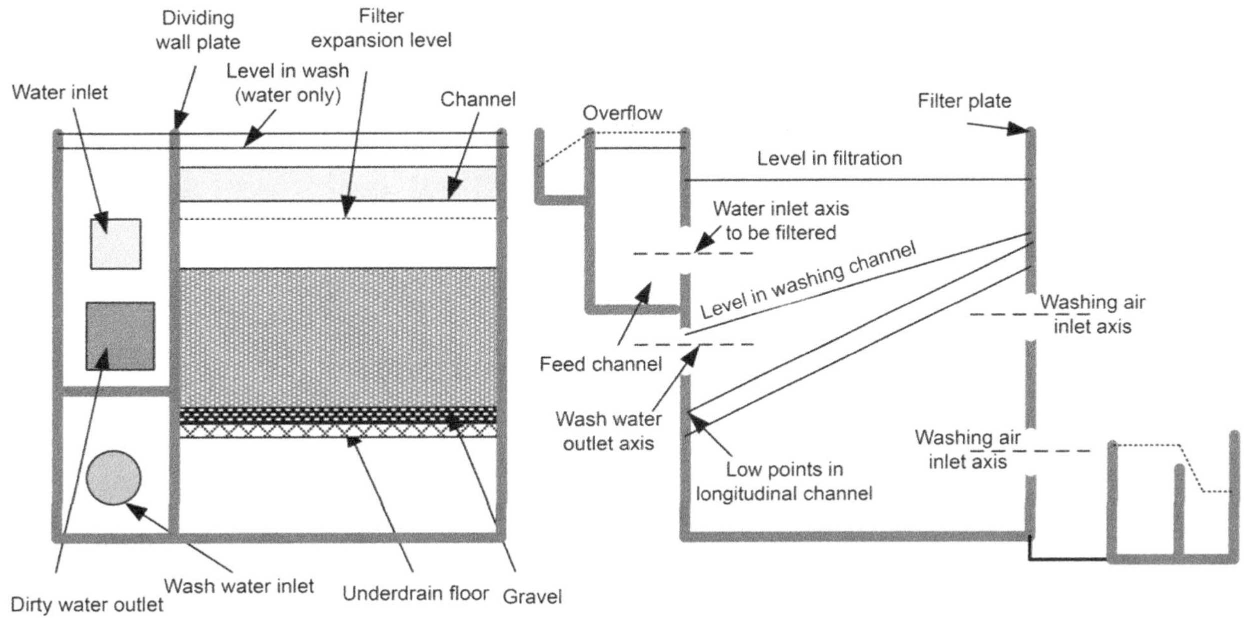

Figure 8.37. *Filtraflo® F-TGV in operation and in wash mode. For a color version of this figure, see www.iste.co.uk/gaid/watertreatment2.zip*

Figure 8.38. *Filtraflo® F-TGV in (a) Chengdu (China) and (b) Huachipa (Peru). For a color version of this figure, see www.iste.co.uk/gaid/watertreatment2.zip*

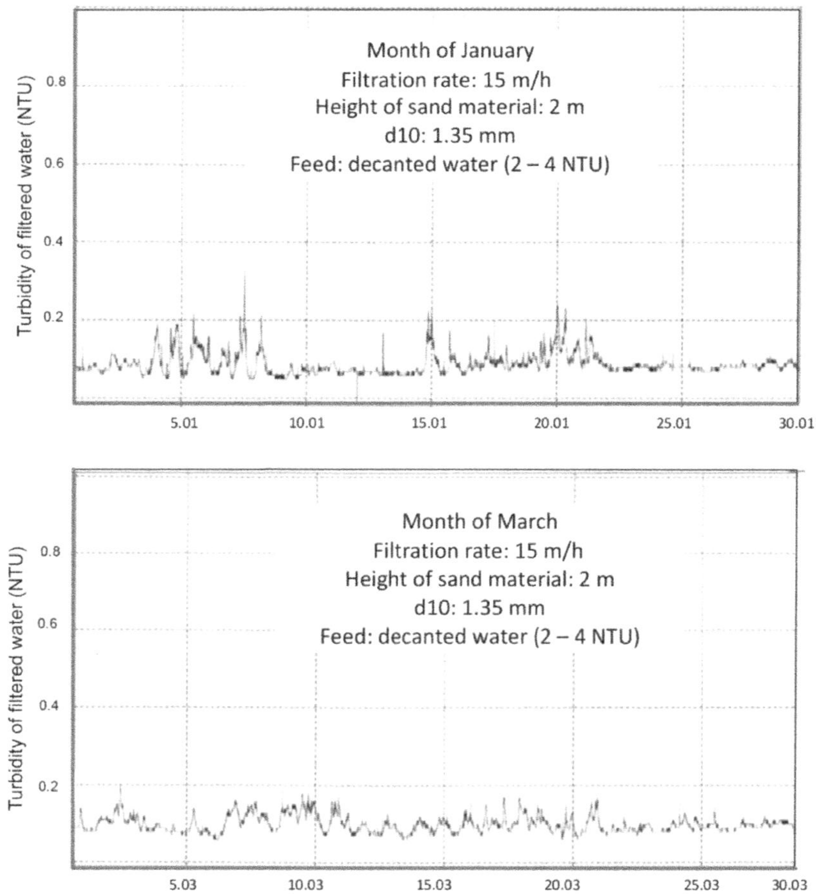

Figure 8.39. *Turbidity recordings of filtered water on Filtraflo® F-TGV for 2 months (January and March) at the Chengdu plant (China). For a color version of this figure, see www.iste.co.uk/gaid/watertreatment2.zip*

The Filtraflo® F-TGV filtered water turbidity recordings at the Chengdu plant (China) during the 20 years of operation (1998–2018) have always indicated a turbidity well below 0.5 NTU (manufacturer's guarantee). On average, it is below 0.2 NTU. However, at the end of the cycle (24–36 h), turbidity rises to 0.4 NTU and requires a washing of the filter.

8.9.1.3. *Surface sweep filters: Filtraflo®SV*

The Filtraflo®SV, also known as a sweeping filter, is a filter made of one or more materials. The Filtraflo®SV is an open gravity flow filter with a rectangular casing, usually constructed of reinforced concrete.

It is characterized by:

– a single filtering layer of homogeneous particle size resting on an underdrain floor that supports the filter media, collects filtered water and distributes backwash water and air scouring;

– a variable material height between 0.9 and 2 m;

– an ES of 0.85–1.0 mm and a UC $\leq$ 1.3;

– a three-phase backwash (air, air/water, and water alone for rinsing), combined with a surface sweep with decanted water or raw water (see Table 8.8).

They are installed after coagulation–flocculation–decantation, regardless of the type of settler. They are also proposed in direct filtration after coagulation (in line or in a specific tank) and flocculation in a specific tank or in line, when water losses must be reduced.

The sweep system can be applied to single-media filters, as well as to multimedia filters.

The main advantage of this system is that it reduces the amount of filtered water needed for rinsing by sweeping the surface with unfiltered water, which facilitates the removal of SS released during the various washing phases. The wash water brought by the sweeping thus makes it possible to accelerate the removal of dirty wash waters.

The intake consists of a spillway and 2 valves: one for the water to be filtered and another for sweeping (30% nominal flow). The second valve can be replaced by a partial and calibrated opening of the main valve.

Filtration rates vary between 8 and 15 $m \cdot h^{-1}$ in N, depending on the application and filter type. The water height above the media varies between 1.10 and 1.50 m.

The pressure loss is calculated according to the height of the material and the height of water above the media.

For example, for a Filtraflo®SV with a media height of 1.4 m and a water height of 1.1 m, the maximum allowable pressure loss is 2.70 m.

The floor of the filter layer consists of reinforced concrete slabs, which are equipped with long-tail nozzles with a density of 40–50 nozzles/m² of floor slab. These nozzles are embedded in a layer of gravel. The slabs are supported by beams.

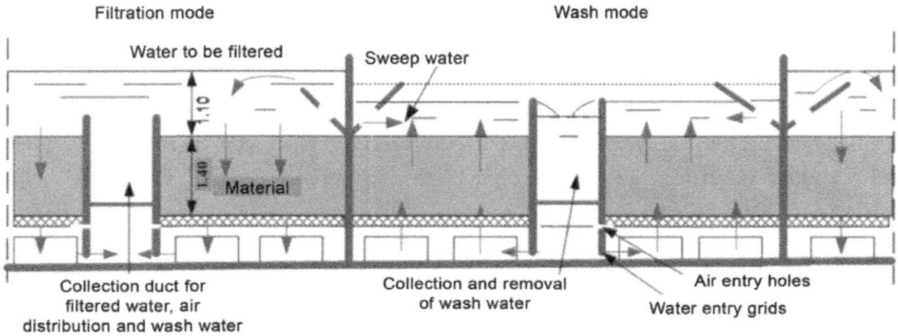

Figure 8.40. *Filtraflo®SV operating diagram. For a color version of this figure, see www.iste.co.uk/gaid/watertreatment2.zip*

The feeder system consists of feeder spillways that allow an even distribution of water to all filters. Depending on their size, the filters have one or more V-shaped side channels, which are either used to ensure a uniform distribution of water over the entire surface (filtration mode) or to create a horizontal sweeping current (wash mode).

The filtered water is collected in a pipe placed under the backwash wastewater reception channel. It is then returned via a regulation system to a reception tank to be evacuated to a thickener or other recovery method. This pipe is also used to ensure the distribution of water and air for the washing.

The cleaning cycle includes:

– a lowering of the water level to the level of the media, followed by the injection of air (50 $Nm^3 \cdot m^{-2} \cdot h^{-1}$) for 1–2 min;

– the air/water phase is then implemented with a water sweeping operation on the surface. The latter is either filtered water or decanted or raw water when it is of good quality (particular and specific conditions);

– the air/water phase is carried out with an air injection (50 $Nm^3 \cdot m^{-2} \cdot h^{-1}$) and backwash filtered water at a flow rate of 8–15 $m^3 \cdot m^{-2} \cdot h^{-1}$. The flow rate of the sweep water is equivalent to the filtration flow rate. The time allotted to this phase is 8 min but can be extended to 12 min if necessary. This phase must not allow the water to overflow above the filter bases, as this could result in loss of media;

– the rinse phase is carried out with filtered water alone (25–50 $m^3 \cdot m^{-2} \cdot h^{-1}$), depending on the type of material and the operating conditions. The surface sweeping is always in operation and in the same flow rates as the second phase. The time allotted to this phase is 8 min but can be extended to 15 min if necessary. Typically, it is recommended to keep the backwash water flowing until the drained water is clear.

The washing fluids (air and water) are distributed under the floor from a single pipe located under the dirty water reception channel. The distribution system consists of rectangular holes that are level with the floor for water and open tubes under the floor for air injection.

As with other filtration technologies, the cleaning cycle is launched on the basis of a pressure loss level, a turbidity of filtered water, a clock or an operator's decision.

Figure 8.41 shows a single-filter or two-filter configuration in the filtration and backwashing phases.

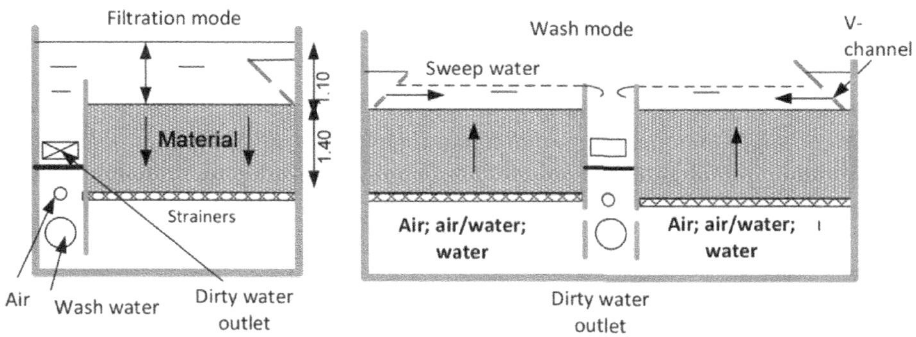

Figure 8.41. *Filtraflo®SV in filtration mode and cleaning mode. For a color version of this figure, see www.iste.co.uk/gaid/watertreatment2.zip*

The advantage of these filters is that they always operate on the total filters (N filters) at a constant flow rate, and there is therefore no increase in flow per filter during the cleaning period, since the settled water or raw water is used as sweep water.

Figure 8.42. *Filtraflo®SV (China). For a color version of this figure, see www.iste.co.uk/gaid/watertreatment2.zip*

8.9.1.4. *Dual media filters: Filtraflo®DC*

Filtraflo®DCs are open gravity flow filters with a rectangular casing, usually constructed of reinforced concrete.

They are characterized by:

– a double filtering layer, resting on an underdrain floor that supports the filter media, collects filtered water and distributes backwash water and air scouring. The floor can be based on a Leopold type of floor, or equivalent;

– a high water height above the media (1.2–1.8 m), making it possible to ensure the degassing of previously aerated or ozonated water, a sufficient flocculation time in case of direct filtration and longer cycle times due to a higher available pressure loss;

– a total media height varying between 1.2 and 2.0 m;

– a three-phase backwash (air, air/water and water alone for rinsing).

They are installed after coagulation–flocculation–settling, regardless of the type of settler. They are also proposed in direct filtration after coagulation (in line or in a specific tank) and flocculation, in a specific tank or in line. They are also proposed when the settler is bypassed, due to the low turbidity of the raw water, which occurs in certain periods of the year.

Dual media filters are used in the same basic configuration as rapid filters. The top layer material has a lighter density than the bottom layer material but a higher porosity.

The porosity of sand is typically 45%, while anthracite has a porosity of 50–55%. As a result, expansion during backwashing and stratification after washing of both media are easy to implement.

Media combinations for this type of Veolia filter include, for example:

– anthracite-sand;

– pumice-sand;

– activated carbon-sand;

– sand-manganese dioxide (MnO_2);

– activated carbon-manganese dioxide (MnO_2).

The filtration rates depend on the total height of the media. They can vary between 8 and 17 $m \cdot h^{-1}$ in N.

The retention capacity of Filtraflo®DC depends on the types of media used. For example, a dual media anthracite-sand filter has a retention capacity of approximately 2 $kg \cdot m^{-2}$. It accepts an inlet turbidity up to 15–20 NTU.

The water is taken into the filters from a general distribution channel through a submerged pipe equipped with an on–off isolation valve. A deflector is placed under the valve to avoid scouring of the material layers. The filtered water is collected in a pipe and placed under the wastewater reception channel. It is then returned through an outlet pipe equipped with a regulation system into a reception tank with an open water level, from where it is evacuated by a spillway.

The cleaning cycle includes:

– a lowering of the water level to the level of the upper layer media, followed by the injection of air (50 $Nm^3 \cdot m^{-2} \cdot h^{-1}$) for 1–2 min;

– the air/water phase is then implemented with filtered water. It is carried out with an injection of air (50 $Nm^3 \cdot m^{-2} \cdot h^{-1}$) and backwash filtered water at a rate of 8–15 $m^3 \cdot m^{-2} \cdot h^{-1}$. The time allotted to this phase is 8 min but can be extended to 12 min if necessary. This phase must not allow the water to overflow the filter bases, as this could result in loss of media;

– the rinse phase is carried out with filtered water alone (35–60 $m^3 \cdot m^{-2} \cdot h^{-1}$), depending on the type of media and the operating conditions. The time allotted to this phase is 8 min but can be extended to 15 min if necessary. Typically, it is recommended to keep the backwash water flowing until the drained water is clear.

The cleaning protocol will be adapted to the material used (depending on the expansion rate of the anthracite).

The support floor for the filter layers consists of reinforced concrete slabs equipped with long-tail nozzles with a density of 50 units/m^2 of floor slab. The heads of the nozzles are embedded in a layer of gravel, which support the layers of media. The floor slabs are supported by reinforced concrete beams.

The washing fluids (air and water) are distributed under the floor from a single pipe located under the dirty water reception channel. The distribution system consists of rectangular holes that are level with the floor for water and open tubes under the floor for air injection.

As with other filtration technologies, the cleaning cycle is launched on the basis of a pressure loss level, a turbidity of the filtered water, a clock or an operator's decision.

The diagram in Figure 8.43 shows a single-filter or two-filter configuration in the filtration and washing phases.

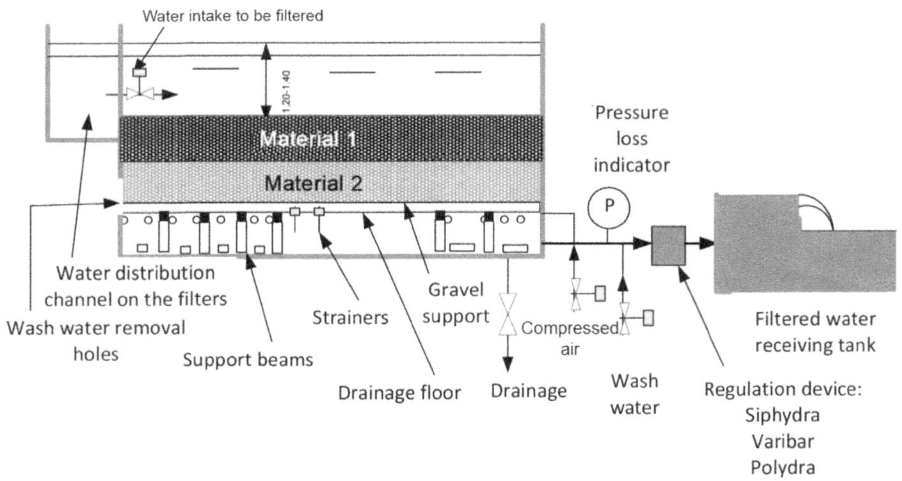

Figure 8.43. *Filtraflo®DC description. For a color version of this figure, see www.iste.co.uk/gaid/watertreatment2.zip*

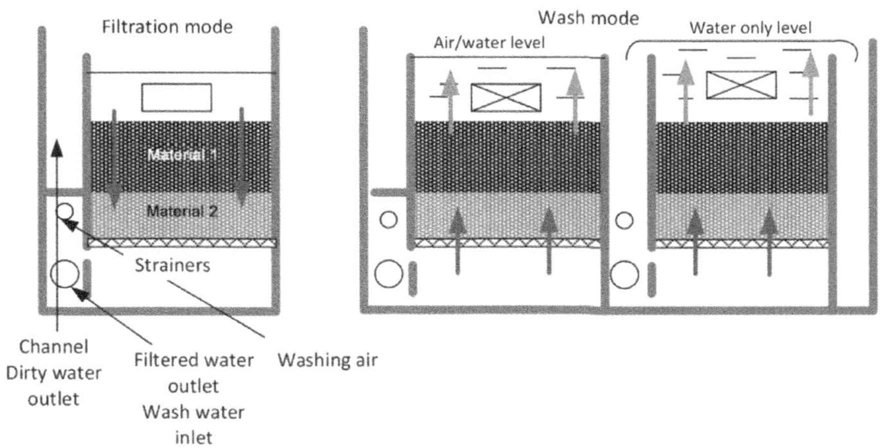

Figure 8.44. *Filtraflo®DC cross-section. For a color version of this figure, see www.iste.co.uk/gaid/watertreatment2.zip*

The turbidity of the filtered water obtained with the Filtraflo®DC is less than 0.2 NTU, as shown in the example of Figure 8.45 on the Tibitoc plant (Colombia).

The total height of the material (anthracite-sand) is 1.8 m, and the filtration speed is 16 m·h^{-1}.

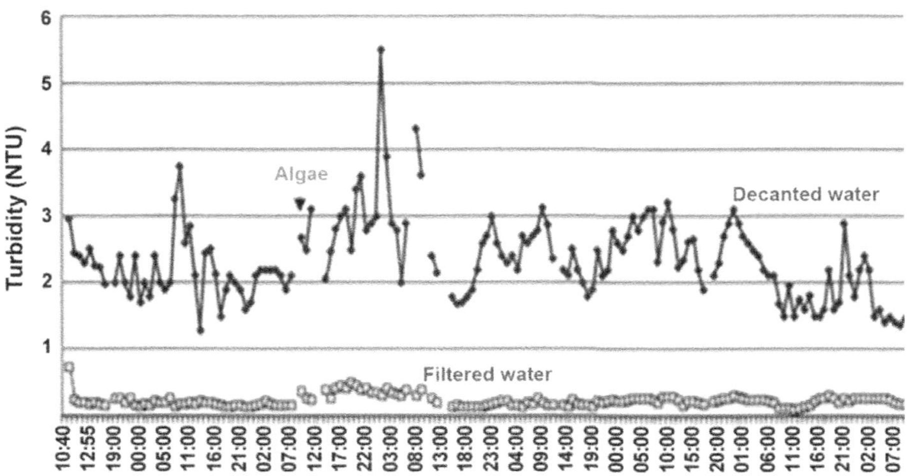

Figure 8.45. *Quality of filtered water at the Tibitoc plant (Colombia). For a color version of this figure, see www.iste.co.uk/gaid/watertreatment2.zip*

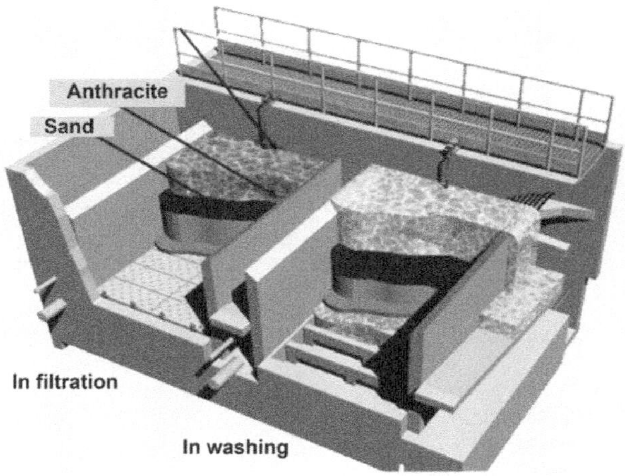

Figure 8.46. *Anthracite-sand Filtraflo®DC. For a color version of this figure, see www.iste.co.uk/gaid/watertreatment2.zip*

Figure 8.47. *Rapid Filtraflo®DC in Villefranche, France (a), and TGV type in Woronora, Australia (b), Oslo, Norway (c) and Tibitoc, Colombia (d). For a color version of this figure, see www.iste.co.uk/gaid/watertreatment2.zip*

8.9.1.5. *Three-media filters: Filtraflo®TC*

Filtraflo®TC are open gravity flow filters with a rectangular casing, usually made of reinforced concrete. They are characterized by:

– a triple layer resting on an underdrain floor that supports the filter media, collects filtered water, and distributes backwash water and air scour. The floor can be based on the Leopold type of floor, or equivalent;

– a high water height above the filtering media (1.2–1.8 m), making it possible to ensure the degassing of previously aerated or ozonated water, a sufficient flocculation time in case of direct filtration, and longer cycle times due to a higher available pressure loss;

– a total height of media varying between 1.5 and 2.0 m;

– a three-phase backwash (air, air/water and water alone for rinsing).

They are installed after coagulation–flocculation–settling, regardless of the type of settler, following softening. They are also proposed in direct filtration after coagulation (in line or in a specific tank) and flocculation, in a specific tank or in line.

Three media filters are used in the same basic configuration as dual media filters. The materials in the three layers must have different densities, going from the lightest (top layer) to the heaviest (bottom layer). The two upper-layer media are particularly affected by porosity, such as between sand and anthracite. The third layer is made of a material whose porosity has little effect on performance. Its density is what is essential. This is why garnet (turbidity and microparticle removal) or manganese dioxide is used to remove iron, manganese and other metals.

Material combinations can, for example, be:

– anthracite-sand-manganese dioxide (MnO_2);

– pumice-sand-manganese dioxide (MnO_2);

– activated carbon-sand-manganese dioxide (MnO_2);

– anthracite-sand-garnet;

– pumice-sand-garnet.

The filtration speeds depend on the total height of the media. They can vary between 8 and 17 $m \cdot h^{-1}$ in N.

The retention capacity of Filtraflo®TC depends on the types of materials used. This is identical to that of a dual media filter because the media in the upper layers remove most of the particles.

Water is fed into the filters from a general distribution channel through a flooded tube equipped with an on-off isolation valve (Figure 8.48). A deflector is placed under the valve to prevent scouring of the media layers. The filtered water is collected in a pipe placed under the dirty water reception channel from the washing process. It is then returned through an outlet pipe equipped with a regulation system into a reception tank with an open water surface, from where it is evacuated by a spillway. The filtered water collection pipe is also used to ensure the distribution of the washing water and air.

The cleaning cycle includes:

– a lowering of the water level to the level of the upper layer media, followed by the injection of air (50 $Nm^3 \cdot m^{-2} \cdot h^{-1}$) for 1–2 min;

– the air/water phase is then implemented with filtered water;

– the air/water phase is carried out with an injection of air (50 $Nm^3 \cdot m^{-2} \cdot h^{-1}$) and backwash filtered water at a flow rate of 8–15 $m^3 \cdot m^{-2} \cdot h^{-1}$. The time allotted to this phase is 8 min but can be extended to 12 min if necessary. This phase must not allow the water to overflow above the filter bases, as this could result in loss of media;

– the rinse phase is carried out with filtered water alone (35–60 $m^3 \cdot m^{-2} \cdot h^{-1}$), depending on the type of material and the operating conditions. The time allotted to this phase is 8 min but can be extended to 15 min if necessary. Typically, it is recommended to keep the backwash water flowing until the drained water is clear. The wash will be adapted to the media used (depending on the expansion rates of the anthracite).

The support floor for the filter layers consists of reinforced concrete slabs equipped with long-tail nozzles with a floor area density of 50 units/m^2. The heads of the nozzles are embedded in a layer of gravel on which the layers of media are laid. The floor slabs are supported by reinforced concrete beams.

The washing fluids (air and water) are distributed under the floor from a single pipe located under the dirty water reception channel. The distribution system consists of rectangular holes that are level with the floor for water and of open tubes under the floor for air injection.

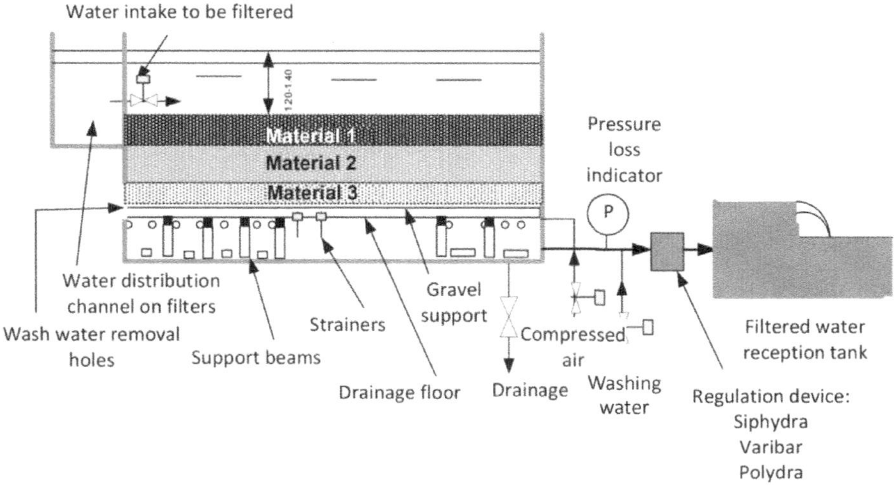

Figure 8.48. *Filtraflo®TC description. For a color version of this figure, see www.iste.co.uk/gaid/watertreatment2.zip*

As with other filtration technologies, the cleaning cycle is launched on the basis of a pressure loss level, a turbidity of the filtered water, a clock or an operator's decision.

Figure 8.49 shows a single-filter or two-filter configuration in the filtration and cleaning phases.

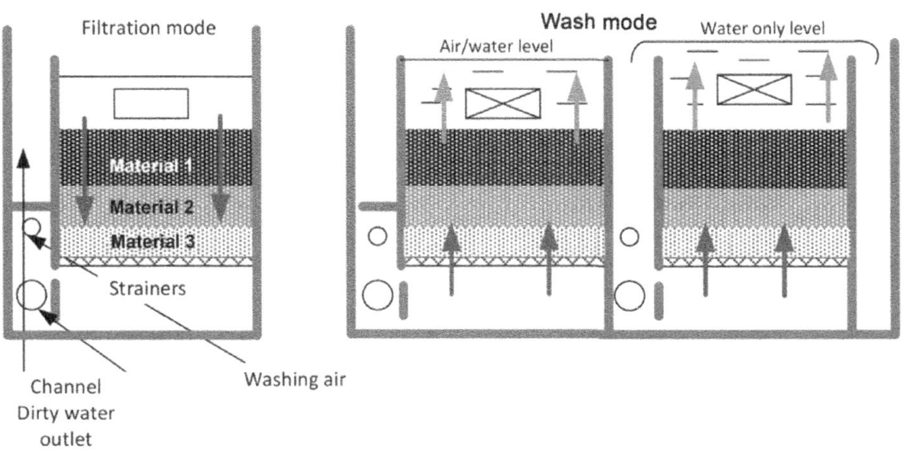

Figure 8.49. *Filtraflo®TC cross-section. For a color version of this figure, see www.iste.co.uk/gaid/watertreatment2.zip*

8.9.1.6. *FV pressure filters*

FV-type filters are closed filters operating under pressure, whose construction enclosure, which is generally metallic, consists of a cylindrical shell that is placed vertically and closed by two dome-shaped ends.

They are characterized by:

– a filtering layer with one or more materials;

– backwash water according to the three-phase protocol;

– a filtration speed: 6–18 m·h^{-1};

– one (or more) filtering layer(s): FV filters can contain one, two or three materials according to the same principle as described for single-layer, dual media or three-layer filters.

Generally, pressurized filters work under a pressure that corresponds to the upper level of the filtering layer. The pressure loss of a filter is checked by measuring the pressure difference between the entry pipe of the water to be filtered and the exit pipe of the filtered water (upstream and downstream pressure gauges).

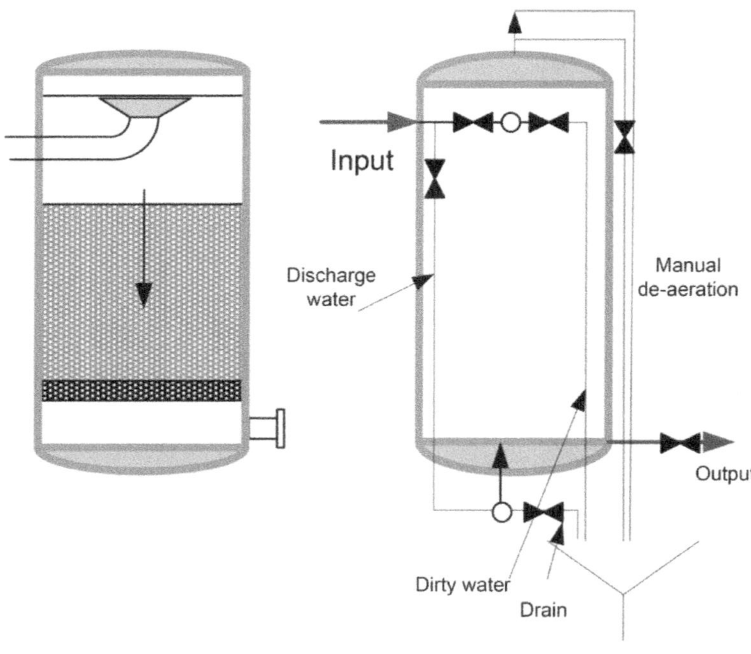

Figure 8.50. *Example of a Veolia pressure filter. For a color version of this figure, see www.iste.co.uk/gaid/watertreatment2.zip*

The pressure loss of a filter file can be controlled by a single set of pressure gauges (one on the upstream collector and one on the downstream collector).

Since the filters in the same file are generally in different states of clogging, the total flow rate processed is unevenly distributed among them, with clean filters being more heavily used than clogged ones.

To keep the flow rates of the filters equal at all times, it is advisable to either:

– equip them with flow regulation devices;

– achieve a complete equalization of the total flow treated on the units in service;

– clean them consecutively, which is only possible if there is a large supply of wash water.

Figure 8.51. *Veolia pressure filters. For a color version of this figure, see www.iste.co.uk/gaid/watertreatment2.zip*

The cleaning conditions for FV filters are identical to those for Filtraflo® and are adapted according to the media present in the filters.

Pressure filters are used:

– for a final clarification of water that was previously flocculated and decanted (traditional way);

– after decanters, on all types that are normally sized;

– in direct filtration, after simple bonding of raw water that is not very loaded and not very colored;

– in traditional Fe and Mn removal and neutralization processes (limestone filters).

8.10. Regulation systems

The role of regulation is to compensate for clogging by ensuring a constant level in the filter and/or a constant filtration rate in the filters. Maintaining a constant level ensures equalization in the case of orifice feeding and prevents scouring due to water fall at the beginning of the filtration cycle.

Veolia has at least two types of regulators:

– hydraulic regulators;

– control valves.

Hydraulic regulators such as Varibar, Siphydra or Polhydra are used in many plants, although the current trend is toward control valves.

8.10.1. *Hydraulic regulators*

We distinguish between Siphydra, Varibar and Polhydra systems.

8.10.1.1. *Siphydra*

The Siphydra's principle of operation is to regulate the flow of water exiting a filter to maintain a constant level on the filter, despite the variation in the pressure loss caused by the progressive clogging of the filter. The Siphydra is generally used on filters with a sand height of 0.9 m, such as Filtraflo®F or Filtraflo®MD.

8.10.1.2. *Varibar*

The Varibar ensures equilibrium between the filters and a stable operation. Three types of Varibar regulation can be used for filter control:

– simplified regulation with maintenance of only one water body, whether upstream or downstream;

– upstream regulation for all filters with a single measurement of the water level in the water channel to be filtered;

– upstream regulation by filter with a water level measurement on each filter.

The Varibar regulation of the filtered water flow is designed so that the regulation automatically absorbs the entirety of the flow entering the filters, with an equal distribution on the filters, and so that the filtration speed remains stable.

The purpose of the Varibar regulation is to maintain the water level at a constant level above the filtering mass, regardless of the flow rate and the variation in pressure loss caused by the progressive clogging of the filtering material, and to ensure that the total flow rate to the reservoirs is equally distributed.

A Varibar regulator essentially consists of a butterfly valve actuated by a long-stroke, double-acting, pivoting cylinder. The regulating assembly is initially set to obtain approximately 0.3 bar in the upstream network and a minimum of 1.8 in the downstream network. There is an electronic version of the Varibar that makes it possible to display the desired flow rate/filter, ensure a slow start after washing (during the desired time), regulate an upstream level that can have several set points, obtain an indicative flow rate of each filter, have a low consumption of driving water less than 50 $L \cdot h^{-1}$ per Varibar, and suppress the float valves that require regular maintenance.

8.10.1.3. *Polhydra*

The Polhydra is a siphon flow regulator whose operation is based on the automatic control of a vacuum tilting blade by means of a hydropneumatic servo device.

The height of the blade is maintained at the desired value by balancing between:

– a depression caused by a vacuum tube placed inside the hood;

– a compensating air inlet, modulated by the variable opening of a shutter.

In general, the Polhydra can be used as a flow regulator at any point in any hydraulic system, provided there is an open water body immediately upstream.

The electronic version of the Polhydra has a pressure sensor that measures the water level in the channel, a regulation valve placed on the air inlet in the Polhydra hood, and an automatic controller to regulate the whole. It preferably equips large filters (>200 $m^3 \cdot h^{-1}$) when the recovery is carried out by an open channel.

8.10.2. *Control valves*

In most cases, control valves require a preliminary distribution of the flow admitted on each filter, preferably by a wall-mounted motorized valve or a butterfly valve.

The "clogging" that occurs during filtration increases the loss in pressure due to the operation of the filter. This results in a gradual decrease in the flow of filtered water and an increase in the height of water above the filter. Conversely, when the filter is clean, the water passes through the filter layer at a rate that is too high. For these reasons, a control valve creates a compensating pressure loss that decreases as the clogging increases.

Each filter is equipped with a compensating pressure loss valve. Its opening/closing is controlled by a flow measurement on the outlet pipe of each filter.

Two types of regulation are possible, according to the liquid level on each filter or according to the flow of filtered water:

– regulation according to the water level in the filter: this type of regulation ensures the compensation of the clogging while always keeping the same water level. A level sensor must be placed in each filter to ensure the control of the regulating valve and the triggering of each washing phase;

– regulation based on the flow of filtered water: the regulator consists of a meter linked with each regulating valve. It allows the opening of the valve to be continuously corrected according to the difference observed between the flow rate read and the set point defined for all the filters by a level indicator located in the upstream channel. This type of regulation implicitly leads to an equal flow rate between the filters in service and thus to an identical level and filtration speed, independent of each clogging level. The filter inlets are controlled by a motorized valve (wall-mounted or butterfly).

8.11. Recycling and microbiological risks

The recycling of specific waters within a plant can be considered to limit water losses.

The main sources of recycling are as follows:

– the first filtered water after cleaning;

– the water used for rinsing;

– the returns from the sludge treatment (thickener, drying beds, etc.).

The presence of pathogenic microorganisms such as *Cryptosporidium* or *Giardia* in drinking water can cause prolonged diarrhea and vomiting in humans and even lead to a fatal prognosis in infants or immunocompromised subjects.

Cryptosporidium and *Giardia* can pass through sand filters that are not preceded by a flocculation stage.

The risk of concentration of these germs following recycling and, consequently, the increase in the probability of leakage into the treated water leads to a necessity for specific treatment of recycled water, such as:

– ozone sterilization;

– ultrafiltration;

– UV and/or microfiltration treatment.

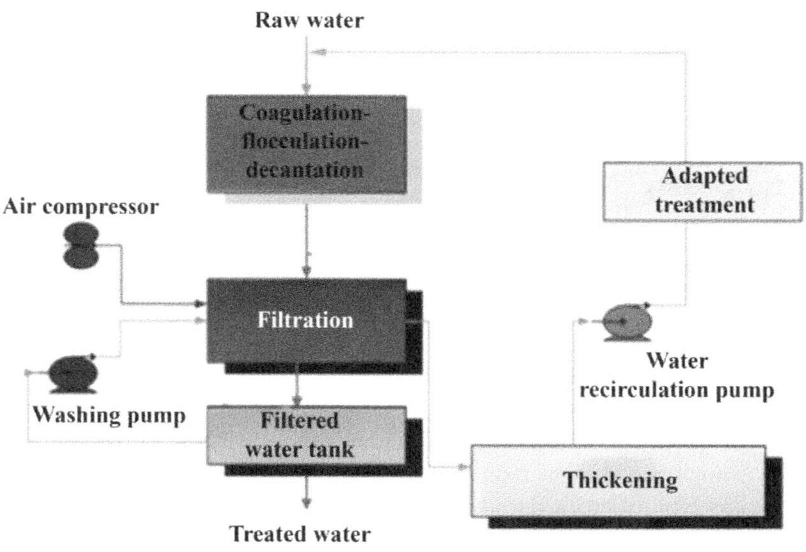

Figure 8.52. *Recycling of dirty water from filters. For a color version of this figure, see www.iste.co.uk/gaid/watertreatment2.zip*

8.12. Monitoring the operation and performance of filters

Many pathogens are often associated with turbidity, so the additional removal of turbidity through filtration is therefore an essential step in allowing produced water quality to meet the standard value for this parameter. Periodic bouts of high turbidity have been associated with increased concentrations of oocysts, SS and spore-forming bacteria in distribution systems. Filters are often the ultimate barrier to particles in drinking water plants.

The main procedures that are considered essential to ensure proper operation of a filtration plant are monitoring of instrumentation, monitoring of filter performance during the filtration phase, management of pretreatment, optimization of filter backwash and inspection of filter materials.

There are various means of controlling the filter operation. These include reducing the filtration rate due to clogging, which is rarely used; measuring the pressure loss in the filter by measuring the height of the water above the media; and operating with a downstream control valve to maintain constant production while controlling the maximum allowable pressure loss.

Sometimes, the turbidity measured in-line on the filtered water is used as an indicator, not just for pressure loss but also particularly for water quality control, especially when a low turbidity is continuously required (e.g. $\leq$0.2 NTU). The pressure loss may not be at its maximum level while the output turbidity (per filter) reaches the required limit.

8.12.1. *Turbidity*

Drinking water plant operators must continuously monitor the turbidity of the water leaving the filters (individually or in combination) to judge the performance of the treatment process, as this continuous monitoring is necessary to ensure that each filter is working properly, to decide when a cycle is complete and to detect any changes in the turbidity that are typically recorded.

The normal operating cycle of a filter includes a maturation period, in which turbidity is maintained at the required level. Beyond the time allocated to this cycle, turbidity can rapidly break through the filter. In addition, the turbidities of filtered water obtained after a wash are high and can reach up to 10 NTU during a period of 5–15 min. This is especially true if the filter has been put back into operation at its nominal rate.

For this reason, whenever possible, refiltration after washing should be performed at a progressive rate.

In most cases, filter performance is assessed through the turbidity of the filtered water. It can be measured continuously on each filter or alternatively by a set of valves if a single device is used.

Treatment objectives in countries:

– 1 NTU 100% of the time;

– 0.5 NTU 95% of the time.

Expected values or treatment goals in North America (not to be confused with guaranteed values):

– NTU in stabilized operation;

– NTU 15 min after washing.

Nevertheless, low turbidity water quality (<0.5 NTU) does not guarantee that it is free of cysts or oocysts. When the turbidity of the filtered water is equal to or less than 0.1 NTU, cyst and oocyst removal becomes more reliable.

Turbidity measurement is the most common control method in treatment plants. It is carried out in line, at each filter outlet or on the mixture of filtered water, depending on the case.

For conventional filtration, continuous monitoring of turbidity at the output of each filter may be required by the client. A multichannel turbidity meter can be used to monitor several filters, and a turbidity meter is used to monitor the quality of the filtered water after mixing.

On the other hand, in the case of in-line or direct filtration, control should preferably be carried out continuously at the outlet of each of the filters because, in this case, filtration is the only clarification stage in the treatment process.

This is particularly true for direct filtration on dual media filters. These filters are indeed designed to work at a high rate and ensure deep clogging. Under these conditions, the risk of breakthrough cannot be neglected, and continuous monitoring of turbidity at the outlet of each filter allows rapid intervention in the event of a problem. In some cases, an additional turbidity meter is installed between the two filter materials, but this is not necessary.

Filter type	Removal log				Turbidity of filtered water (NTU)
	Giardia	*Cryptosporidium*	Bacteria	MS2 Phage	
Rapid filter	1.5	–1.5	2	1.0–1.5	<0.5
	2–2.5	2–2.5	3	1.5–2	<0.2
Dual media filter	2.5–3.0	2.5–3.0	3	2.5	<0.2
TGV filter	2–2.5	2–2.5	2.0	1.0–1.5	<0.3
Direct filtration without reagents	1.0–1.5	1.0	1.5	1.0	<0.2

Table 8.11. *Removal log of microorganisms during filtration under turbidity conditions of filtered water*

8.12.2. *Particle count*

The particle count provides information on the number of particles of a given size that are present at the filter outlet and possibly at the inlet. This is particularly interesting for the monitoring of *Cryptosporidium* (3–5 μm) or *Giardia* (10–15 μm).

For turbidity, the measurement can be carried out continuously on each filter or, alternatively, through a single device and an adapted set of valves.

Despite its interest, particle counting is a rather underdeveloped control technique in drinking water treatment plants, and there are notably two reasons for this:

– manual counts based on microscopic observations are time consuming, and electronic automatic counting systems are relatively expensive;

– there are currently no standards or recommendations for the number and size of particles that should be in the treated water. There is also no standard analysis protocol.

Automatic monitoring devices can detect particles larger than 0.5–1 μm. Therefore, there is no direct and universal correlation between the particle count and the turbidity measurement. Indeed, very small particles smaller than 1 μm will be included in the turbidity measurement and will not be reliably detected by the counting system.

8.12.3. *Problems with the operation of filters*

The operating problems encountered in the filtration stage in drinking water plants can be due to a lower efficiency against turbidity, a short duration of the filtration cycles due to a too fast increase in pressure loss, a compaction of the material, etc.

These problems disrupt the filtration process and often result in a reduction in water production. The main reasons are as follows:

– an insufficient dosage of coagulant and flocculant in the previous stage, which leads to a high delivery of turbidity in the settled water. In this case, the particle flow becomes too high compared to the filter capacity. The filter then needs to be washed too often, which reduces water production;

– sudden changes in feed rate can lead to turbidity leakage through unexpected particle dropouts. If the filtration rate is increased too quickly to a value that is more than 30% of the nominal rate, the balance between particle retention forces and hydraulic shear forces, which tend to dislodge the retained particles, is disrupted. This results in a temporary dislodging of the particles until the nominal filtration speed is restored;

– poor filter cleaning allows some of the particles to accumulate in the filter media. The consequence is a shorter cycle time or an overly fast breakthrough of the filter.

This accumulation of particles can lead to the formation of mud balls that grow over time and can no longer be removed because the washing conditions are insufficient to suspend or separate them from the mass. Cases of floor lifting have been observed. Mud balls are formed by the accumulation of coagulated particles (coagulated flocs), as well as organic matter, which causes fermentation and therefore bacterial growth within the filtering mass. They have not been eliminated during the cleaning process and create a compact layer around the material grains over time. In this situation, if the wash water velocity or the duration of the rinse phase is insufficient, some particles, although detached from the filter grains, cannot be removed from the filter. They will accumulate in the filter.

When an issue such as mud ball development is identified on a filtration installation, the possible solutions are to act on the cleaning procedure by increasing the water velocity and/or the duration of the rinsing and to act on the sand by subjecting it to a chemical treatment with chlorine, potassium permanganate or soda ash. If these solutions are ineffective, the filter media can be changed completely. This has happened many times in plants.

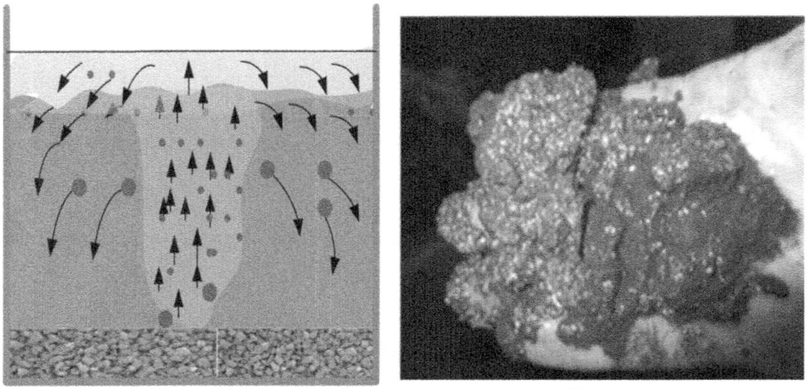

a) b)

Figure 8.53. *Filter with mud balls (a) and example of mud balls in a plant in India (b). For a color version of this figure, see www.iste.co.uk/gaid/watertreatment2.zip*

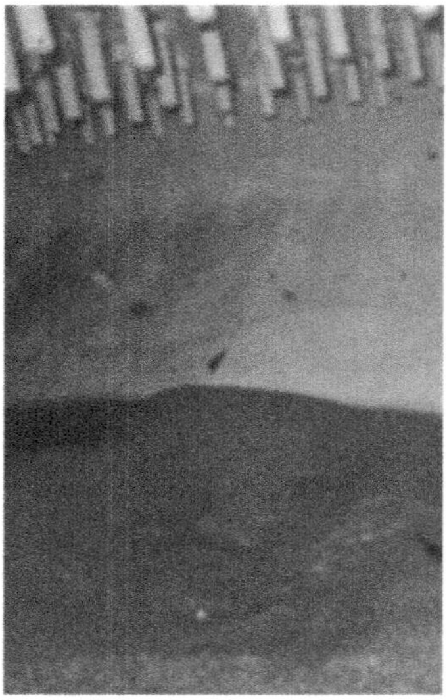

Figure 8.54. *Presence of sand at the bottom of the filter due to the deterioration of the strainers. For a color version of this figure, see www.iste.co.uk/gaid/watertreatment2.zip*

– The control of nozzles is an important monitoring operation because their deterioration leads to leakage of SS and sand. In both cases, a degradation of the quality of the filtered water is observed, often with high turbidities in the filtered water.

– Too much clogging can cause uneven distribution of the backwash water, creating areas with a high washing rate. This can lead to the gravel being lifted and cause damage to the floor.

– Air binding is a common problem observed in plants treating oxygen-supersaturated water or water that has undergone ozone pretreatment or aeration. The retention of dissolved O_2 within the filter material creates pockets and thus occupies volume within the material. This results in a rapid increase in pressure loss and a poor quality of filtrate because the flow rate is higher around the pockets, as well as a reduction in filtration cycles. This effect is observed during washings by the presence of large air bubbles coming out from the media and bubbling on the water surface.

Problems seen by the operator	**Heterogeneity of the cleaning**	**Increased turbidity of the filtered water**	**Increased filter volume**	**Filter compaction**	**Short cycle time**
Recommendations	Filtering floor (nozzles)	Expansion and cleaning conditions	Expansion and cleaning conditions	Expansion and cleaning conditions	Cleaning conditions
	Aeration	Turbidity of dirty water	Turbidity of dirty water	Turbidity of dirty water	Turbidity of dirty water
	Cleaning conditions	Filtration rate	Air binding	Pressure losses	Turbidity analyzer
	Core drilling (mud balls), deposits, etc.	Media height	Core drilling (mud balls), deposits, etc.		Pressure losses
		Turbidity analyzer			Air binding
		Tracer (salt)			Core drilling (mud balls), deposits, etc.

Table 8.12. *Some recommendations to implement according to the problems encountered*

– Material loss during washes is a common phenomenon if the washes are not controlled. This results in a progressive decrease in media height with degraded operating conditions and performance.

– Short filter cycles due to rapid pressure loss, or early particle breakthrough, are usually indicators of filter malfunction, problems with the drainage system, excessive polyelectrolyte dosage, the presence of algae blocking the filter, improper selection of one or both sizes and depths of the media filter, or simply ineffective coagulation.

8.13. References

Adin, A., Baumann, E.R., Cleasby, J.L. (1979). The application of filtration theory to pilot-plant design. *J. Am. Water Works Assoc.*, 71(1), 17–27.

Amburgey, J.E. and Brouckaert, M. (2005). Practical and theoretical guidelines for implementing the extended terminal subfluidization wash (ETSW) backwashing procedure. *Journal of Water Supply: Research and Technology. AQUA*, 54(5), 329–337.

Amirtharajah, A. (1977). Optimum backwashing of sand filters. *J. Environ. Eng. Div. ASCE*, 104(EE5), 917–938.

Amirtharajah, A. (1988). Some theoretical and conceptual views of filtration. *J. Am. Water Works Assoc.*, December, 36–39.

Bai, R. and Tien, C. (1997). Particle detachment in deep bed filtration. *Journal of Colloid and Interface Science*, 186, 307–317.

Bergendahl, J. and Grasso, D. (2000). Prediction of colloid detachment in a model porous media: Hydrodynamics. *Chemical Engineering Science*, 55, 1523–1532.

Camp, T.R. (1961). Discussion of Conley – Experiences with anthracite filters. *J. Am. Water Works Assoc.*, 53(12), 1453–1478.

Camp, T.R. (1964). Theory of water filtration. *J. San. Eng. Div. ASCE*, 90(SA4), 1–17.

Camp, T.R., Graber S.D., Conklin, G.F. (1971). Backwashing of granular water filters. *J. Environ. Eng. Div. Proc. ASCE*, 97(EE6), 903–919.

Cartwright, O. (2001). Removal of Giardia and cryptosporidium in drinking water, Kinetico incorporated SW224 backwashable macrolite pressure filtration system. Environmental Technology Verification Report, NSF International, Ann Arbor.

Cleasby, J.L. (1993). Declining rate filtration. *Water Science and Technology*, 27(7–8), 11–22.

Cleasby, J.L. and Woods, C.W. (1975). Intermixing of dual and multi-media granular filters. *J. Am. Water Works Assoc.*, 67(4), 187–197.

Cleasby, J.L., Williamson, M.M., Baumann, E.R. (1963). Effect of rate changes on filtered water quality. *J. Am. Water Works Assoc.*, 55(7), 851–869.

Coad, M.A. and Ives, K.J. (1981). Investigation of deep bed filters using tracers. *Filtration Society*, 131–136.

Conley, W.R. and Hsiung, K. (1969). Design and application of multimedia filters. *J. Am. Water Works Assoc.*, 61(2), 97–115.

Crittenden, J.C., Trussel, R.R., Hand, D.W., Howe, K.J., Tchobanoglous, G. (2005). *Water Treatment Principles and Design*. John Wiley & Sons, New York.

Culp, R.L. (1977). Direct filtration. *J. Am. Water Works Assoc.*, 69, 375–391.

Edwards, M. and Scardina, P. (2007). Addressing problems with gas supersaturation at drinking water utilities. *J. Am. Water Works Assoc.*, 99(5), 137–147.

Ergun, S. (1952a). Fluid flow through packed columns. *Chemical Engineering Progress*, 48(2), 78–89.

Ergun, S. (1952b). Determination of geometric surface area of crushed porous solids. *Analytical Chemistry*, 24, 369–388.

Fitzpatrick, C.S.B. (1990). Detachment of deposits by fluid shear during filter backwashing. *Water Supply*, 8, 177–183.

Gadkari, S.K., Raman, V., Gadkari, A.S. (1980). Studies of direct filtration of raw water. *Indian J. Environ. Health*, 22(1), 575–591.

Graham, N.J.D. (1988). Filter pore flocculation as a mechanism in rapid filtration. *Water Research*, 22(10), 1229–1238.

Han, S.J., Fitzpatrick, C.S.B., Wetherill, A. (2009). The impact of flow surges on rapid gravity filtration. *Water Research*, 43(5), 1171–1178.

Hamann, C.L. and McKinny, R.E. (1968). Upflow filtration process. *J. Am. Water Works Assoc.*, 60(9), 1023–1039.

Hewitt, S.R. and Amirtharajah, A. (1984). Air dynamics through filter media during air scour. *J. Environ. Eng. Div. ASCE*, 110(3), 184–195.

Hutchinson, W. (1976). High-rate direct filtration. *J. Am. Water Works Assoc.*, June, 292–298.

Hutchison, W. and Foley, P.D. (1974). Operational and experimental results of direct filtration. *J. Am. Water Works Assoc.*, 66(1), 145–173.

Ives, K.J. (1978). A new concept of filterability. *Prog. Water Technol.*, 10(5/6), 123–137.

Ives, K.J. (1990). Testing of filter media. *Aqua*, 39(3), 144–151.

Jegatheesana, V. (1999). Mathematical modeling of deep bed filtration. PhD Thesis, University of Technology, Sydney.

Kawamura, S. (1975). Design and operation of high-rate filters – Part 1. *J. Am. Water Works Assoc.*, 67(10), 535–549.

Kim, J.K. and Lawler, D.F. (2006). Particle detachment during hydraulic shock loads in granular media filtration. *Water Science and Technology*, 53(7), 177–184.

Logsdon, G.S. (1970). Direct filtration – Past, present, future. *Civ. Eng. ASCE*, 70(7), 68–81.

Logsdon, G.S., Hess, A.F., Chipps, M.J., Rachwal, A.J. (2002). *Filter Maintenance and Operations Guidance Manual*. American Water Works Association, Denver, CO.

Lopato, L., Galaj, Z., Binning, P.J., Arvin, E. (2009). *Influence of Inhomogeneities on Rapid Sand Filter Efficiency*. IWA, Kobe.

Lopato, L., Röttgers, N., Binning, P.J., Arvin, E. (2010). Monitoring of gas bubbles in rapid sand filters. In *Nordic Drinking Water Conference*, Copenhagen.

Nix, D.K. and Taylor, J.S. (2003). *Filter Evaluation Procedures for Granular Media*. American Waterworks Association, Denver, CO.

Payatakes, A.C., Tien, C., Turian, R.M. (1973). A new model for granular porous media: Part I. Model formulation. *American Institute of Chemical Engineers Journal*, 19, 58–66.

Raguenes, I. (1994). *Filtration des eaux potables, cours de l'eau*. OTV-Veolia, Saint-Maurice.

Raveendran, P. and Amirtharajah, A. (1995). Role of short-range forces in particle detachment during backwashing. *Journal of Environmental Engineering*, 121, 860–868.

Rijnaarts, H.M., Norde, W., Bouwer, E.J., Lyklema, J., Zehnder, A.J.B. (1993). Bacterial adhesion under static and dynamic conditions. *Applied Environmental Microbiology*, 59, 3255–3265.

Scardina, P. and Edwards, M. (2002). Practical implications of bubble formation in conventional treatment. *J. Am. Water Works Assoc.*, 94(8), 85–94.

Sweeney, G.E. and Prendiville, P.W. (1974). Direct filtration: An economic answer to a city's water needs. *J. Am. Water Works Assoc.*, 66(2), 65–82.

Tate, C.H. and Trussel, R.R. (1980). Recent developments in direct filtration. *J. Am. Water Works Assoc.*, 72(3), 165–181.

Tate, C.H., Lang, J.S, Hutchinson, H.L. (1977). Pilot plant tests of direct filtration. *J. Am. Water Works Assoc.*, 69(7), 379–392.

Wen, C.Y. and Yu, Y.H. (1966). Mechanics of fluidization. *Chemical Engineering Progress*, 62, 100–111.

Yao, K.M., Habibian, M.T., O'Melia, C.R. (1971). Water and wastewater filtration: Concepts and applications. *Environ. Sci. Technol.*, 5, 1105–1112.

9

Adsorption on Activated Carbon

Activated carbon is a microcrystalline, non-graphic form of carbon. X-ray analyses of activated carbons show a much more disordered structure than graphite, with crystals in the layers less than 10 nm thick. The spaces between the crystals constitute the porous structure, with an internal surface of 500–2,000 $m^2 \cdot g^{-1}$.

Due to its chemical structure, activated carbon preferentially adsorbs organic matter and other non-polar compounds from gaseous or liquid streams. This makes it the material of choice for the treatment of drinking water, municipal wastewater and industrial water.

The removal of organic pollutants and micropollutants is achieved by adsorption. Adsorption is a mechanism that describes the existence of a substance concentration that is higher at the point between the fluid and the solid than the concentration of that substance in the fluid.

9.1. Activation processes of activated carbon

The manufacturing principle of activated carbons is based on the selective removal of groups of compounds from a material made of carbon. This removal is achieved by carbonization and/or activation. The product generates a matrix with a large internal porosity.

Various natural raw materials, such as coconut shells, wood, bamboo, peat, coal, pits of various fruits (apricots, dates, etc.), olive pits, bones, and so on, are the origin of this adsorbent material. These raw materials are transformed into activated carbon by means of two complementary steps: carbonization and activation.

Carbonization is the thermal decomposition of carbonaceous materials: any species other than carbon are removed. Its structure is formed of irregularly arranged carbon sheets, leaving interstices between them. The material thus only has a basic porosity known as primary porosity (pores that are too small or in limited numbers) and cannot be used as an adsorbent without an additional treatment.

Activation consists of developing the pore structure and creating surface functions, generally oxidized, which are at the origin of the interactions between the solid and the adsorbed molecules. The adsorption capacity of the carbon is created by removing the tars and other disorganized carbonaceous matter through an activation process, which results from the decomposition of the organic compounds that obstruct the pores.

The porous structure and chemical nature of an activated carbon depend on the raw materials used for its manufacture, as well as on the activation method.

9.1.1. *Chemical activation*

The material is carbonized by chemical activation after the addition of reagents to limit tar formation. The most commonly used activating agents are phosphoric acid, zinc chloride and potassium sulfide. In some cases, alkali metal hydroxides, magnesium and calcium chloride are used. These products are powerful dehydrating agents that promote the formation of aromatic rings of the carbon skeleton by creating an internal pore structure and active sites on the surface.

The temperature required for pyrolysis is lower than that required for physical activation using gaseous fluids. This favors the development of a porous structure and, in particular, micropores and mesopores. The chemical activation leads to an activated carbon with very good sorption properties.

Chemical activation is most often used on raw materials such as lignocellulosic material. Cellulose is composed of macromolecules up to 1,800–2,000 nm in length, oriented along their longitudinal axes, which form agglomerates called micelles. The molecule chains oriented in this manner are delineated by lateral bonds of different strengths. The activating agent causes cellulose to undergo a transformation known as swelling, during which the arrangement of the molecules remains unchanged, but the side bonds are broken. This results in the formation of inter- and intramicellar voids, which increase until the cellulose is broken down. Simultaneously, other hydrolytic or oxidizing reactions occur, by which macromolecules are progressively depolymerized. The processes lead to the formation of a homogeneous mass consisting of depolymerized material that is uniformly saturated with the activating agent. Chemical activating agents promote dehydration, which helps the decomposition of organic substances through the action of heat, thus preventing the formation of non-carbonized degradation products (tar).

Structural reorganization allows the development of microporosity and mesoporosity by enlarging the pore diameter. An important factor in chemical activation is the impregnation coefficient, which is the weight ratio between the anhydrous activating agent and the dry matter. The effect of the impregnation coefficient on porosity is significant because the volume of salts extracted into the carbonized material is equal to the volume of pores released. Low impregnation coefficients lead to an increase in the total pore volume, as well as a high micropore volume.

High impregnation coefficients imply a greater number of macropores, and the volume of micropores decreases. Generally, activated carbons produced by this chemical activation have an open porous structure (the pores open on the surface of the grain), which is ideal for the adsorption of large molecules. This process is, for example, used to produce wood-based activated carbon.

9.1.2. *Physical activation*

The main purpose of carbonization is to reduce the volatile content of the source material so that it can be converted into a suitable form for activation. During this phase, the carbon content is approximately 80%. Most of the non-carbon elements, such as hydrogen and oxygen, are removed in gaseous form by pyrolytic decomposition of the material. The released atoms are grouped and organized into a crystallographic formation known as elemental graphite crystals. Physical activation allows existing pores to expand and new ones to be created. The material is calcined in the absence of oxygen to obtain a carbon-rich amorphous solid.

The carbonization of the lignocellulosic material takes place between 500 and 700°C. In the resulting product, the mutual arrangement of the crystals is irregular. However, the interstices are filled or saturated due to the deposition of tarry substances. The carbonized product has a low adsorption capacity.

The carbonized product is then partially activated, either by removing the tarry substances by heating them in the flow of an inert gas or by extracting them with a suitable solvent or by a chemical reaction (e.g. heating in a sulfur vapor atmosphere at lower temperatures, at which reactions with carbon occur).

An activated carbon with high adsorption capacity is then prepared by activating the carbonized material using a chemical agent that reacts with the carbon. The most common activating agents are water vapor, carbon dioxide and oxygen (air).

The activation stage typically takes place at temperatures between 800 and 1100°C. The reaction of oxygen with carbon is exothermic, and oxygen essentially burns the most reactive part of the carbon skeleton, producing CO and CO_2.

At lower temperatures, the reactions are too slow. The temperature must be carefully chosen so that the reaction between the carbon and the gaseous agent is optimal, and the reactions take place on the internal surface of the carbon. The removal of carbon from the pore walls causes the formation of new pores, as well as the development of the macroporous structure.

This process is used to produce activated carbons from coconut shells, coal, peat, wood and other plant material.

9.2. Physicochemical properties of activated carbon

The basic structure of activated carbon is close to the structure of pure graphite. However, the structure of activated carbon is quite disorganized compared to graphite due to the random oxidation of the graphite layers. The structure that develops depends on the carbonization and activation temperatures. Thus, during the carbonization process, several aromatic rings with a structure similar to graphite are formed.

Activated carbon is considered to be a skeleton made of interconnected microcrystallites consisting of fused hexagonal rings of carbon atoms. For reference, graphite crystal consists of flat layers of carbon atoms ordered in regular hexagons, similar to aromatic rings.

Each microcrystallite includes an irregular stack of sheets (5–20) with a spacing of 0.36 nm. Functional groups are present on these sheets. The diameter of the microcrystallites, as well as the stacking height, has been estimated to be between 2 and 5 nm.

The hexagonal layers are held within approximately 0.335 nm of each other by Van der Waals forces.

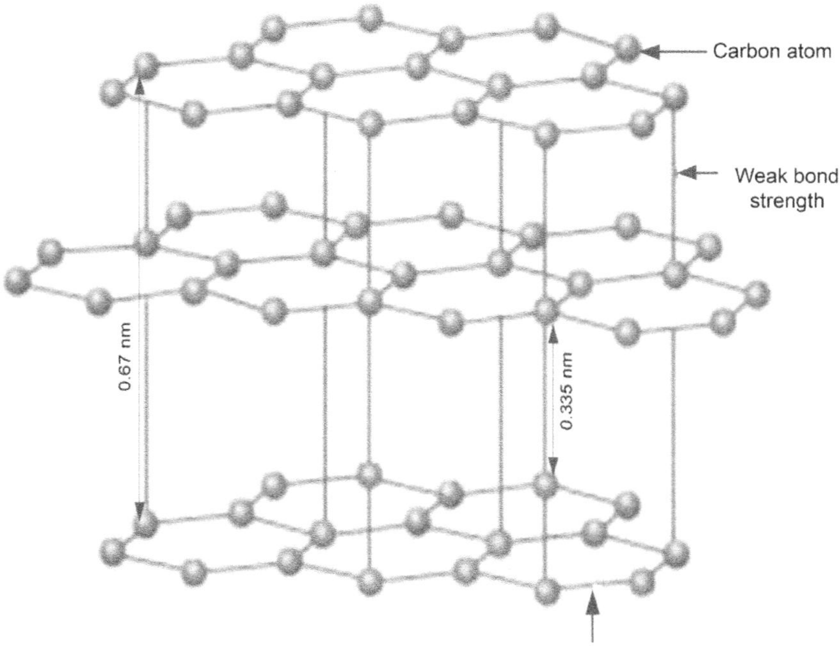

Figure 9.1. *Crystal structure of graphite. For a color version of this figure, see www.iste.co.uk/gaid/watertreatment2.zip*

9.2.1. *Parameters influencing adsorption*

The adsorption process can be considered chemisorption or physical adsorption. In physical adsorption, substances are retained on the carbon surface by Van der Waals forces. In chemisorption, the adhesion forces are stronger, and adsorption occurs on active sites that are present on the carbon surface. Therefore, the efficiency of carbon depends on the accessibility of its surface, as well as on the significance of these active sites on the surface.

In physical adsorption, the surface accessibility mechanism and the presence of active sites are less significant.

The factors affecting the adsorption process are the specific surface, the nature and the initial concentration of the organic molecules, the pH of the water, the temperature, the physicochemical characteristics of the activated carbon and the injected dosage. A large number of parameters linked to both the activated carbon and the molecules to be removed therefore participate in the global mechanism.

The physical characteristics of the activated carbon particles make it possible to determine the conditions of use of the activated carbon.

9.2.1.1. *Pore structure of activated carbon and pore volume*

Activated carbon is a wide range of carbonaceous substances with different properties (porosity, specific surface, adsorption capacity). These characteristics are mainly due to the origin of the material and its manufacturing process. This is why the surface or pore volume of activated carbons can vary considerably from one type to another.

Pore volume ($cm^3 \cdot g^{-1}$): A high volume of micropores allows a better adsorption capacity. However, the distribution of pores is related to the size of the molecules to be removed. The presence of humic and fulvic acid molecules favors the use of a carbon that has a high volume of macropores and mesopores.

A porous solid can be defined based on the volume of adsorbed substance that is necessary to saturate all the open pores of one gram of this solid. An open pore is a pore whose access is located at the surface of the grain and is thus accessible to the fluid. The formation of pores occurs during the activation process.

	Pore characteristics		
	Macropores	**Mesopores**	**Micropores**
Pore diameter (nm)	>50	2–50	<2
Pore volume ($cm^3 \cdot g^{-1}$)	0.2–0.8	0.02–0.10	0.2–0.6
Specific surface ($m^2 \cdot g^{-1}$)	0.5–2.0	20–70	400–1,500

Table 9.1. *Pore characteristics*

The free space between the sheets constitutes the microporosity of the material. The porosity can limit the adsorption capacity of the material toward high molecular weight organic molecules and influence the diffusion rate of adsorbates inside the activated carbon particle. Let us also note that the size of the bacteria (>200 nm) does not allow them to penetrate the micropores and mesopores. The estimation of the pore diameters (assumed to be cylindrical) leads to the characterization of several groups (Figure 9.2).

Pores with a diameter greater than approximately 50 nm are classified as macropores. Their volume in activated carbon is generally between 0.2 and 0.5 $cm^3 \cdot g^{-1}$, and their specific surface is 0.5–2 $m^2 \cdot g^{-1}$. Macropores are essentially open externally and allow the fluid to access the internal surface of the activated carbon. Their contribution to the specific surface is low and therefore do not participate much in adsorption.

Mesopores are intermediates between macros and micropores. They are transitional pores with a diameter between 2 and 50 nm. Their specific surface is generally approximately 5% of the total surface of the activated carbon. They favor the transport of fluid toward the micropores and influence adsorption kinetics.

Pores with a diameter of less than approximately 2 nm are called micropores. The volume of micropores is usually approximately 0.15–0.50 $cm^3 \cdot g^{-1}$. The micropore region is sometimes subdivided into primary (d < 0.8 nm) and secondary (0.8 nm < d < 2 nm) micropores. The micropores alone can determine the adsorption capacity of an activated carbon. The specific surface of the micropores can represent more than 80% of the total specific surface. Thus, they play an important role in adsorption.

A **high specific surface** ($m^2 \cdot g^{-1}$) favors the adsorption capacity. It is the total surface per unit mass of adsorbent accessible to the molecules. The specific surface includes the external surface and the internal surface of an adsorbent. Its estimation is conventionally based on measurements of the single-layer adsorption capacity "qm" of the adsorbent for a given adsorbate of a known and acceptable surface. The specific surface Ss is calculated by measuring the quantity of a gas adsorbed at low temperature (generally nitrogen at 77 °K) with a gas porosimeter and by assuming that the surface is covered by a single layer of molecules. Knowing the diameter of a molecule and thus its surface, we deduce the adsorption surface. The larger the adsorption surface is, the higher the quantity of adsorbable matter. It is related to the pore volume Vp, ($cm^3 \cdot g^{-1}$) and to the width of the pore lp (nm):

$$Ss = \frac{2.10^3 Vp}{lp}$$

The adsorption capacity particularly depends on the specific surface and the pore size distribution. For example, adsorption of small molecules such as phenol occurs in micropores, takes place in mesopores for natural organics and dyes, and in macropores for bacteria. The adsorption capacity is significantly influenced by the surface chemistry of the carbon. This surface is usually divided into three zones: the basal plane, surface functions (mainly oxygen functions) and inorganic ash. Different studies have shown that the oxygenated acid functions of the surface participate in the adsorption of organic compounds in aqueous solution. The forces of attraction between the hydrophobic organic compounds and the hydrophobic surface of the solid are strong.

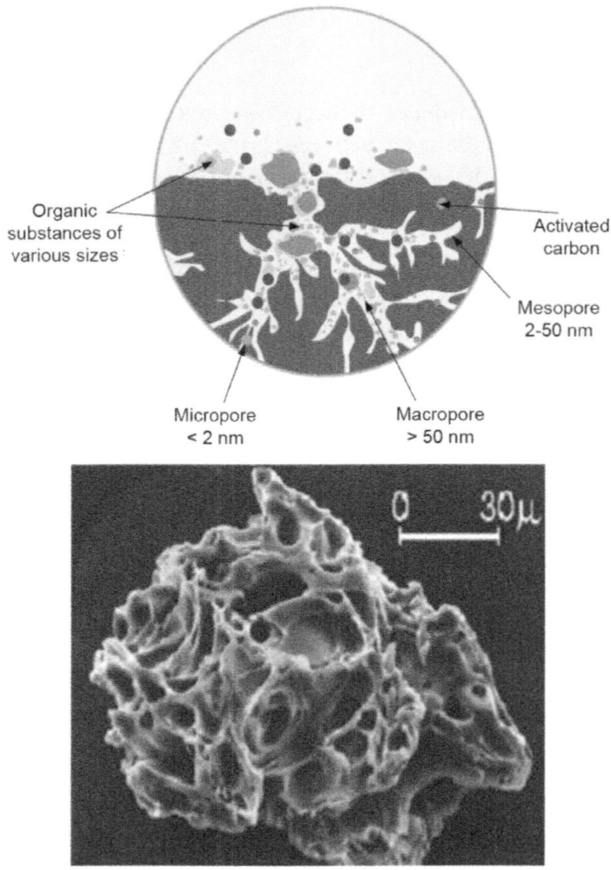

Figure 9.2. *Pore size observed on activated carbon. For a color version of this figure, see www.iste.co.uk/gaid/watertreatment2.zip*

A fine **particle size** (powdered activated carbon [PAC]) accelerates the transfer of molecules within the porosity and makes it possible to obtain faster kinetics. Granulometry thus conditions the adsorption speed (the smaller the grain is, the faster the diffusional transfer toward the center because the path for the molecules is smaller).

The **hardness** of the material expresses the resistance of the activated carbon to abrasion, settling, attrition and vibration. It is an important factor in the design of the treatment system and its life span, given that it allows assessment of the formation of fine particles that are harmful to the operation of the installations (clogging the bed, deposit in valves and pipes, etc.). It depends on the raw material and the level of activation. The hardness of the material must be higher than 3 mhos to avoid attrition of the granular activated carbon (GAC, during filter washing) or of the micrograin applied in fluidization.

The **density** of the activated carbon ($g \cdot cm^{-3}$) must be sufficient to obtain a rapid settling of the PAC (settling mode) or its settling after washing the GAC filters.

The **surface functions** ($meq \cdot g^{-1}$) participate in the adsorption of organic molecules but are not indicated by the suppliers. This parameter is used when the manufacturer implements an oxidation of the material in their treatment process.

The surface of activated carbon contains a wide variety of functional groups (surface functions). The type and number of these functional groups depend on the activation method, as well as the type of raw material used. The irregularities of the microcrystalline structure of activated carbon and the presence of heteroatoms (oxygen, hydrogen, metals, etc.) allow the formation of functional groups on the surface of the material during the activation stage, the nature of which essentially depends on the initial material and the activation processes.

There are three types of these functions: acidic, basic, or neutral. However, there is a majority of acidic groups that predominate on the edges of the surface of the pores. They are mainly carboxylic acid, ether, amine, lactone, phenol and carbonyl type functions.

These functional groups give activated carbon acid–base properties. Their presence has a marked effect on the adsorption of polar molecules. In addition, these heteroatoms can affect the catalytic capacities of the material and physically hinder the adsorption of non-polar molecules.

The Boehm dosage method is used to determine the basic and acidic surface functions that are present.

9.2.1.1.1. Functional groups containing oxygen

The oxygen in the source material has a great influence on the arrangement and size of the crystals formed. In activated carbons prepared from materials with high oxygen content, the distance between the parallel graphite layers is significantly smaller. Oxygen is chemisorbed and bound as surface oxides on the edge of the plan-graphite layer. These surface oxides have important effects on the adsorption capacity of the activated carbon because they participate in the binding of organic molecules.

Activated carbons prepared at low temperatures of 200–500°C, which are called L-activated carbons, generally develop acidic surface oxides. The acidic surface oxides mainly include phenolic hydroxyl groups.

Activated carbons prepared by chemical treatment in aqueous solutions with oxidizing agents such as permanganate, persulfate, hydrogen peroxide and nitric acid or another acid give identical characteristics to L-type activated carbons.

Activated carbons prepared at high temperatures of 800–1,000°C, called H-activated carbons, develop basic surface oxides.

The presence of oxygen-containing functional groups gives a polar nature to the surface of the activated carbon, which implies a preferential adsorption of polar organic substances.

9.2.1.1.2. Functional groups containing hydrogen

The hydrogen present in the source material is in the form of hydrocarbon chains. Most of the hydrogen is removed during activation at temperatures below 950°C, but some remains after activation and is only released at higher temperatures. This is because hydrogen (in the form of aromatic and aliphatic molecules) is more strongly chemisorbed than oxygen. The hydrogen in aromatic molecules is covalently bonded to the carbon atoms at the edge of the aromatic planes. Hydrogen in aliphatic molecules is present as aliphatic chains and alicyclic rings attached to the surrounding aromatic rings.

The carboxylic, phenolic, carbonyl (quinone type) and lactone groups are the most common (Figure 9.3). These surface groups are particularly involved in the hydrophilic nature of activated carbon and in its electrokinetic potential.

In general, the analyses focus on the quantification of the four main surface functions:

– Group I: relatively strong acid functions (carboxylic acids).

– Group II: lactone functions.

– Group III: phenolic functions.

– Group IV: carbonyl functions.

As shown in Figure 9.4, it is the combination of porosity properties and surface functions that influence the more or less selective adsorption of organic compounds.

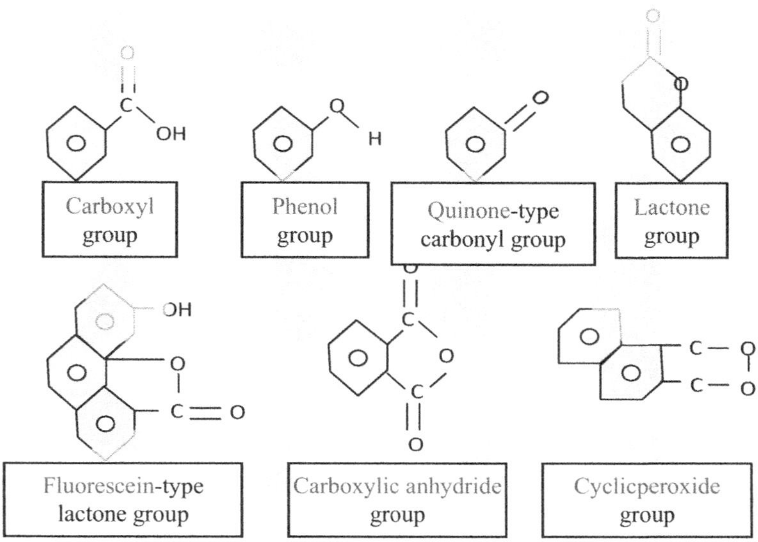

Figure 9.3. *Main functional groups of the activated carbon surface*

Functional groups	Identification reactions
Carboxylic function	With $NaHCO_3$ $RCOOH + NaHCO_3 \rightarrow RCOONa + CO_2 + H_2O$ With Na_2CO_3 $2RCOOH + Na_2CO_3 \rightarrow 2RCOONa + CO_2 + H_2O$ With NaOH $RCOOH + NaOH \rightarrow RCOONa + H_2O$ With acetate $RCOOH + CH_3COONa \rightarrow RCOONa + CH_3COOH$
Phenol function	Difference in neutralization between NaOH and Na_2CO_3 With diazomethane $ROH + CH_2N_2 \rightarrow ROCH_3 + N_2$
Carbonyl function	Difference in neutralization between C_2H_5ONa and NaOH
Lactone and weak carboxylic functions	Difference in neutralization by Na_2CO_3 and $NaHCO_3$

Table 9.2. *Methods of analysis of functional groups*

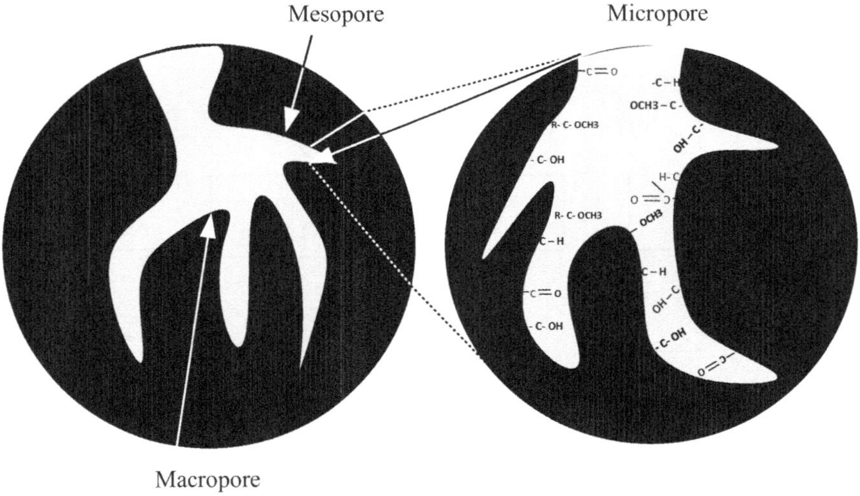

Figure 9.4. *Surface functions in micropores*

Thus, in the case of a non-dissociated molecule, adsorption is influenced by the porous carbon structure, whereas in the case of an ionized molecule, the surface functional groups are predominant (Figure 9.5). Similarly, it is suggested that high molecular weight molecules, such as humic substances, are preferentially removed by a macroporous carbon with many surface functions. In contrast, according to the same authors, the removal of small molecules requires mesoporous carbon and is supported by a large number of functional groups.

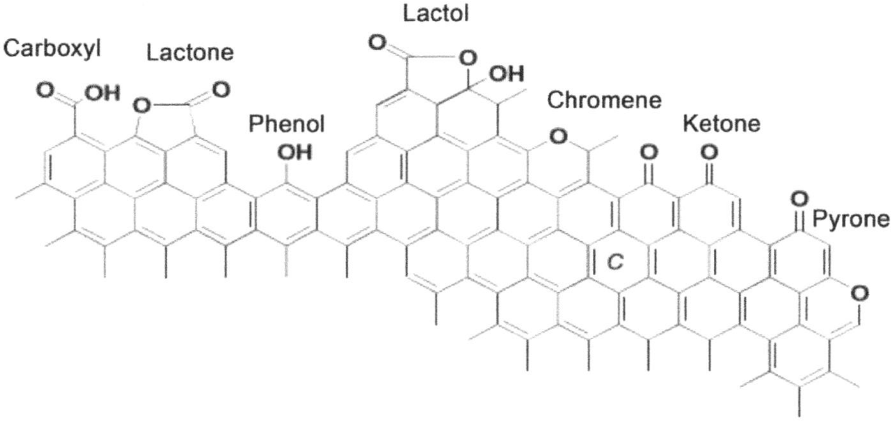

Figure 9.5. *Different surface functions distributed on the surface of the activated carbon*

The **pH at the point of zero charge** (PZC) is defined as the pH of the aqueous solution in which the solid exists under a neutral electrical potential. If both the carbon and the molecule are charged, electrostatic interactions may exist and explain preferential or, on the contrary, unfavorable adsorption.

– If pH < PZC, then the surface is positively charged.

– If pH > PZC, then the surface is negatively charged.

A negative charge on the surface of the activated carbon results from the presence of carboxylic (Ar-COOH), phenolic (Ar-OH) and carbonyl (Ar-C=O) groups. These functions give the carbon an acidic nature and hydrophilic properties. The carboxylic and lactone groups tend to polarize the surface, decreasing the adsorption of non-polar organic compounds in aqueous solution. The positive charge is generally attributed to basic surface complexes such as pyrones or chromenes, nitrogenous functionalities, or the existence of electron-rich regions in the graphene layers that accept protons from the solution.

Electron-rich sites act as Lewis bases by accepting protons. The basic nature of some carbons is due to the chemisorption of oxygen on the surface, forming a peroxide (O_2^{2-}). The peroxides on the surface thus constitute basic active sites.

The iodine index is a measure of the pore volume present for the diameter range of 1 to 2.8 nm. This is related to the microporosity of the activated carbon. This index refers to the amount, in milligrams, of iodine adsorbed per gram of activated carbon when the equilibrium concentration is typically 0.02. Activated carbons with a high iodine value would be suitable for the adsorption of low molecular weight substances. The iodine molecule has a surface area of 0.44 nm^2, which corresponds to the specific surface present in pores with diameters of less than 1 nm. This index is very important for the characterization of activated carbons because it gives an essential idea on the surface available for small substances (micropores). It is an indicator of GAC filter saturation, which operators use to estimate the height of the adsorption front and to anticipate activated carbon replacement.

The **molasses index** (indicated by the supplier) is related to the large pore volume. This ensures that there is a sufficient proportion of these pores to adsorb the large molecules, without being too reliable. It generally varies between 95 and 600 mg molasses per g of AC.

The **methylene blue index** is the number of milligrams of methylene blue adsorbed by one gram of activated carbon when the equilibrium concentration of methylene blue is 1 $mg \cdot L^{-1}$. The methylene blue index gives an idea of the available surface for microporous and mesoporous adsorbents. The surface occupied by a molecule of methylene blue is 2.08 nm^2.

The **phenol index** is the number of milligrams of phenol adsorbed per gram of activated carbon when the equilibrium concentration of phenol is equal to 1 $mg \cdot L^{-1}$. Phenol is often used as the first molecule to simulate certain toxic chemicals in the study of solid–liquid phase adsorption and to measure the porosity of activated carbon.

The **ash content** determines the inorganic, inert, amorphous and unusable parts present in the activated carbon. The ash often consists of calcium salts and metal oxides. The lower the ash content is, the better the activated carbon. The elemental composition of activated carbon generally consists of 85–90% C, 0.5% H, 0.5% N, 5% O and 1% S. The rest (5-6%) represents inorganic constituents (ash).

Table 9.3 summarizes the overall porosity of some activated carbons from their source material. It also indicates the compounds targeted for application in drinking water.

Raw materials	Activation types	Porosity	Targeted compounds
Coconut	Physical	Microporous	Pesticides Taste and odor Phenol Micropollutants Toxins
Coal **Lignite**	Physical	Microporous Mesoporous	Pesticides Taste and odor Micropollutants Humic substances
Wood	Chemical	Macroporous Mesoporous	Humic substances Color
Wood	Physical	Microporous Mesoporous	Pesticides Taste and odor Micropollutants Humic substances

Table 9.3. *Main characteristics of some activated carbons*

Figure 9.6. *Origin and manufacturing method of activated carbon*

9.2.1.2. *Parameters related to the characteristics of water and organic molecules*

The **pH** conditions the form in which the molecule is found (ionized or neutral). Retention is at its maximum when the pH is equal to the pKa. The effect of the pH is negligible for molecules that are not in ionized form at the pH of the solution.

The optimal pH for adsorption is one that neutralizes the organic molecules, in other words, the isoelectric point of a contaminant. A variation in the pH modifies the charge of the molecules and increases the affinity with water. Generally, the adsorption of organic pollutants increases when the pH decreases. In this case, the negative charges on the surface of the carbon are neutralized, which increases the diffusion capacity of a pollutant and, consequently, the number of available adsorption sites.

Temperature affects the affinity between the carbon and the pollutant. As the temperature of the solution decreases, adsorption increases due to its exothermic reaction. However, the diffusion rate increases as the temperature increases, which improves adsorption.

The **viscosity** of the solution is a factor affecting the adsorption rate, and it is likely that decreasing the viscosity will increase the rate.

The **polarity** of the molecules is a parameter that is involved in the adsorption mechanism. Indeed, activated carbons are non-polar materials and will therefore adsorb polar molecules well. As water is strongly polar, the non-polar molecules (hydrophobic) are easily adsorbed on the activated carbon, whereas the polar molecules (hydrophilic) have a greater affinity with water and are relatively less adsorbable.

The **size** of the molecules can strongly affect the adsorption process. In general, the size of the molecule must be smaller than the pore diameter of the adsorbent for it to diffuse rapidly into the pore volume and reach the adsorption site.

Solubility and **hydrophobicity** are used to estimate the feasibility of molecule adsorption. Activated carbon adsorption is mainly governed by hydrophobicity and can therefore be predicted using log Kow.

The octanol/water partition coefficient **(Kow)** is defined as the ratio of the concentration of a chemical in the octanol phase to its concentration in the water phase of a two-phase octanol/water system. The log Kow is equal to the logarithm of the ratio of the concentrations of the substance in octanol and in water. This parameter makes it possible to apprehend the hydrophilic or hydrophobic nature of a molecule. Indeed, if the log Kow is positive and very high, it shows that the molecule considered is much more soluble in octanol than in water, which reflects its hydrophobic nature, and conversely, if the log Kow is negative, it means that the molecule considered is hydrophilic. The values of Kow have no unit. In terms of log Kow, the range varies between –3 and 7. The log Kow is generally inversely related to the solubility in water and directly proportional to the molecular weight of a substance. Given its increasing use in estimating the adsorption of organic molecules on activated carbon, Kow is considered a necessary parameter.

Log D is a more refined measure that considers the modification of log D according to the pH of the aqueous phase and the pKa of the compound studied. For neutral (uncharged) compounds, log D and log Kow are equal. A positively or negatively charged molecule has a lower log D value than log Kow. Table 9.4 shows log Kow values for drugs, endocrine disruptors, industrial products, pesticides, disinfection by-products, etc. Substances with a log Kow greater than 3 will be very well adsorbed by the activated carbon. Those with a log Kow between 1 and 3 will be fairly well adsorbed. Molecules with a log Kow between 0 and 0.5 will be poorly adsorbed, and those with a negative log Kow will not be adsorbed.

The structure of the molecule influences the adsorption kinetics. When a molecule is not symmetrical, the adsorption is less good than with a symmetrical molecule. The OH, NH_2, SO_2 and COOH functional groups tend to reduce the adsorption mechanism, as well as the ionized molecules. However, the NO_2, NOH, COOR and X functional groups, which are molecules with a double bond or containing delocalized electrons, facilitate the adsorption mechanisms.

Figure 9.7 shows the adsorbability of some organic molecules according to the porous distribution of activated carbon.

9.3. Transport process in activated carbon: mass transfer

Mass transfer occurs from the fluid phase to the center of the activated carbon. This process takes place in three stages (Figure 9.8):

– **Stage 1: External diffusion.** The first stage is the external mass transfer of the compound from the liquid phase through the hydrodynamic layer that surrounds the activated carbon particle. The molecules must thus pass through a layer, known as the diffusion layer, around the activated carbon grains to reach their surface. This stage depends on the hydrodynamic and physicochemical conditions of the medium. External mass transfer, or film diffusion, is significant and controls the overall kinetics of molecule adsorption during filtration on GAC or in a mixed reactor (PAC).

– **Stage 2: Internal diffusion.** Once external diffusion has taken place, the internal mass transfer transports the molecules to an adsorption site. This adsorption happens on the surface (surface diffusion) and in the micropores. Indeed, the internal mass transfer of the molecules is carried out toward the mesopores and then the micropores. At the level of the microporous system, the adsorbed molecules diffuse through the micropore networks. This diffusion inside the carbon is performed under the influence of the concentration gradient. Global diffusion includes diffusion in the pores and diffusion on the surface of the pores. Surface diffusion (microporous transfer) is recognized as the limiting stage in kinetics, although pore diffusion (macropore transfer) also controls kinetics. The adsorption reaction of the compound on the surface of the activated carbon is very rapid and is not considered to limit the speed. Thus, overall adsorption is considered a two-stage mass transfer process: external and internal mass transfer. Internal mass transfer is often the slowest and thus controls the overall adsorption of molecules. Internal mass transfer can occur through the porous fluid or along the pore walls in parallel mechanisms. Both pore and surface diffusion are controlled by the pore structure of the activated carbon, as well as the properties of the molecules. Given that the processes take place at the same time, the fastest stage is usually pore diffusion. The latter controls the overall adsorption of an adsorbate.

– **Stage 3: Adsorption.** The molecules are adsorbed on the surface according to kinetics depending on the type of molecules, the surrounding medium and the characteristics of the activated carbon. This stage is very fast for physical adsorption.

Compounds and log Kow	Compounds and log Kow	Compounds and log Kow
Acenaphthene 3.90	Diazinon 3.01	Metolachlor 3.25
Acetochlor 3.17	Dibromochloromethane 2.16	Metolachlor EZA –0.03
Acetochlor EZA 1.82	Dibutyltin cation 4.44	Metolachlor OSA –0.03
Acetochlor OSA 1.82	Dicamba -1.88	Metamitron 0.85
Aclonifen 3.09	Diclofenac 0.70	Metazachlor 2.50
Alachlor 2.64	1.2 Dichlorobenzene 3.38	Metsulfuron –1.70
Aldicarb 1.13	1.4 Dichlorobenzene 3.38	MIB 3.31
Aldrin 3.01	1.2 Dichloroethane 1.45	Methyl parathion 3.32
Ametryn 2.98	Dichloromethane 1.25	Methoxychlor 5.08
Amino triazole –0.97	Dichlorprop 2.29	2.4 MCPA –0.81
AMPA –2.17	Dichlorvos 1.90	Monuron 1.46
Anthracene 4.54	Dicofol 4.30	Naphthalene 3.36
Atrazine 2.33	Dieldrin 5.40	Napropamide 3.36
Azithromycin 4.02	Diflufenicanil 4.20	Nicosulfuron 1.00
Azoxystrobin 2.50	Dimethenamid 2.47	Nonylphenol 5.00
Benazolin 1.34	Dimethachlor 2.17	Octyl phenols 5.30
Bentazone –0.46	Dinoseb 3.15	Oxadiazon 5.30
Benzene 2.11	Diquat –4.60	Parathion 3.81
Benzo (a) pyrene 6.35	Diuron 2.81	Paracetamol 0.46
Benzo (b) fluoranthene 6.67	Endosulfan 3.62	2.4'.PCB 5.10
Benzo (k) fluoranthene 6.30	Endrin 3.20	4.4'.PCB 5.58
Benzo (g,h,j) perylene 6.61	Erythromycin 4.51	PCB 118 4.4-5.0
Benzothiazole 2.10	Estrone 3.13	Pendimethalin 5.20
Bezafibrate 4.25	Estriol E3 2.45	Pentachlorobenzene 5.20
BDE 047 6.77	Ethephon -1.89	Pentachlorophenol 3.00
Bifenox 3.64	Ethofumesate 2.70	PEFOS 3.5-5.8
Biphenyl 4.09	Epoxiconazole 3.30	Phenanthrene 4.57
Boscalid 2.96	17- alpha-ethinyl estradiol 3.67	Phenmedipham 3.59
Bromacil 2.02	17- beta-estradiol 4.01	Prochloraz 4.30
Bromobenzene 2.99	Fenuron 1.00	Prometon 2.99
Monobutyltin cation 3.31	Fenoprop 3.80	Propamocarb HCl -1.20
Carbamazepine 2.45	Flufenacet 3.20	Propyzamide 3.00

Compounds and log Kow	Compounds and log Kow	Compounds and log Kow
Carbendazim 1.52	Fluorobenzene 2.27	Propazine 2.94
Carbaryl 2.81	Fluoranthene 5.10	Prosulfocarb 4.48
Carbofuran 2.07	Fluorene 4.20	Prothioconazole 3.82
Clarithromycin 3.16	Fluoxetine 4.05	Pyrene 5.18
Chlordane 3.32	Gemfibrozil 4.77	Quinoxyfen 4.70
Chloroalkanes (C10-13) 7.00	Genistein 2.84	Quinmerac 1.76
Chlorobenzene 2.84	Geosmin 3.10	Simazine 2.16
Chloroform 1.97	Glyphosate -3.20	Spiroxamine 2.80
Chlormequat chloride -1.56	Heptachlor 4.41	Sulcotrione -1.70
Chlortoluron 2.41	Heptachlor epoxide 2.70	Sulfamethazine 3.11
Chlorpyrifos 3.31	Hexachlorobenzene 5.73	Sulfamethoxazole 0.89
Chlopropham 3.06	Hexachlorobutadiene 4.78	Tebuconazole 3.70
Chlorsulfuron -0.99	ϒ HCH (lindane) 3.72	Terbacil 1.89
Chlorthalonil 2.94	Hydroxy alachlor 2.74	Terbutryne 3.65
Chlorfenvinphos 4.22	Hydroxy metolachlor 2.81	1.2.4 Trichlorobenzene 4.02
Chlorpyrifos 4.23	Ibuprofen 3.97	Trietazine 3.35
Ciprofloxacin 0.28	Indeno (1,2,3-cd) pyrene 7.60	1.1.2.2 Tetrachloroethane 2.39
Cyanazine 2.24	Iodobenzene 3.25	1.1.1. Trichloroethane 2.47
Cybutryne 2.70-4.00	Iodosulfuron 1.59	Triclosan 4.76
Cypermethrin 5.50	Isodrine 6.0	Tetrachloroethylene 2.53
Cyprodinil 4.00	Ioxynil 2.20	Carbon tetrachloride 2.60
DDT 6.90	Imazapyr 0.11	Thirame 1.73
2.4 D 2.81	Imidacloprid 0.57	Tributyl phosphate 4.00
Deethylatrazine 1.51	Isoproturon 2.50	Triclopyr -0.45
DEHP 7.60	Linuron 2.19	Trichlorobenzene 4.05
Deisopropylatrazine 1.15	Malathion 2.89	Trichloroethylene 2.99
Deschloroalachlor 3.08	Mancozeb 1.30	Trifuran 5.07
Deschlorometolachlor 3.10	Mecoprop -0.19	Trifluralin 5.34
Deethylhydroxyatrazine -2.70	Mesotrione 0.11	Tylosin 0.14
	Metaldehyde 0.12	

Table 9.4. *Log Kow of micropollutants: drugs, endocrine disruptors, industrial products, pesticides, disinfection by-products, etc.*

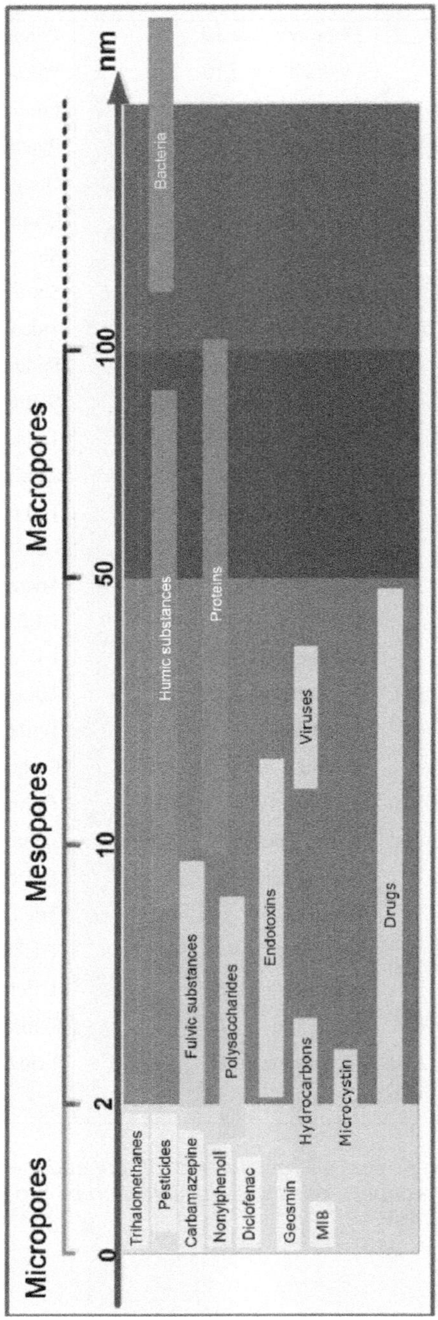

Figure 9.7. *Porous distribution and adsorbability of some organic molecules. For a color version of this figure, see www.iste.co.uk/gaid/watertreatment2.zip*

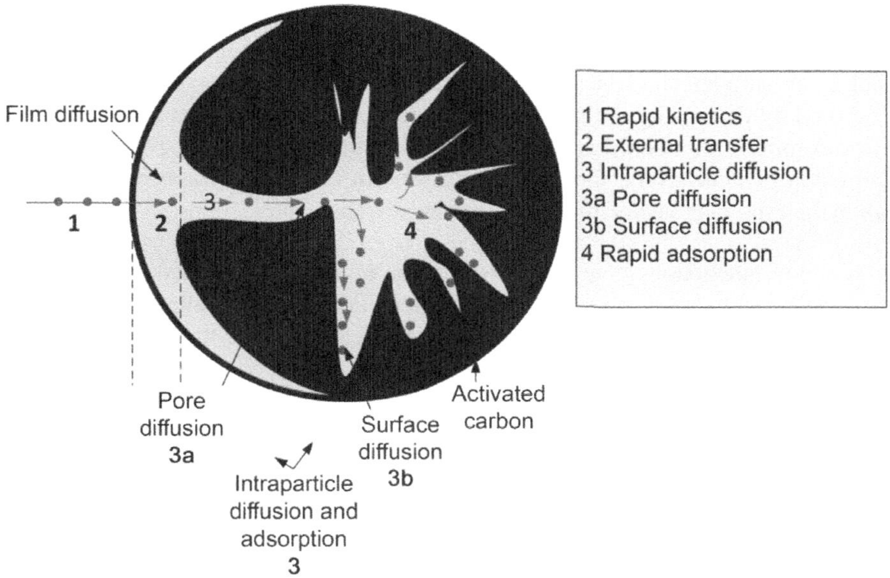

Figure 9.8. *Different stages of adsorption on a grain of activated carbon. For a color version of this figure, see www.iste.co.uk/gaid/watertreatment2.zip*

9.3.1. *Adsorption mechanisms and isotherms*

Adsorption is a surface phenomenon by which gas or liquid molecules attach themselves to the solid surfaces of adsorbents according to various more or less intense processes. This is explained by the intervention of intermolecular forces of attraction of various natures and intensities, which are responsible for the cohesion of the condensed phases, liquid or solid. A molecule that is unequally attracted by other two-phase molecules finds an energetically favorable position on the surface of the phase that attracts it the most. This latter is called the adsorbent, and the molecules thus adsorbed constitute the adsorbate.

The principle of adsorption includes physical and chemical adsorption. Physical adsorption involves attractive dispersion forces and, at very short distances, repulsive forces.

Physical adsorption is the most common type of adsorption. Adsorption is said to be physical adsorption when it is due to physical interaction forces between the atoms or groups of atoms of the solid and the molecules. The term *surface* corresponds to the totality of the surface of the solid, the geometrical surface for a non-porous solid, to which the internal surface of the pores is added for a porous solid, being accessible to the molecules. It occurs without modification of the molecular structure and is reversible (i.e. the adsorbed molecules can be easily desorbed by decreasing the pressure or by increasing the temperature). Contrary to chemisorption, physisorption is a non-specific phenomenon that gives rise to the formation of multilayers: the first layer is due to adsorbate/adsorbent interactions, while the following layers involve adsorbate/adsorbate interactions.

London interactions give rise to bonds that hold together molecules that do not have permanent electrical charges. These interactions are also called Van der Waals forces. They are caused by the movement of electrons inside the molecules, which can generate small instantaneous dipole moments. A local dipole can induce another instantaneous dipole on another molecule, oriented in such a manner that the interaction between the two dipoles is attractive.

Thus, this physical adsorption process involves relatively weak intermolecular forces (less than 100 $kJ \cdot mol^{-1}$). There are three types of physisorption forces:

– the dispersion forces (Van der Waals, London), which are always present;

– polar forces resulting from the presence of an electric field in the micropores;

– hydrogen bonds due to hydroxyl or amine groups.

The interaction between the adsorbent and the adsorbed molecule depends on the suitability between the properties of the two entities (polar–polar, non-polar–non-polar), the molecular weight and the shape of the molecule.

	Physical adsorption	Chemical adsorption
Type of bonds	Van der Waals, London	Ionic, covalent
Bond energy	Weak	Strong
Reversibility	Easy	Difficult, if not irreversible
Layer type	Polymolecular	Monomolecular
Adsorption heat ($kJ \cdot mol^{-1}$)	<100	100–500

Table 9.5. *Physical and chemical adsorption parameters*

Chemical adsorption results from a chemical reaction with the formation of chemical bonds between the adsorbate molecules and the adsorbent surface. The bond energy is much higher (between 100 and 500 $kJ \cdot mol^{-1}$), and the process is much less reversible and sometimes even irreversible. Given that just a single layer of adsorbate can chemically bind to the surface of the adsorbent, chemisorption is necessarily restricted to a single layer. This type of adsorption is notably encountered during catalytic reactions on the surface of the solid (such as in situ ozonation).

9.3.1.1. *Adsorption isotherms*

Adsorption isotherms are experimental curves that represent the variation (mass or volume) of adsorbed substrate per weight of adsorbent, according to concentration (in the liquid phase) or pressure (in the gas phase). They can be used to select the most suitable type of carbon and to determine the adsorption capacity for a continuous or batch process. They can also be used to approximate the number of solids required to achieve treatment efficiency.

Adsorption isotherms are curves relating the quantity of pollutants adsorbed per unit mass of material (qm) to the concentration of pollutants residing in solution (Ce). Adsorption isotherms are classified into six main classes (Figure 9.9):

– Type I isotherms are typical of single-layer adsorption or correspond to the filling of micropores with saturation when the available volume is completely filled. This type of isotherm is characteristic of adsorption on microporous carbons.

– Type II isotherms generally correspond to multilayer adsorption on open surfaces. A Type II isotherm can also result from a sum of isotherms I + II (filling of micropores followed by multilayer adsorption on an external surface).

– Type III isotherms reflect a lack of affinity between the adsorbate and the adsorbent and relatively strong adsorbate–adsorbate interactions.

– Type IV isotherms can result from the combination of Type I and Type V isotherms. This is the case for water on oxygen-rich carbons.

– Type V isotherms also reflect a strong interaction between the adsorbates. Moreover, the existence of hysteresis during desorption reflects the presence of mesopores, in which the vapor condenses, forming a meniscus of strong curvature.

– Type VI isotherms show characteristic stages of multilayer adsorption on a very homogeneous non-porous surface.

The isotherms commonly used are the Langmuir isotherm and the Freundlich isotherm.

Thus, for different initial concentrations of the molecules to be removed and for different masses of activated carbon in a given volume of water, we obtain a curve representing the adsorption efficiency for the tested molecules.

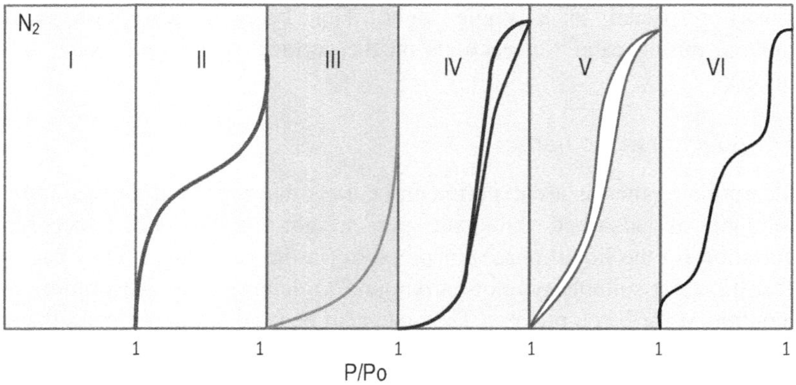

Figure 9.9. *Classification of isotherms. For a color version of this figure, see www.iste.co.uk/gaid/watertreatment2.zip*

However, adsorption isotherms do not explain the adsorption mechanisms, but they do provide access to the sizing parameters of activated carbon structures.

Absorbent capacity, in relation to the different compounds of a mixture, is the most important factor for the performance of the activated carbon toward the pollutants and micropollutants present in water. They are therefore important for the design of activated carbon reactors and the associated performance. The adsorption capacity of an adsorbent is defined as the quantity of substrate (generally in mass) adsorbed per unit weight of activated carbon at a given temperature. It requires the consideration of many parameters, both for the adsorbate (size of the molecules, solubility in water, etc.) and the adsorbent (specific surface, structure and type of particles, constituent, etc.).

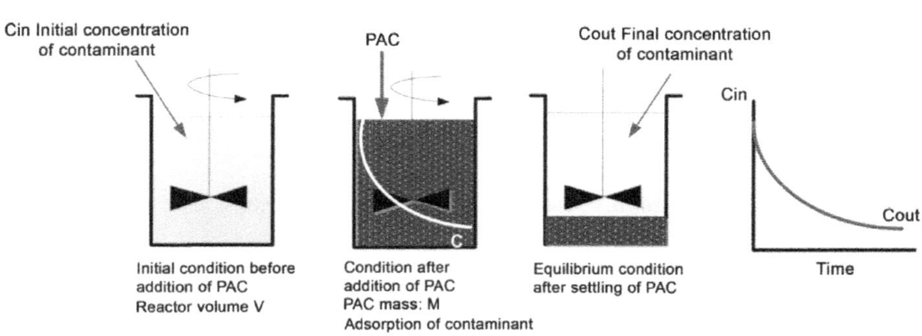

Figure 9.10. *Adsorption of a contaminant with powdered activated carbon in an integral mixing reactor*

At low concentrations of organic substances, it is generally observed that the adsorption rate, according to the concentration of the dissolved substance, follows the Freundlich law.

Thus, for a given equilibrium concentration Ceq (which is most often the required concentration at the outlet of a carbon reactor, in other words, Cout = C outlet), the adsorption capacity is given by:

$$qm = \frac{(Cin - Ceq)\ V}{M_{AC}}$$

with:

– *qm*: quantity of molecules adsorbed per g of activated carbon (mol (or g)·g^{-1} AC);

– *Ceq*: remaining concentration of molecules at equilibrium (mol (or g)·L^{-1});

– *Cin*: initial concentration of molecules (mol (or g)·L^{-1});

– *V*: volume of solution (L);

– M_{AC}: mass of activated carbon (g).

9.3.1.1.1. Adsorption models

Adsorption isotherms are most often modeled by the Langmuir and Freundlich models.

Langmuir model

The Langmuir model is characterized by the fact that adsorption is carried out in a single layer in energetically equivalent adsorption sites. These sites can only contain one molecule per site, and there are no interactions between the adsorbed molecules. The saturation of the adsorption sites occurs progressively at low concentrations. The number of adsorption sites on the surface is limited. The shape of the isotherm is characterized by a long plateau, indicating weak multilayer formation. In this case, we consider the adsorption sites to be equivalent: the adsorption is homogeneous. An L-type isotherm is characteristic of an adsorbent whose micropore volume is particularly high; the pores are too small to allow the formation of an additional layer (Figure 9.9).

The dynamic equilibrium between the molecules that reach the surface (adsorbed molecules) and those that leave it (desorbed molecules) makes it possible to assess the adsorption.

The isotherm is modeled by the following equation:

$$qm = \frac{qmax\ K\ Ceq}{1+KCeq} \text{ or even } \frac{1}{qm} = \frac{1}{qmax\ KCeq} + \frac{1}{qmax}$$

with:

– *qm*: adsorbed quantity per mass of activated carbon (mol or g per kg AC);

– *Ceq* = C out: concentration at equilibrium in the fluid phase in (mol or $g \cdot m^{-3}$);

– *K*: the constant that describes the energy of adsorption; the Langmuir constant ($m^3 \cdot g^{-1}$ or mol^{-1}). The latter depends on the temperature and gives an indication of the adsorbate affinity for the adsorbent: the higher it is, the stronger the affinity;

– *qmax*: maximum quantity that can be adsorbed on a given mass of solid, meaning a total occupation of the adsorption sites (mol or g per kg AC).

The values of *qmax* and *K* are obtained by plotting $1/qm$ against $1/Ceq$.

The **RL ratio** indicates favorable adsorption when RL approaches zero and unfavorable adsorption if R approaches 1:

RL = 1/K. Cin

Distribution coefficient

The distribution coefficient is a special case of the Langmuir relation. It is defined for low concentrations of adsorbed species. It is the ratio of the quantities fixed per gram of solid to the quantity of solute remaining in solution by volume of solution. The distribution coefficient characterizes the affinity of the molecules for the active carbon:

$$Kd = \frac{(Cin - Ceq)V}{M\,.Ceq.1000}$$

with:

– *Kd*: distribution coefficient ($L \cdot g^{-1}$);

– *Cin*: initial concentration of molecules ($mg \cdot L^{-1}$);

– *Ceq* = C out: remaining concentration of molecules ($mg \cdot L^{-1}$);

– *V*: volume of solution (L).

Freundlich model

In the Freundlich model, adsorption first occurs in a single layer and then in multilayer. It is represented by a two-parameter equation (k and n) and consists of an exponential distribution of the energies of the adsorption sites on the surface of the support. It is characterized by adsorption in localized sites. In the case of an adsorbent with heterogeneous surfaces, one can use the Freundlich model:

$$qm = k.Ceq^1/n \text{ or even } Ln\,qm = Ln\,k + \frac{1}{n}\,Ln\,Ceq$$

qm and Ceq are, respectively, the adsorption capacity in $mg \cdot g^{-1}$ and the residual concentration of the molecules at equilibrium in mg (or mol)$\cdot L^{-1} \cdot$k; n are Freundlich constants to be determined for each adsorbent/adsorbate couple at each temperature; k depends on the total number of adsorption sites.

The (dimensionless) constant n gives an indication of the adsorption intensity. It is generally accepted that low values of 1/n ($0.1 < 1/n < 0.5$) are characteristic of good adsorption, while higher values reveal moderate ($0.5 < 1/n < 1$) or weak ($1/n > 1$) adsorption. The constant "1/n" is very often replaced by "n". Note that if n (or 1/n) tends toward 1, the isotherm becomes linear.

The passing, in a neperian logarithmic scale, of the Freundlich isotherms by plotting Ln qmax according to Ln Ceq, makes it possible to linearize them, with a slope of 1/n and ordinate ln k, facilitating the exploitation and comparison of the various equilibria. For low values of equilibrium concentrations, the Freundlich and Langmuir equations lead to equivalent adsorbed quantities.

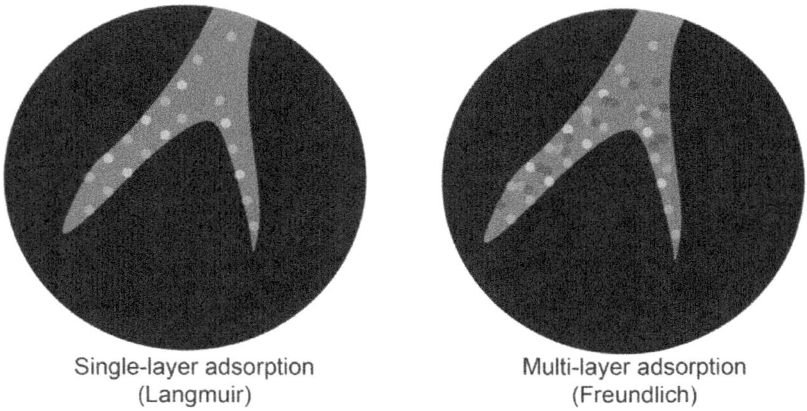

Figure 9.11. *Schematic diagram of adsorption modes. For a color version of this figure, see www.iste.co.uk/gaid/watertreatment2.zip*

For high values of Ceq, K Ceq (Langmuir isotherm) becomes greater than 1, and the denominator of this isotherm approaches K Ceq. When it reaches this value, the terms K Ceq of the numerator and denominator of the equation cancel, and we obtain qm = qmax.

In the case of mixtures of organic substances, as is often the case in the water to be treated, the different compounds do not necessarily have the same adsorption capacities and kinetics. Some molecules are adsorbed first, and others are more difficult to adsorb, giving rise to competitive adsorption. This is the reason why adsorption kinetics are carried out on the water to be treated when possible. Each adsorption constant per molecule gives an indication of their adsorption potential.

The Freundlich model is well suited to define activated carbon behavior toward organic molecules in water.

Temkin model

The Temkin empirical model describes chemisorption mechanisms by considering the variation in adsorption energy. This is assuming that the heat of adsorption varies linearly with the degree of recovery of the carbon surface by adsorbed molecules:

$$qm = bT\ (Ln\ KT\ Cout)$$

with:

– *qm*: adsorption capacity.

– *bT* ($mg \cdot g^{-1}$) and *KT* ($mg \cdot L^{-1}$): Temkin constants.

– *Cout*: remaining concentration of molecules ($mg \cdot L^{-1}$).

When the water to be treated is available, adsorption kinetics are performed to determine the optimal contact time between the activated carbon and the organic molecules present in the water. These curves are obtained by plotting the evolution of organic substance concentrations (expressed in DOC, TOC, Ox. $KMnO_4$, etc.) according to time for different initial concentrations and/or for different masses of carbon. Generally, they tend toward a maximum adsorbed quantity, which represents the ideal adsorption capacity of activated carbon. Nevertheless, the diversity of the molecules to be treated (humic matter and/or micropollutants) can lead to quite long contact times before reaching the required final concentration. Contact times can vary between 5 and 120 min (or even 24 hours). This can lead to very large structures, especially when the flow to be treated is high.

Manufacturers prefer to intervene on the mass of carbon to be used in the case of reactors with powdered or micrograin carbon to limit the contact time to 10 min. This leads to acceptable and easily exploitable works.

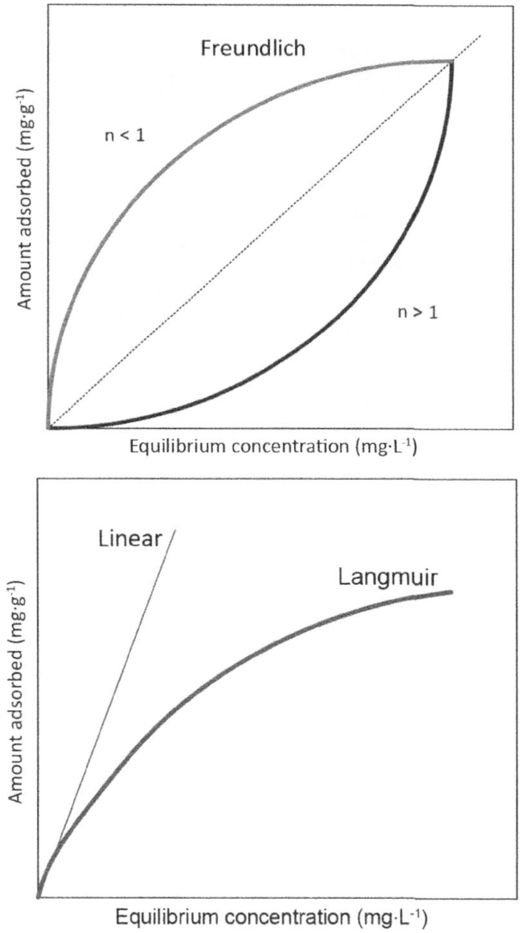

Figure 9.12. *Adsorption isotherms with Freundlich and Langmuir models. For a color version of this figure, see www.iste.co.uk/gaid/watertreatment2.zip*

9.3.1.1.2. Adsorption kinetics

Adsorption kinetics is the second parameter indicating the removal performance of an adsorbent (activated carbon). It allows us to estimate the quantity of pollutants adsorbed as a function of time. The kinetics provide information on the adsorption mechanism, as well as on the mode of transfer of solutes from the liquid phase to the solid phase.

The mass balance gives the following:

– Quantity adsorbed: initial quantity – final quantity;

– qm M = V (Cin – Cout);

with:

– qm: mass of contaminants adsorbed by the activated carbon $mg \cdot g^{-1}$;

– M: mass of activated carbon injected in g;

– V: volume of the reactor in L;

– Cin and Cout: initial and final concentrations of the contaminant in $mg \cdot L^{-1}$.

$$qm = \frac{V}{M}(Cin - Cout)$$

With the Freundlich equation, we can write:

$$qm = \frac{V}{M}(Cin - Cout) = kCout^{1/n}$$

Knowing the values of parameters *k* and 1/*n*, we can deduce the mass of activated carbon required for the adsorption of a contaminant.

For example, the removal of a pesticide with a concentration in water of 1 $\mu g \cdot L^{-1}$ requires the following mass of PAC:

Freundlich parameters: k: 74.5 and 1/n: 0.32;

– flow to be treated: 1,200 $m^3 \cdot d^{-1}$;

– Cout: 0.1 $\mu g \cdot L^{-1}$

$$\frac{V}{M} = \frac{kCout^{\frac{1}{n}}}{Cin - Cout} = 74.5\frac{(0.1)^{0.32}}{1 - 0.1} = 35.65\ L \cdot g^{-1}$$

Therefore, M/V = 0.028 $g \cdot L^{-1}$ = 0.028 $kg \cdot m^{-3}$.

The quantity of carbon to be injected is:

$$PAC = 0.028\ (\mathrm{kg} \cdot \mathrm{m}^{-3})1{,}200(\mathrm{m}^3 \cdot \mathrm{d}^{-1}) = 33.6\ \mathrm{kg} \cdot \mathrm{d}^{-1}$$

The adsorption kinetics of a material can be modeled. The molecule-activated carbon adsorption, with a localized reaction in the pores of the activated carbon, kinetically includes three important stages:

– diffusion through the film surrounding the activated carbon particles, called "external diffusion";

– diffusion in the pores of the activated carbon, called "intraparticle diffusion";

– the adsorption reaction.

The first two stages govern the adsorption kinetics.

External diffusion

The kinetic expression used to model external diffusion, such as the transfer of molecules from water to activated carbon, is:

$$-\frac{dCt}{dt} = k\frac{a}{V}\,(Ct - Ceq)$$

The integrated form gives:

$$Ln\frac{(Cin - Ceq)}{(Ct - Ceq)} = k\frac{A}{V}t = k't$$

with:

– Ceq = Cout: concentration of molecules at equilibrium ($g \cdot m^{-3}$);

– Cin: concentration of molecules at the inlet ($g \cdot m^{-3}$);

– Ct: concentration of molecules at time t ($g \cdot m^{-3}$);

– t: reaction time (min);

– A: specific surface of the activated carbon/solution interface (m);

– V: volume of solution (m^3);

– k: reaction constant ($m \cdot min^{-1}$).

A simple plot of $Ln\frac{(Cin-Ceq)}{(Ct-Ceq)}$ according to the reaction time makes it possible to assess whether the external diffusion stage is crucial for the whole reaction. The value of k' can be deduced from this.

Intraparticle diffusion

Intraparticle diffusion models are applied in three different forms.

The quantity adsorbed (qt) at time t is plotted according to $t^{1/2}$. We can obtain a line passing through the origin, whose slope indicates the transfer coefficient.

However, the same plot can lead to two or three lines with different slopes: this is a multilinear reaction in which external surface adsorption is produced in the first stage, and intraparticle diffusion is controlled in the second stage.

The third stage is the final equilibrium stage, for which the adsorption rate is slow. This reflects the passage of molecules from the mesopores to the micropores.

The fractional approach of the equilibrium change is done according to a function of $(Dt/r^2)^{1/2}$, where r is the radius of the activated carbon particle and D is the diffusion coefficient of the molecules in the carbon.

The intraparticle diffusion constant kp ($mg \cdot g^{-1} \cdot min^{-1/2}$) is obtained from the line $qt = f(t^{1/2})$, which is rather expressed according to a formulation $qt = kp\, t^{1/2} + B$.

B is a constant ($mg \cdot g^{-1}$).

The adsorption reaction

The pseudo-first order expression is of the following form:

$$\frac{dqt}{dt} = k_1\,(qm - qt)$$

which gives:

$$Ln(qm - qt) = Ln(qm) - k_1 t$$

This is a global pseudo-first-order kinetic expression for the adsorption reaction. With a partial order of 1 with respect to the concentration of unavailable sites on the activated carbon and a partial order of 0 with respect to the concentration of the molecules in water.

A pseudo-second-order expression is also possible:

$$\frac{dqt}{dt} = k_2\,(qm - qt)^2$$

which leads to:

$$\frac{1}{(qm - qt)} = \frac{1}{qm} + k_2 t$$

This is a pseudo-second-order kinetic expression for the adsorption reaction. With a partial order of 2 with respect to the concentration of unavailable sites on the activated carbon and a partial order of 0 with respect to the concentration of the molecules in water.

At high concentrations of pollutants, the pseudo-first-order equation is generally a better predictive model. However, at low initial adsorbate concentrations, the pseudo-second-order model is more suitable.

The rate of adsorption is greater as follows:

- the concentration of the solute is high;
- the granulometry of the adsorbent is low;
- the specific surface of the adsorbent is high;
- the size of the adsorbable molecules is small.

Some factors accelerate extraparticle diffusion kinetics, such as agitation and solute concentration. Other factors accelerate intraparticle diffusion and fixation kinetics, such as small particle size and high affinity of the solute for the adsorbent.

9.4. The different forms of conditioning of activated carbons

Depending on their applications, activated carbons are available in powder, micrograin or granular form. The size of each of these activated carbons varies according to the type of activated carbon.

9.4.1. *Powdered activated carbon*

PAC, which consists of very fine particles with a diameter of 5–35 μm, is only applied during short periods of degradation of raw water quality (presence of toxins, accidental pollution, momentary increase in the concentration of organic matter, momentary increase in the concentration of pesticides, etc.). In this case, it is dosed directly into the treated water, usually before or at the same time as the coagulant dosage.

It is also used continuously in PAC reactors for the removal of organic substrates, micropollutants, color, etc. Activated carbon is also used in combination with membranes in a multibarrier approach integrating PAC and ultrafiltration (UF).

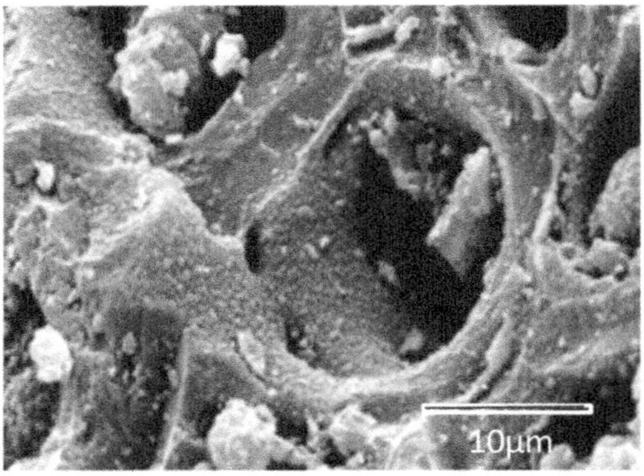

Figure 9.13. *Powdered activated carbon made from date pits*

9.4.2. *Micrograin activated carbon (µgrain)*

Micrograin activated carbon has a particle size that is generally between 0.3 and 0.8 mm and a specific surface of approximately 900–1,000 $m^2 \cdot g^{-1}$.

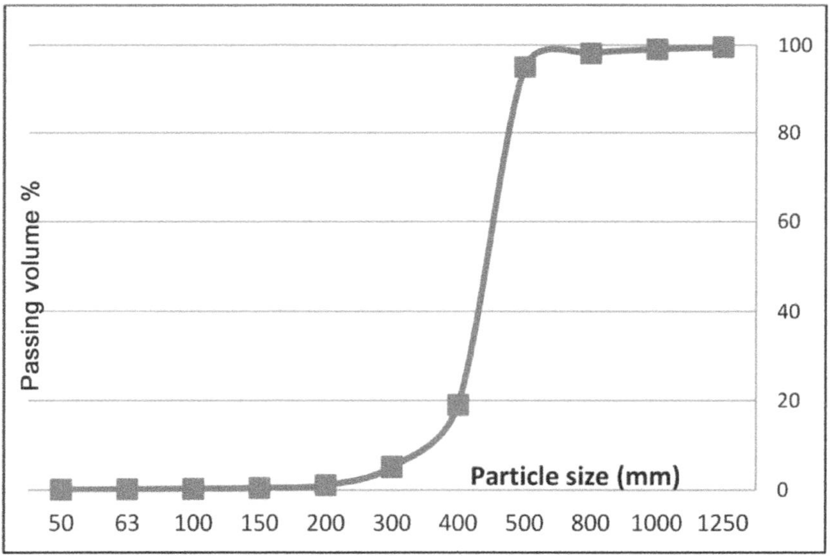

Figure 9.14. *Example of the particle size of a micrograin activated carbon*

Re-agglomerated activated carbons are produced from a mixture of raw materials (mineral and preferably bituminous coal) to obtain a constant product after steam activation. The two most important stages in the production of agglomerated activated carbon are carbonization (baking) and activation. During carbonization, the raw material is oxidized and carbonized: all the volatiles of the raw material are removed to obtain the purest possible product. This stage is very exothermic, and temperature control is critical. The product goes from 150 to 500°C.

During the pulverization stage, a carbon binder is added to the powdered carbon. It is then compacted, crushed and sent to a carbon furnace. The advantage of this manufacturing process is the creation of macropores and mesopores, known as "transport", to the micropores. The creation of the final porosity takes place in the activation stage. The steam, in an oven heated at very high temperature, will oxidize the carbonated surface. The point is to create a porosity that will be involved in any reaction kinetics that the activated carbon granule will undergo:

– in the activation phases, when the steam must oxidize the carbon grain in a uniform manner;

– in the adsorption phase to promote the transport of pollutants;

– in the reactivation phases, when the pollutants have to be pyrolyzed and gasified.

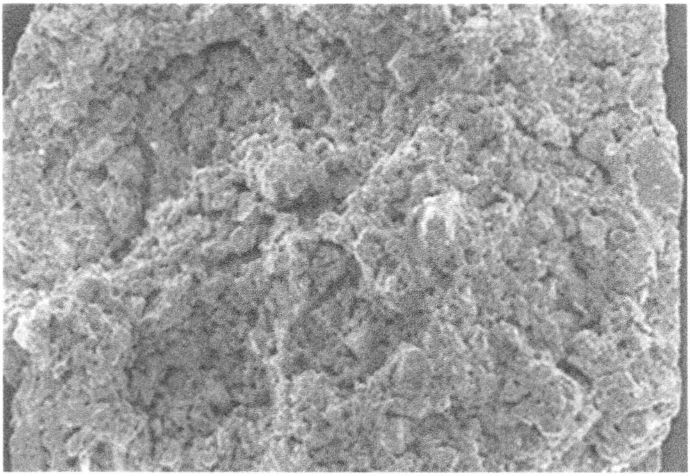

Figure 9.15. *Electron microscope views of the Microsorb 400 R*

The pore distribution of the activated carbon and its performance are then determined by the retention time in the activator and the conditions in the furnace. The activated carbon is then ground and sieved to obtain different particle size distributions specific to each application. The result is a micrograin activated carbon, which is used for the adsorption of organic substances. The activated carbon is easily wet and has a very low fine particle content of 0.1%. The µgrain is regenerative.

It is used in continuous fluidized bed reactors for the removal of organic substances, toxins, molecules responsible for tastes and odors, color and micropollutants.

9.4.3. *Granular activated carbon*

GAC comes in irregularly formed particles, in a range of sizes from 0.4 to 2.5 mm. It is used in continuous fixed bed reactors for the removal of organic substances, toxins, taste and odor molecules, color and micropollutants. It acts as a physical filter for particles, as well as a biological support for the treatment of ammonia or organic substances.

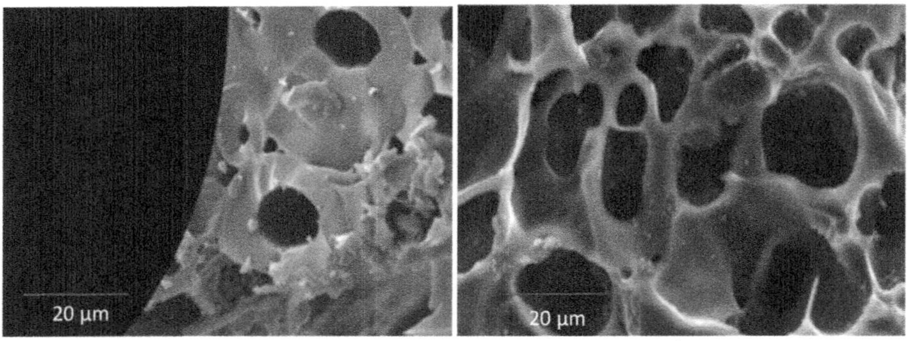

a) b)

Figure 9.16. *(a) Picabiol activated carbon with a plant base (Jacobi) and (b) F400 (Chemviron) with a coal base*

9.5. Adsorption reactors on activated carbon: removal process

The main mechanisms involved in the removal of organic substances depend on the reactor used.

The adsorption mechanism prevails for PAC and fluid bed reactors. For fixed bed reactors, biological activity can complement the adsorption mechanism.

The importance of these processes in the removal depends on the residence time of the activated carbon in the water (carbon age).

In a PAC reactor, kinetics is not a limiting parameter for adsorption, since the homogenization of the reactor allows rapid external diffusion kinetics, and the small size of the PAC particles limits the amount of internal diffusion.

A fixed bed reactor has a high carbon age; the lifetime is calculated for a replacement between 1 and 4 years. For example, the development of microorganisms that are adapted to the conditions of the water to be treated is possible.

Activated carbon can therefore be used in adsorption or adsorption-biological mode. However, adsorption mode is most often used, which works with residence times that are too low to allow the development of microorganisms. The PAC or the micrograin are constantly renewed by the addition of fresh material. This makes it possible to obtain stable performances.

An activated carbon with a life span of more than 30 days goes into biological mode (if the water temperature is favorable), while a carbon with a life span of less than 10 days is always in adsorption mode.

It is, however, possible that under certain operating conditions, some adsorption reactors can operate in a hybrid mode, with an adsorption part (persistent pollutants) and a biological part (biodegradable pollutants). This is particularly the case in reactors that have not yet reached saturation and have started to develop a biofilm (age of the intermediate carbon). Similarly, in a reactor with continuous injection of PAC, the two processes can, in theory, coexist if the injected carbon is maintained for a sufficient amount of time in the reactor before being withdrawn. The newly injected PAC performs adsorption, while the used PAC can be covered with a biofilm and be responsible for biodegradation.

However, this situation does not imply stable performance because the kinetics of biological degradation vary greatly with water temperature (variable during the year for surface waters), and a presence of biodegradable dissolved organic carbon (BDOC) is always available to maintain this biological activity. This implies rigorous control and activated carbon replenishment, which can be very frequent and large depending on the water quality and operating conditions.

Fixed bed processes differ from PAC reactors by the retention of particles. Indeed, the passage of water within the bed traps the particles. Moreover, the accumulation of this phenomenon, coupled with the development of the biofilm, can lead to clogging of the bed.

In all cases, it is a question of treating organic substances, which are often classified according to the following categories (non-exhaustive list):

– Category I: organic compounds that cause taste, odor and color problems;

– Category II: synthetic organic chemicals including the following micropollutants: endocrine disruptors, drug residues, industrial products, pesticides;

– Category III: organic compounds (precursors) that react with disinfectants to produce "disinfection by-products".

Furthermore, the efficiency of disinfection is improved when activated carbon processes are used to reduce the organic content of the water before this final step. Thus, during chlorination, fewer side reactions take place, and the formation of chlorinated organics is reduced. It should be possible to achieve a given level of inactivation of unwanted microorganisms with a lower dosage of disinfectant for a given period of time or a shorter contact time for the same dosage of disinfectant to achieve the same level of inactivation.

9.6. PAC reactors: description of PAC reactors

The use of PAC differs from that of GAC through its point of application, the type of reactor used, the relative importance of adsorption kinetics and the nature of the competitive adsorption effects. The general objective of any PAC application is a fast and optimal loading of the material, to avoid long hydraulic contact times in large reactors, and a high dosage of activated carbon for the removal of organic substances and micropollutants.

The kinetics of the overall adsorption of organic molecules can be illustrated using a perfectly mixed reactor (Figure 9.17), where V is the volume of the reactor, with an amount of PAC injected to move from an initial concentration (Cin) of organic molecules to a final concentration (Cout) at the outlet.

Agitation in the reactor is an important parameter because the external mass transfer resistance is reduced by increasing the agitation power. This reduces the thickness of the film surrounding the activated carbon particles.

After the addition of the activated carbon, the concentration of the molecules in solution decreases according to the reaction time until a very low final concentration is reached. The activated carbon-organic molecule couple is at equilibrium (Figure 9.17). This is because the mixing conditions in the PAC reactor make the concentration of the target compound equal to the concentration of the effluent, C out, at all points in the reactor.

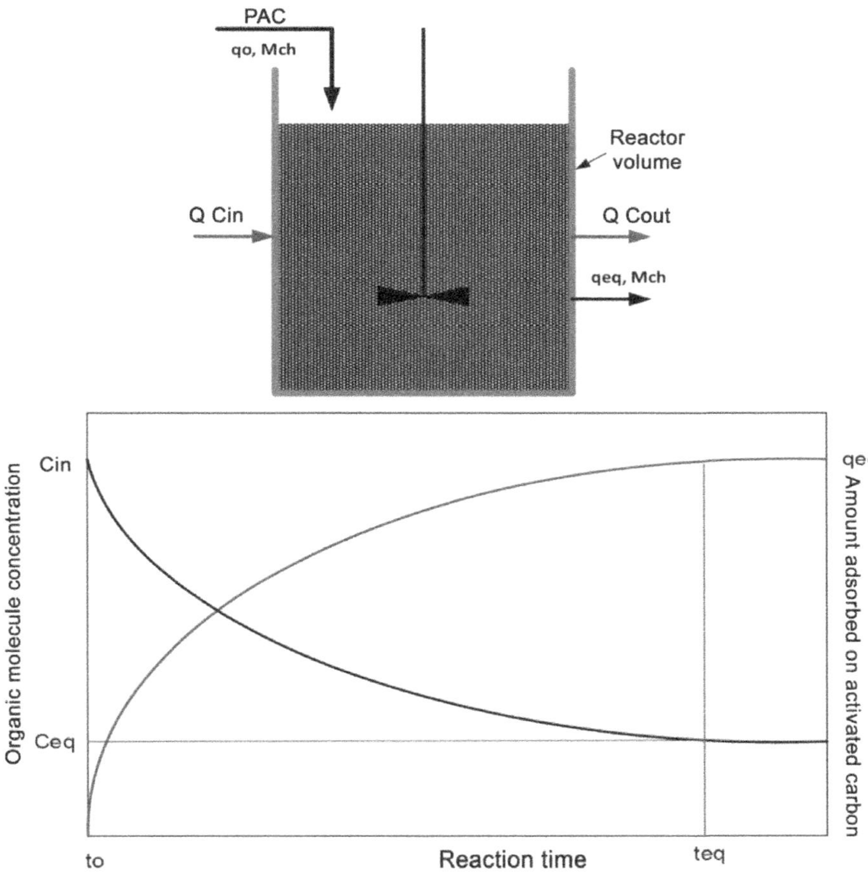

Figure 9.17. *PAC reactor and equilibrium kinetics. For a color version of this figure, see www.iste.co.uk/gaid/watertreatment2.zip*

The concentration at this time is called the equilibrium concentration, Ceq. It is most of the time assimilated to the desired outlet concentration (Cout) by the designers of drinking water treatment plants. The time corresponding to Ceq is the equilibrium time, teq.

Other parameters affect the kinetics or equilibrium time, such as the size of the activated carbon, the diffusion coefficient of the molecules, the initial concentration of the molecules, and the mode and power of agitation (Figure 9.18).

In the case of PAC, where the particles are very small, the distance of the internal diffusion path that the molecules must travel to reach an adsorption site decreases, and the adsorption sites are then reached more quickly. The same mass of fine activated carbon particles reaches a state of equilibrium quicker than large particles of the same carbon. As the diffusion coefficient and the concentration of the molecules increase, the flow in the carbon increases, and the adsorption sites are reached faster. This also decreases the reaction time to reach equilibrium.

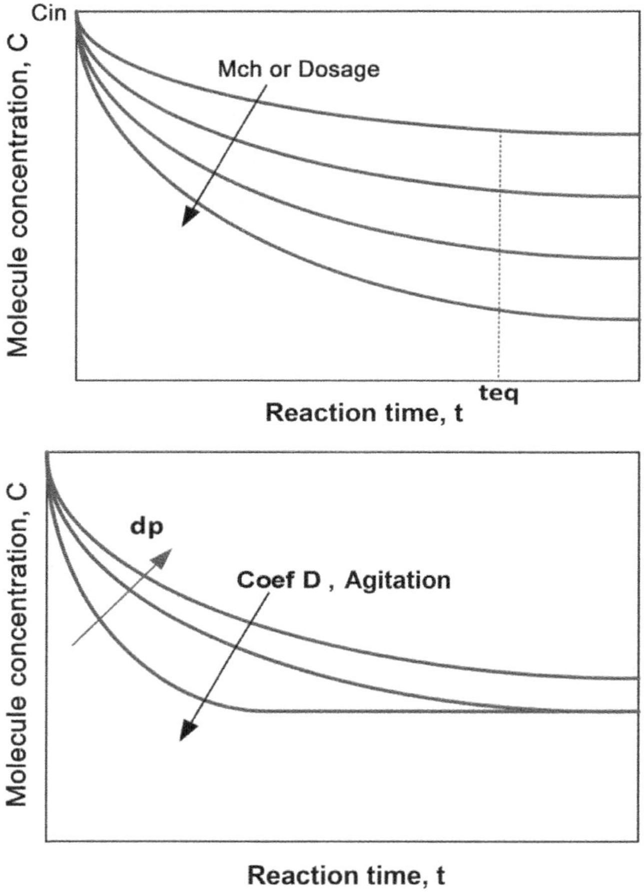

Figure 9.18. *Influence of the parameters: mass of carbon or dosage, particle diameter or agitation. For a color version of this figure, see www.iste.co.uk/gaid/watertreatment2.zip*

The factors affecting the equilibrium time are summarized as follows:

$$teq = f\ (dp^2, D^{-1}, Cin^{-1})$$

9.6.1. *Mass balance*

Increasing the dosage of activated carbon injected ($g \cdot m^{-3}$) in a PAC reactor decreases the final concentration of organic molecules (Cout), and the concentration adsorbed in the activated carbon increases.

9.6.1.1. *Integral mixing*

In a fully mixing adsorption reactor, a first-order reaction gives:

$$\frac{dC}{dt} = -kC$$

By integrating between C = Cin and C = Cout, and t = 0 and t, we obtain:

$$\int_{Cin}^{Cout} \frac{dC}{C} = -k \int_{o}^{tf} dt \;\; \text{or even } Ln\frac{Cout}{Cin} = kt$$

k depends on the mass of the adsorbent, the temperature and the type of compound.

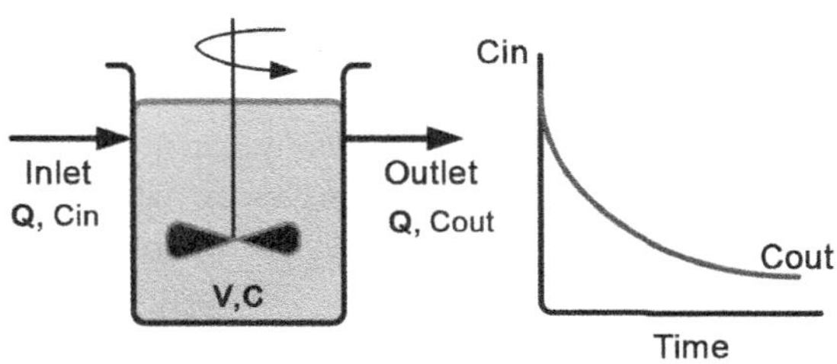

Figure 9.19. *Operating diagram of the integral mixing reactor. For a color version of this figure, see www.iste.co.uk/gaid/watertreatment2.zip*

With qo being the quantity adsorbed at time to, and qm being the quantity adsorbed at time t for the mass of activated carbon (M_{AC}) present in the PAC reactor:

$$\mathrm{V.Cin + M.qo = V\,Cout + M.qm}$$

For qo = 0, we derive the following dosage:

$$Dosage = \frac{M}{V} = \frac{(Cin - Cout)}{qm} = \frac{(Cin - Cout)}{kCout^{\frac{1}{n}}}$$

By varying the dosage of activated carbon or the initial concentration of gold-organic molecules, the amount adsorbed can be calculated. This makes it possible to determine the adsorption capacity of the activated carbon at any initial concentration Cin.

However, the surface water matrix contains many organic molecules, such as humic and fulvic substances, pesticides and other micropollutants.

Therefore, it is important to establish adsorption isotherms for the water to be treated (when possible) to determine the real adsorption capacity associated with this matrix. This establishment should be made because there is significant adsorption competition between the different molecules, and it can favor some molecules over others.

Thus, when molecule 1 is more strongly adsorbed than molecule 2 (in other words, $k1 > k2$), weaker adsorption of molecule 2 occurs.

To remedy this situation when isotherms are not accessible, manufacturers intervene by increasing the carbon dosages to compensate for the decrease in adsorption of the weakest compounds. This is especially the case given that, in a surface water matrix, the concentrations of natural organic matter vary between 1 and 20 $mg \cdot L^{-1}$, while those of micropollutants vary from 0.2 to 10 $\mu g \cdot L^{-1}$.

Even if the size of the molecules and their polarity are different, adsorption competition exists. This is because humic substances, which are large molecules, cannot access the micropores due to size exclusion, whereas small molecules, such as pesticides, can. However, size exclusion can result in partial or complete blockage of a pore. Partial blockage can slow down the kinetics of small molecules, while complete blockage results in a decrease in the adsorption surface area.

Lastly, in treatment processes with PAC injection, a coagulant is generally added to stabilize the activated carbon suspension by weighing down the particles and facilitating their settling in the case of separation by decanter. The addition of this reagent does not modify the adsorption mechanism.

If the concentrations of the substances to be removed are high or constant, the PAC is placed in a contact tank with a contact time of 10–15 min, which is located upstream of coagulation–flocculation–settling. It is then removed during the settling process.

When concentrations of substances in the water are seasonal or occasional, PAC is only introduced upstream of the coagulation tank or directly into it (Figure 9.20).

The coagulation and flocculation contact times are the reaction times for adsorption. It is then removed during settling.

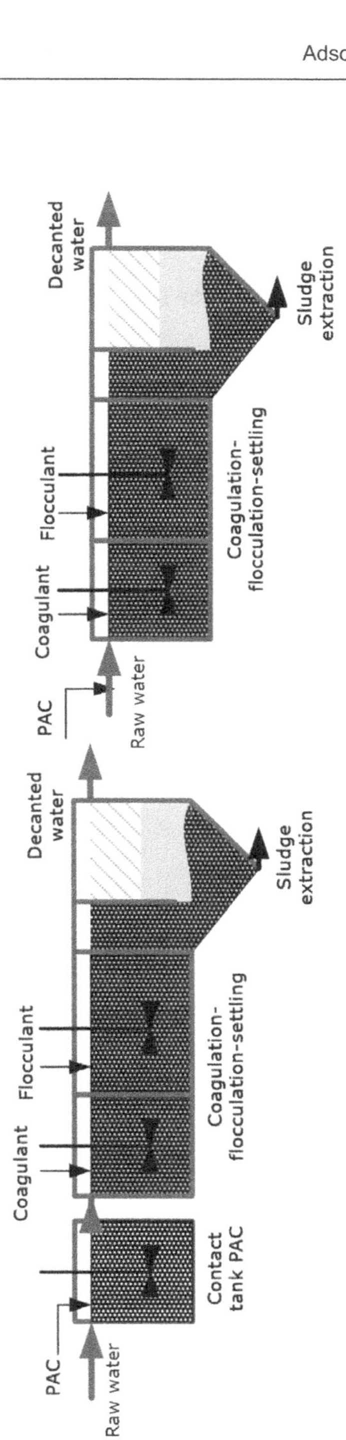

Figure 9.20. *Injection of PAC in a contact tank or in line in the coagulation tank. For a color version of this figure, see www.iste.co.uk/gaid/watertreatment2.zip*

The same provisions are applied with a flotation process.

In recent years, processes with continuous PAC recirculation (Figure 9.21) have been developed, which make it possible to maintain a constant concentration in the contact tank and to improve the reaction kinetics. These reactors operate with lower PAC dosages than those of direct injection for the same contact time and better efficiency.

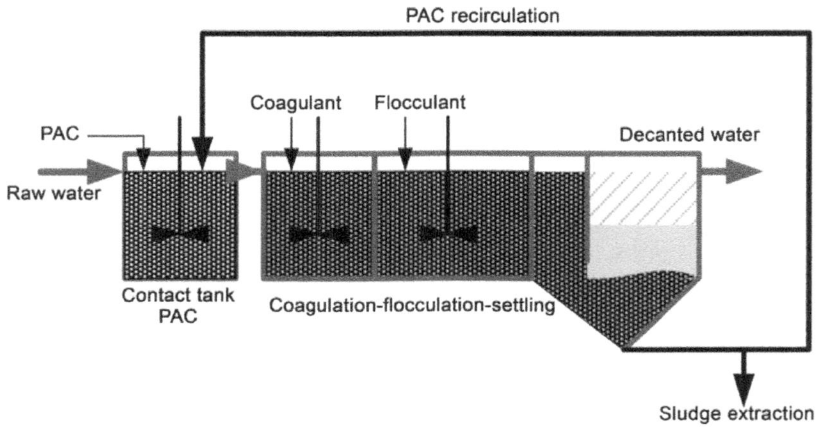

Figure 9.21. *Process with PAC recirculation. For a color version of this figure, see www.iste.co.uk/gaid/watertreatment2.zip*

The principle of this type of reactor is to maintain a high concentration of PAC in the contact tank (1–5 $g \cdot L^{-1}$), which permanently mobilizes sufficient available adsorption sites to remove the organic substances and micropollutants. This PAC concentration is maintained by recirculating the PAC extracted from the decanter. Fresh PAC is injected continuously or at a controlled frequency, and used PAC is removed from the decanter. There is a permanent balance between fresh and used PAC. To maintain a sufficiently long carbon age (7–10 days) in the contact tank, it is important that there are as few suspended particles as possible in the inlet water (turbidity < 5 NTU). If this is not the case, this results in an increase in the mass of non-active SS, which is recirculated together with the activated carbon, resulting in a reduction in the age of the carbon and the need to top up with fresh PAC more frequently. The advantage of these processes is that they react with water that is less loaded with natural organic substances, since a large part of the latter will have been removed in a first stage involving coagulation–flocculation–settling with injection of coagulant and polymer (Figure 9.22). The same provisions are applied with a flotation device in the first stage.

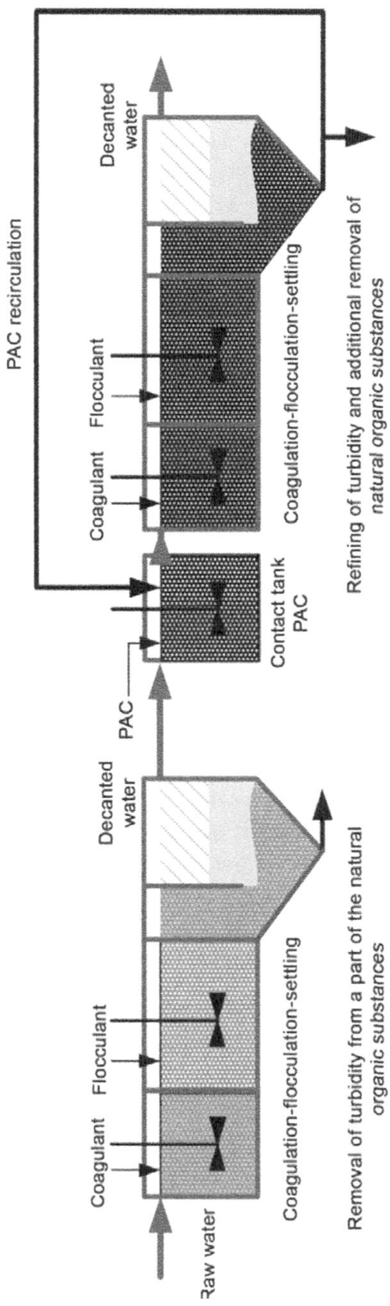

Figure 9.22. *PAC reactor located in the second stage. For a color version of this figure, see www.iste.co.uk/gaid/watertreatment2.zip*

Different types of PAC are commercially available due to variations in raw materials and manufacturing processes. Since the types of organic contaminants vary considerably from one water to another, activated carbon may be less effective for one type of water and very good for another. Therefore, comparative tests between various activated carbons are recommended to select the most suitable one for the water to be treated. This is due to the influence of the water characteristics on the adsorption kinetics: low molecular weight organic substances diffuse quickly and reach equilibrium. High molecular weight organic substances, such as humic substances, diffuse more slowly, and the adsorption equilibrium is only reached by increasing the dosages of activated carbon for the same reaction time.

The addition of PAC to water containing a chemical oxidant (NaOCl, Cl_2, ClO_2, O_3, $KMnO_4$) resulting from chemical preoxidation reacts as a reducer and neutralizes the oxidant. This can affect its reaction sites.

Biological growth on the carbon present in the contact tank in recirculation mode is unlikely if the carbon is less than 10 days old. The frequent renewal of the activated carbon leaves little room for bacterial growth.

In PAC reactors, the addition of coagulants, such as $FeCl_3$, has several functions. It neutralizes the colloids that are still present and refines the turbidity of the treated water. The hydroxide flocs, associated with the injected polymer, coat the activated carbon particles and accelerate their settling. The high molecular weight molecules can form metallic colloids with the iron, which adsorb on the activated carbon, creating a bridge with the activated carbon particles.

The same is true with low molar mass organic molecules, which form complexes (ligands) with metal ions (establishment of covalent bonds). These organometallic complexes adsorb onto the carbon, either directly or indirectly, through a bridge created by another metal cation.

9.6.1.2. *Sizing and operating parameters*

In an equilibrium system, the decrease in pollutant concentration as a function of time and activated carbon mass is written according to first-order mass transfer kinetics:

$$-\frac{dC}{dt} = \frac{M}{V}\frac{dq}{dt}$$

So:

$$\frac{(Cin - Cout)V}{M} t = (qm - qo)t$$

with:

– Cin and Cout are the concentrations of pollutants at the inlet and outlet of the reactor ($g \cdot m^{-3}$), respectively;

– V: volume of the reactor (m^3);

– M: mass of carbon (kg) in the reactor;

– t: contact time (h);

– qm: maximum adsorption capacity at equilibrium;

– qo: adsorbed capacity at *to*.

Considering that qo = 0 at time *to* since the carbon is new, we get:

$$\frac{(Cin - Cout)V}{M} = qm$$

By replacing qmax with the Freundlich equation parameters:

$$qm = kCout^{\frac{1}{n}}$$

The carbon dosages (g) are then given by:

$$\frac{(Cin - Cout)V}{qm} = \frac{M}{V} = dosage\left(\frac{kg}{m3}\right) \quad \text{or even} \quad \frac{(Cin - Cout)V}{kCout^{\frac{1}{n}}} = dosage\left(\frac{kg}{m3}\right)$$

It is therefore possible to adapt the dosage of PAC to be injected according to the variation in the concentration of the organic substances to be removed.

Knowing the values of parameters k and 1/n, we can deduce the mass of activated carbon required for the adsorption of a contaminant.

For example, the removal of a pesticide with a concentration of 1 $\mu g \cdot L^{-1}$ in water requires the following mass of PAC:

– Freundlich parameters: k: 74.5 and 1/n: 0.32;

– flow rate to be treated: 1,200 $m^3 \cdot d^{-1}$.

$$\frac{V}{M} = \frac{kCout^{\frac{1}{n}}}{Cin - Cout} = 74.5\frac{(0.1)^{0.32}}{1 - 0.1} = 35.65 \text{ L} \cdot \text{g}^{-1}$$

Therefore, M/V = 0.028 $g \cdot L^{-1}$ = 0.028 $kg \cdot m^{-3}$.

The quantity of coal to be injected is:

$$PAC = 0.028\ (\text{kg} \cdot \text{m}^{-3})1{,}200(\text{m}^3 \cdot \text{d}^{-1}) = 33.6\ \text{kg} \cdot \text{d}^{-1}$$

The parameters that are regularly monitored in the PAC reactors are as follows:

– A contact time that is too short (<8 min) prevents molecules with slow adsorption kinetics from being properly adsorbed. The PAC dosage conditions the performance of organic substance removal. They can vary between 5 and 60 $mg \cdot L^{-1}$, depending on the type and concentration of organic compounds present. Dosages of 5–20 $mg \cdot L^{-1}$ are typical for the nominal control of micropollutants, tastes and odors, as well as the removal of 1–2 $mg \cdot L^{-1}$ of DOC.

– The age of the PAC is useful in recirculating PAC reactors because it determines whether the adsorption mechanism is predominant with a renewal of the material that allows active sites to be permanently available. If this age is too high, that is to say, an aged and therefore likely saturated PAC, the efficiency will be impacted. The age of the PAC therefore corresponds to the "average" age of the particles in the reactor and can be controlled according to the renewal rate of fresh PAC. It is therefore possible to associate it with an equivalent dosage of PAC; in other words, the PAC dosage corresponding to the quantity of fresh PAC to be injected.

Considering the mass of PAC in the reactor and the flow rate of water to be treated, the age of the carbon is obtained by:

$$Age = \frac{M_{AC}}{D\ Q}$$

with:

– *Age*: average age of the PAC particles in the reactor (d);

– *M*: total mass of PAC in the reactor (g);

– *D*: dosage of PAC ($g \cdot m^{-3}$);

– *Q*: average daily flow rate ($m^3 \cdot d^{-1}$).

It is therefore easy to determine the age of the carbon by quantifying the carbon mass present in the reactor by multiplying the concentration by the volume of the reactor. Usually, a carbon concentration between 1.2 and 5 $g \cdot L^{-1}$ is maintained in the reactor.

However, depending on the variation in organic substances in the water during the year, the dosages of injected activated carbon are different. In the presence of low concentrations of organic substances, the tendency is to reduce the carbon dosages. As the extractions are adapted to the frequency of the additions, an increase in the age of the carbon is obtained as the carbon mass must remain constant.

If, on the other hand, the carbon mass is reduced by extractions to maintain a standard carbon age (<10 d), the operator must be ready to inject a very large quantity of carbon instantaneously when the concentration of organic substances increases to obtain the desired efficiency with regard to organic substances.

9.6.2. *GAC reactors in adsorption mode: GAC filters*

The GAC filter is mainly used as an adsorbing filter for various kinds of organic substances (humic substances, toxins, molecules responsible for taste and odor, color, etc.) and micropollutants. The operation of a GAC filter is generally based on water flow by gravity through the adsorbent material. Figure 9.23 shows the progression of the adsorption front within the reactor over time. The feed water has an organic concentration that is equal to Co. Since the activated carbon is new, the filtered water has an outlet concentration equal to 0 for a certain period of filter operation. As the water flows through the filter, the various pollutants are removed by adsorption.

After a certain period of operation, the activated carbon begins to saturate at the top (due to the downward flow). This creates an adsorption front that moves from top to bottom. The adsorption front defines the boundary between saturated and unsaturated carbon. Thus, there is a gradually expanding saturation zone and an active adsorption zone, in which mass transfer takes place.

This adsorption zone that moves through the filter is the mass transfer zone. The depth of the mass transfer zone varies with the GAC adsorption capacity and the diffusion rates of the contaminants. It is affected by the physicochemical factors that influence the adsorption of pollutants. When the adsorption front reaches the bottom of the filter, signifying that the carbon is completely saturated, it is replaced by new carbon or regenerated activated carbon.

At the same time, the so-called "breakthrough curve", which represents the concentration of pollutants at the bed outlet, is plotted. It represents the quantity of pollutants (micropollutants, organic substances, toxins, etc.) removed by the activated carbon at the maximum outlet concentration acceptable to the operator.

The relation between the concentration of a substance at the outlet of a GAC filter, the operating time and the volume of treated water, or bed volume (volume of water treated/volume of activated carbon) is used to determine the breakthrough curve.

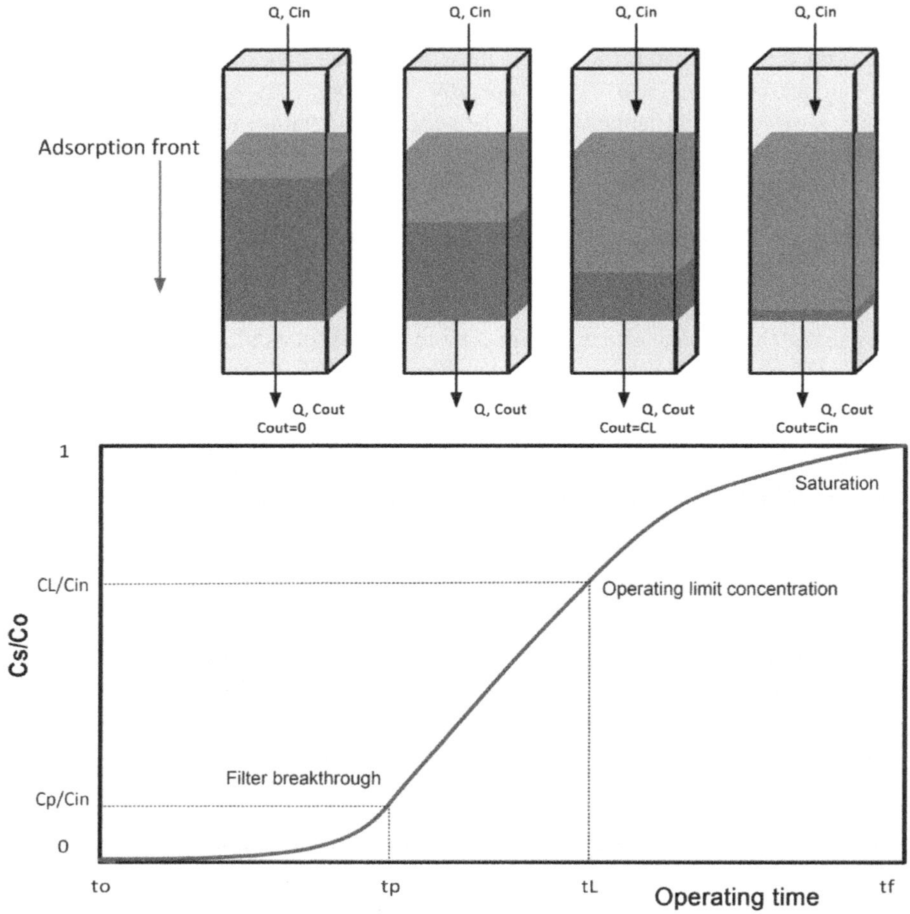

Figure 9.23. *Evolution of the adsorption front and of the outlet concentration during an operating cycle. For a color version of this figure, see www.iste.co.uk/gaid/watertreatment2.zip*

The breakthrough point (Cp) corresponds to the evolution of the concentration at the GAC filter outlet. The breakthrough is therefore the ratio between the total flow of treated water (until the concentration of the pollutant at the filter outlet has reached the desired limit concentration) and the mass of activated carbon. The higher the concentration of the pollutant at the outlet, the greater the volume of treated water/carbon mass.

Breakthrough is the time, tp, at which a concentration of pollutants above the target is detected. For a heterogeneous mixture of pollutants, the non-adsorbable and biodegradable compounds affect the appearance of the breakthrough curve. Indeed, in the presence of non-adsorbable material, the residual pollutant concentration is never zero.

$$Breakthrough\ (m3.kg - 1) = \frac{Vt\ treated\ water}{M} = \frac{tp}{tc\ \rho_{ac}}$$

with:

– *M*: mass of activated carbon in the GAC filter (kg);

– *Vt*: volume of treated water (m^3);

– *tp*: breakthrough time (t);

– *tc*: contact time in the GAC filter (t);

– *ρac*: apparent density of activated carbon ($kg \cdot m^{-3}$).

The breakthrough time is very short for pollutants with low or medium adsorbability. On the other hand, it can be very long (more than 2–3 years) for highly adsorbable molecules.

After a new operating time, an operating limit concentration (C_L) is obtained, which corresponds to the maximum output concentration set by the operator. It is related to the breakthrough time (t_L) linked to the limit concentration (C_L). The activated carbon is moderately saturated but sufficiently saturated so that it no longer complies with the concentration limit set at the outlet.

In the case of potabilization, the operating limit and the breakthrough point are often mixed for pesticides to respect the maximum allowable concentration of 0.1 $\mu g \cdot L^{-1}$ in individual substances, and 0.5 $\mu g \cdot L^{-1}$ in total substances.

For total organic carbon (TOC), the quality reference limit of 2 $mg \cdot L^{-1}$ can be slightly exceeded if the THM concentration is respected, knowing that THM formation depends on various factors, including TOC and operating conditions (see Chapter 22).

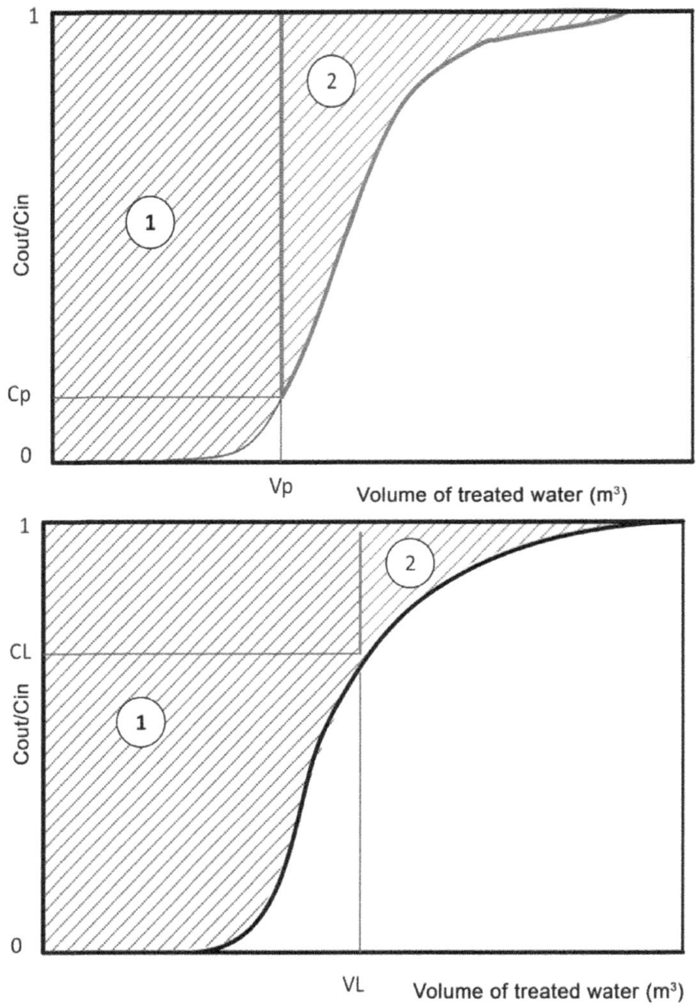

Figure 9.24. *Adsorption on GAC filter with breakthrough (Cp) or concentration limit curve (CL). For a color version of this figure, see www.iste.co.uk/gaid/watertreatment2.zip*

The percentage of activated carbon mass use is the ratio of the amount adsorbed at breakthrough (or concentration limit) to the total amount obtained after saturation of the activated carbon (Figure 9.24). The saturated carbon flow (kg·m^{-3}) represents the mass of activated carbon consumed per unit volume of treated water. It is given by:

$$sat.flow = \frac{M}{Vp} = \frac{\rho ac}{BVp}$$

with:

– M: mass of activated carbon in the filter (kg);

– Vp: volume of treated water at breakthrough (m^3);

– ρac: apparent density of activated carbon;

– BVp: volume of treated water/volume of activated carbon up to breakthrough (or concentration limit).

Figure 9.24 shows, for example, that the GAC capacity at breakthrough (Cp) is: $area_1 \times Cin$. The percentage of the activated carbon mass use is:

$$saturation\ \% = \frac{area1}{area1 + area2}$$

The performance of a GAC filter consists of defining the adsorption capacity beforehand to deduce the operating time and therefore the lifetime.

As the operation of drinking water plants is usually continuous, natural organic matter (NOM: humic and fulvic substances) adsorbs onto the activated carbon, even during periods when micropollutants are absent from the feed water. Therefore, GAC is "preloaded" by humic substances.

Figure 9.25 illustrates the "preloading" phenomenon by showing the concentration profiles of a micropollutant and two NOM fractions: a small, well-adsorbed NOM fraction that competes directly with micropollutant adsorption sites and a larger, less adsorbed fraction that restricts access pores and slows adsorption kinetics.

The concentration profile of small, well-adsorbed NOM is similar to the micropollutant profile but generally occurs earlier due to the initial concentration of NOM being higher than the micropollutant.

The adsorption of micropollutants is rapid in the upper zone and remains fairly stable in the lower zone. A GAC filter is usually installed at a second filtration stage after filtration on inert material (sand, anthracite-sand, pumice-sand). As a result, the feed water is practically free of suspended solids (turbidity < 0.5 NTU). The absence of suspended solids avoids an accumulation of these particles in the filter and considerably reduces the frequencies of the backwash. The GAC filter operates in adsorption mode at first and then in adsorption-biology mode, the importance of which depends on the concentration of BDOC.

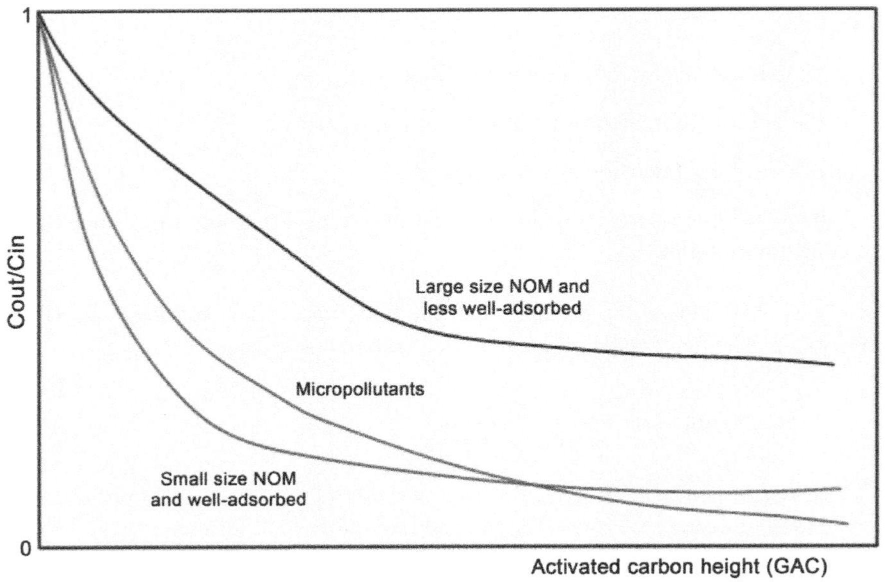

Figure 9.25. *Principle of natural organic matter (NOM) removal in a GAC filter. For a color version of this figure, see www.iste.co.uk/gaid/watertreatment2.zip*

9.6.2.1. *Operating parameters*

The generally laminar flow in a GAC filter maintains the concentration gradient within the reactor and associates the performance with a plug flow reactor.

In a plug flow adsorption reactor, the mass balance leads to:

Accumulation/removal: inlet load – outlet load

To account for dynamic adsorption on GAC, it is assumed that the adsorption curve is "symmetrical"; in other words, that equilibrium is achieved on each grain or dx slice of activated carbon. The decrease in the residual adsorption capacity (qr) of the carbon, as a function of time, is proportional to the adsorption rate:

$$\frac{\partial qr}{\partial t} = -KqrC$$

The concentration of molecules in the water at any distance x from the filter diminishes according to:

$$\frac{\partial C}{\partial H} = -\frac{K}{v}qrC$$

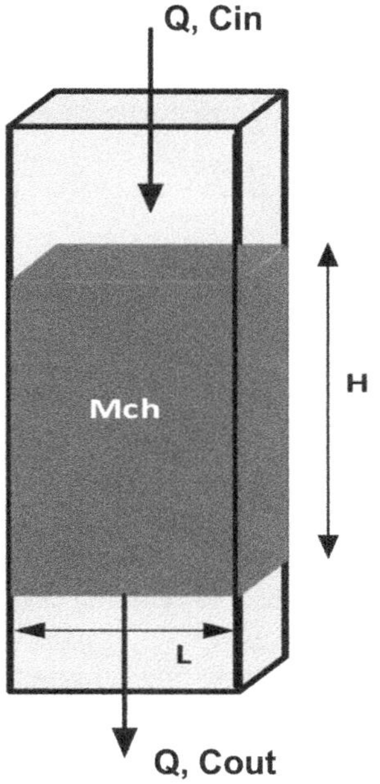

Figure 9.26. *How a GAC filter works*

Solving this system of differential equations leads to the following equation:

$$Ln\frac{qm\ Cout}{qr\ Cin} = KCin\ tf\ - \frac{KqmH}{v}$$

with:

– *qr*: residual carbon adsorption capacity ($mg{\cdot}g^{-1}$);

– *qm*: maximum adsorption capacity ($mg{\cdot}g^{-1}$);

– *H*: height of activated carbon in the filter (m);

– *tf*: operating time (h);

– *Cin*: initial molecule concentration ($mg.L^{-1}$);

– *Cout*: final molecule concentration ($mg{\cdot}L^{-1}$);

– *K*: speed constant ($h^{-1}{\cdot}mg^{-1}{\cdot}L^{-3}$);

– *v*: filtration rate ($m{\cdot}h^{-1}$).

This equation cannot be used directly because the term qr (residual carbon adsorption capacity) is not known. At the beginning of the leak curve when Cout « Cin, we consider that qr = qm, that is:

$$Ln\frac{Cout}{Cin} = KCin\,tf\ - \frac{KqmH}{V}$$

With In C/Cin = f (t), the linear part of the curve can be used to determine the speed constant K and the adsorption capacity qe with:

$$Cout\ = Cin\exp(KCin\,tf)\ - (\frac{KqmH}{V})$$

Due to the mass of activated carbon present in the filter and a sufficient contact time (10–20 min), the GAC balances out with the inlet concentration Cin. In a PAC reactor, the activated carbon balances out with the final concentration.

Since the inlet concentration Cin is much higher than the outlet concentration Cout, the x/m value in a filter is higher than the x/m value for a PAC reactor. This results in less activated carbon consumption to treat the same volume of water.

An organic substance of concentration Cin enters the GAC reactor of height H, side L and flow rate Q. The outlet concentration is Cout. The difference in concentration between the inlet and outlet (Cin-Cout) is adsorbed onto the activated carbon and increases the carbon load q (Figure 9.24).

Dm is the diffusion coefficient in the x direction ($m^2{\cdot}s^{-1}$) with:

$$Dm = \frac{kT}{6\pi\eta r} = \frac{RT}{6\pi\eta rN}$$

with:

– *k*: Boltzmann constant 1.3805×10^{-23} $J{\cdot}K^{-1}$;

– *T*: in K (273.15+°C);

– *R*: 8.3145 $J{\cdot}mol^{-1}{\cdot}K^{-1}$;

– η: dynamic viscosity $Ns{\cdot}m^{-2}$;

– *r*: radius of the adsorbent material in m;

– *N*: Avogadro number 6.02×10^{23} molecules $g^{-1} \cdot mol^{-1}$.

For example, the diffusion coefficient at 20°C for a GAC particle r of 1 mm, or 10^{-3} m, is:

$$Dm = \frac{(8.3145)\,(293.15)}{6(3.14)(1.002 \times 10^{-3})(10^{-3})(6.02 \times 10^{23})} = 21.43 \times 10^{-17} m^2.s^{-1}$$

η : 1.002×10^{-3} $Ns \cdot m^{-2}$

9.6.2.2. *Design parameters: contact time (tc)*

The GAC filter is considered as a reactor. The contact time is therefore the design parameter. There is a minimum GAC height and a minimum contact time that must be implemented to have a sufficient mass transfer zone to avoid having an overly fast breakthrough. An increase in contact time often results in an increase in operating time for a given initial pollutant concentration.

However, the relation tc according to operating time is not linear:

$$tc = \frac{H}{v} = \frac{SH}{Q}$$

with:

– tc: contact time (h);

– H: activated carbon height (m);

– v: filtration rate ($m \cdot h^{-1}$);

– Q: flow rate ($m^3 \cdot h^{-1}$);

– S: filter surface (m^2).

For the adsorption of natural organic matter such as humic substances (NOM), a longer tc (10–20 min) is more effective, with a higher saturation flow. A tc of less than 10 min should be avoided for the removal of NOM and micropollutants. This should especially be the case in the presence of micropollutants, whose adsorption capacity can be reduced in the presence of NOM, which are known to diffuse deeply along the filter, thus shortening the adsorption rates of micropollutants.

Conversely, for the molecules responsible for tastes and odors, particularly geosmin and methyl isoborneol (MIB), a contact time between 6 and 8 min is acceptable.

The **filtration rate** is related to the flow rate and the filtration surface. It can vary between 5 and 20 m·h^{-1}. The higher the speed, the higher the activated carbon height. For a tc of 10 min and a filtration rate of 5 m·h^{-1}, the height of the carbon is 0.83 m, whereas for a speed of 15 m·h^{-1}, it is 2.5 m.

The **filtration height** is related to the contact time and the filtration rate. Generally, heights below 0.8 m are avoided.

The effect of the activated carbon **particle size** impacts the adsorption rate if it is controlled by intraparticle diffusion. A fine particle size decreases the time required to reach equilibrium in GAC filters and reduces the mass transfer area. This recommendation is particularly important when a removal of more than 90% of organic substances and micropollutants is desired. This is in contrast to PAC, where a fine particle size accelerates the diffusion rates.

In addition, a fine particle size increases the pressure loss and therefore requires that the average diameter is between 0.8 and 1.4 mm and a uniformity coefficient < 1.5.

The effect of pore distribution shows that NOM molecules are preferentially adsorbed in macropores and mesopores. A high portion of these pores in an activated carbon reduces access to the micropores.

9.6.2.3. *Lifetime of the activated carbon*

The lifetime of activated carbon can be calculated as follows:

$$Quantity\ of\ organic\ substance\ removed\ (kg \cdot h - 1) = (Cin - Cout) \cdot Q$$

$$Volume\ of\ carbon\ (m3) = H \cdot S$$

$$Mass\ of\ activated\ carbon\ in\ the\ filter\ (kg) = \rho ac \cdot\ Volume\ of\ carbon$$

$$Maximum\ adsorption\ capacity\ qmax\ (kg \cdot kg - 1) = kCin^{\frac{1}{n}}$$

$$GAC\ consumption\ in\ relation\ to\ the\ substance\ removed\ (kg \cdot h - 1) = (Cin - Cout)Q \cdot qm$$

$$operating\ time\ (h) = \frac{mass\ of\ carbon\ in\ the\ filter\ (kg)}{GAC\ consumption\ (kg \cdot h - 1)}$$

with:

– *Cin*: concentration of organic substance in the inlet water ($kg \cdot m^{-3}$);

– *Cout*: concentration of organic substance in the treated water ($kg \cdot m^{-3}$);

– *Q*: water flow rate ($m^3 \cdot h^{-1}$);

– *H*: height of activated carbon in the filter (m);

– *S*: filtration surface (m^2);

– *ρac*: density of activated carbon ($kg \cdot m^{-3}$);

– *k and 1/n*: Freundlich constants.

The lifetime of the activated carbon before replacement can be estimated by the following equation when all parameters are available:

$$tf = \frac{M\,n}{Q\,\rho ac} - \frac{M}{Q\,\rho ac}\big(Kd\,\rho s(1-n)\big)$$
$$(2arcerfc\,\frac{2Cin\,r^2Q\rho ac}{15Cs\,DhKd\,\rho s(1-n)M} - 1)$$

with:

– *tf (h)*: operating time to pass from an inlet concentration Cin ($mg \cdot L^{-1}$) to an outlet concentration Cout ($mg \cdot L^{-1}$);

– *Q*: operating flow rate ($m^3 \cdot h^{-1}$);

– *M*: mass of carbon in the GAC filter (kg);

– *ρac*: apparent density of the activated carbon ($kg \cdot m^{-3}$);

– *ρs*: density of the activated carbon ($g \cdot cm^{-3}$);

– *n*: macropore volume of the activated carbon ($cm^{-3} \cdot g$);

– *Kd*: distribution coefficient ($cm^3 \cdot g^{-1}$);

– *r*: radius of an activated carbon grain (cm);

– *Dh*: intrapore diffusion coefficient (10–12 $m^2 \cdot s^{-1}$).

The value of Kd is deduced from the Freundlich parameters

with $Kd = k\ Cout^{\ n-1}$.

9.6.2.4. *Washing conditions*

When the GAC filter is placed in the second stage of filtration (after filtration on inert material), the backwashing frequency is once a month in cold periods and every 15 days in hot periods because of the risks of bacterial development on the material. In all cases, the operators adapt the backwashing frequencies according to the local conditions.

When the GAC filter is placed in the first stage of filtration after coagulation–flocculation–settling or flotation, the washing frequency is every 24–48 h, depending on the quality of the settled water.

The general washing of a GAC filter (second stage) is performed with water alone at 25 $m \cdot h^{-1}$ for 10 min. The required expansion is 25–30%. It is recommended to perform a longer wash every 3 months, with times of approximately 30 min. However, an air/water wash is recommended every 15–30 days to remove preexisting particles or bacteria that have developed, especially in hot weather. The air/water phase includes air at 50 $Nm^3 \cdot m^{-2} \cdot h^{-1}$ and water at 8–10 $m \cdot h^{-1}$ for 8 min.

In all cases, the first filtered water corresponding to one or two volumes of carbon is evacuated. The washes are triggered manually by the operator, by an indication of clogging on the filters, or by a clock.

9.6.2.5. *Release of bacteria and fine particles*

The long operating time of a GAC filter supports the development of bacteria on the surface of the activated carbon. These bacteria are detached during the washes (air/water and water-only phases), but they can also be released periodically with the filtered water. Generally, these bacteria are not pathogenic, and water disinfection is effective.

However, the GAC filter can colonize potentially pathogenic microorganisms, increasing the risk of bacterial contamination of the filtered water. As such, water disinfection, which is the stage following GAC filtration, must be well controlled. This should be the case even if these bacteria have difficulties colonizing the distribution network, because of the competition with the organisms that are already present in the biofilm, that have previously been established in the network.

In addition, the fine particles released after a filter backwash, or even sometimes during operation due to flow surges, can provide protection for bacteria from disinfectants, especially in small plants operating with two parallel filters, where the filtration rate can double during a filter backwash if the total production is maintained.

The fine particles associated with bacteria settle in the reservoirs, as well as in the networks. However, they are resuspended in the water when the hydraulic conditions of the distribution network change.

In this respect, Veolia has set up a protocol for the disposal of filtered water in the plants it operates in (see section 9.7 on Veolia technologies).

The GAC filter is sometimes laid out as a dual media filter with a sand layer of approximately 0.4-0.5 m in height and a grain size < 0.85 mm to retain the passing of fine particles in the filtered water.

9.6.2.6. *Regeneration of GAC*

The complete regeneration cycle for saturated GAC in drinking water plants includes the following stages:

– GAC drying at 120°C: the GAC extracted from the filters is very wet. It is therefore important to remove the excess water;

– thermal regeneration: the GAC is heated to a high temperature (700–1000°C) under a controlled atmosphere to volatilize the molecules fixed within the GAC pores;

– mechanical sieving to refocus the regenerated GAC in the initial granulometric choice during the design;

– addition of new GAC identical to the original carbon to compensate for regeneration losses and possibly operating losses to reach the filter volume.

Based on the analysis of the received and regenerated carbon, the guarantees on the regenerated GAC are mostly related to the iodine value and the grain size. The guaranteed values are determined on the basis of the results of analyses of a carbon sample, taken up to two months before the start of operations:

– iodine index: the iodine value of the carbon after regeneration is at least 350 mg/g of activated carbon higher than the value before regeneration, but not more than 900 $mg \cdot g^{-1}$ AC;

– grain size: the grain size of the regenerated carbon is equal to that of the carbon to be regenerated.

GAC can be regenerated several times as long as the operation is carried out under good conditions; in other words, saturated GAC has the following characteristics: an iodine value higher than 500 $mg \cdot g^{-1}$, mineral content (Ca, Al, Cu, Mn, Fe) with low concentrations, hardness higher than 90%.

9.6.2.7. *The advantages and disadvantages of GAC filters*

Advantages	Disadvantages
The lifetime of the granular activated carbon depends on the concentration of organic substances and the objectives with regard to pesticides. The choice of the type of activated carbon is also decisive for the expected performance. GAC has an adsorption capacity in relation to many organic substances. It retains a large number of organic substances at the same time. It operates well under a wide range of temperatures. It can be made from various sources of vegetable or mineral origin. It is easily found on the market. It can be arranged in single, double or triple layer, depending on the water quality and the treatment objectives. Can be used as a combustion material (under certain conditions with respect to adsorbed substances).	Limited lifetime. They produce fine particles during washing. They promote the growth of bacteria on the substrate, which can be released with the filtered water. Cost: the need to regularly replace saturated carbon can make it costly for high concentrations of contaminants. Regeneration: it can only be regenerated three times when it has adsorbed organic substances. It is not regenerable when it has adsorbed metals, for example. The investment costs are more expensive than those for powdered activated carbon.

Table 9.6. *Advantages and disadvantages of GAC filters*

9.6.3. *Fluidized bed activated carbon reactors*

The last ten years have seen the development of powdered and micrograin activated carbon reactors with the goal of removing organic matter, pesticides and other micropollutants. They have been successfully implemented in many drinking water plants: reactors-decanters, moving bed reactors or fixed bed reactors.

To improve the performance of activated carbon reactors and to optimize operating costs, new reactors have appeared on the market, for which a fluidization of the adsorbent media is realized. These reactors use a large mass of mobile carbon, which makes it possible to smooth the peaks of contamination, to potentially reduce the contact times and to adapt the dosages of activated carbon.

9.6.3.1. *Principle of micrograin fluidization*

When the water flows through a bed of activated carbon particles at a low flow rate, there is no disruption of the bed. This is the process of upflow fixed beds. In the fixed bed, the particles are in direct contact with each other and support each other.

If the flow velocity increases, the bed gradually expands. The resulting phenomenon is called fluidization, and the bed is called a fluidized bed. Liquid–solid fluidization consists of passing the water to be treated through a bed of solid particles at a sufficient speed to put them in suspension. The bed of particles behaves like a fluid that expands according to the velocity of the water. A fluidized bed consists of a liquid–solid mixture that has properties similar to those of a fluid. Thus, the mirror surface of the bed is relatively horizontal, and the bed can be considered a homogeneous mixture of liquid and solids that can be represented by a single apparent density.

In fluidized beds, the contact of solid particles with the dissolved substances that are present in the water is much better than in fixed beds because each particle in suspension offers all its contact surface. The transport and diffusion of molecules within the adsorbent media is improved. In the fluidized bed, the particles have an average free distance between them and are supported by the drag force of the fluid. The expanded bed has some of the properties of a fluid. It is also known as a fluidization reactor.

In addition, due to the high contact surface between water and the high concentration of activated carbon in the reactor, the mechanisms of the reactions between the dissolved substances and the activated carbon are fast in a fluidized bed reactor. Thus, there is uniformity in the adsorption kinetics of the molecules, considering that each particle in suspension will have the same behavior as the other particles present in the bed.

The different stages can be summarized as follows: at low velocity of the ascending flow, the bed remains fixed, and we have a simple percolation. As the speed increases, the drag forces also increase. They become sufficient to balance the weight of the particles, which move slightly and become suspended: the minimum fluidization velocity "Umf" is reached. The solid particles slightly move away from each other, the cross-sectional area of the water passage increases, and the speed between particles decreases. The suspension remains homogeneous, and the particles behave globally as a fluid. By further increasing the flow rate, the bed undergoes a uniform expansion until the rate causes a release, and the particles are carried away by the water and can thus escape the reactor.

9.6.3.2. *Expansion parameters – minimum fluidization velocity*

To determine the expansion of the adsorbent media, the minimum fluidization speed is first calculated and then corrected to an expansion rate that takes the desired expansion into account.

The minimum fluidization speed (Umf) is defined as the surface velocity of water at which the drag force of the ascending flow becomes equal to the weight of the particles present in the reactor. More than a hundred correlations have been proposed for more than fifty years, allowing the prediction of the Umf value. Generally, Umf is a function of the properties of the particles to be fluidized (diameter (dp), density (ρp), sphericity (φ)) and of the characteristics of the liquid used for fluidization (density (ρw), viscosity at a given temperature (μ)). The porosities of the bed at rest and in expansion are also parameters that are considered.

Most of the relationships proposed in the literature (Table 9.7) for the calculation of Umf consider the following dimensionless numbers:

$$Re\,mf = \frac{\rho dpUmf}{\mu} \quad Ar = \frac{g\rho w(\rho ac - \rho w)dp^3}{\mu^2}$$

with:

- ρac: particle density (activated carbon in our case);
- dp: particle diameter;
- μ: viscosity of water at a given T °C;
- ρw: density of water;
- g: gravitational acceleration.

Ergun's semiempirical correlation represents the variation in pressure loss according to the apparent flow velocity.

$$\frac{1.75(1-\varepsilon_{mf})(Umf^4\ \rho w)}{\varepsilon_{mf^3} dp\phi_p} + \frac{150(1-\varepsilon_{mf})^2(\mu\ Umf)}{\varepsilon_{mf^3}(dp\phi_p)^2} = \Delta P/L$$

where ΔP is equal to the apparent weight of the particles per unit area of the bed, and φp is the sphericity of the particles, which is defined as the area of a volume-equivalent sphere divided by the area of the particle.

The first term, in Umf, corresponds to the viscous energy dissipation and predominates for laminar flows ($Re_{mf} < 10$) and follows from the Kozeny equation. The second term, in Umf^2, represents the kinetic energy dissipation and predominates for turbulent flows ($Re_{mf} > 200$). By removing $\Delta P_{mf}/L_{mf}$ between this equation and the fundamental fluidization relation:

$$\frac{\Delta Pmf}{\Delta Lmf} = (1 - \mathcal{E}mf)(\rho ac - \rho w)g$$

We deduce the value of Umf:

$$\frac{1.75(dp\ Umf\,\rho_{ac})}{\varepsilon_{mf^3}\mu} + \frac{150(1-\varepsilon_{mf})(dp\ Umf\ \rho w)}{\varepsilon_{mf^3}\mu} - \frac{dp^3\ \rho w(\rho_{ac} - \rho w)g}{\mu^2} = 0$$

This equation is valid for perfectly spherical particles. As the particles used are never spherical, we bring the form factor into play:

$$\Phi p = \frac{4dp^4}{Sp}$$

where Sp is the surface of the particles.

The bed porosity at minimum fluidization ε_{mf} can be calculated using the following correlation:

$$\mathcal{E}mf = 0.586\ \Psi s^{-0.72}\ Ar^{-0.029}\ (\rho w/\rho ac)^{0.021}$$

For $0.85 < \varphi_p < 1.1$; $Ar < 10^5$; $500 < \rho ac/\rho g < 5 \times 10^4$

9.6.3.2.1. Expansion rate

The expansion rate considers the minimum fluidization speed, as well as other parameters such as the porosity of the expanded bed and experimental constants related to the operating conditions, such as the Kozeny constant.

Table 9.7 summarizes the different equations tested.

Authors	Equations of the expansion rate
Leva	$\frac{22.12 \cdot {d_{60}}^{0.82} \cdot (\rho_w(\rho_{ac} - \rho_w)^{0.94})}{\mu^{0,88}} (\frac{\varepsilon_{exp}}{\varepsilon})\eta$
Kozeny	$\frac{(\rho ac - \rho w) \cdot 1{,}000 \cdot g \cdot \varepsilon^3}{hk \cdot \mu \cdot 0.001(\frac{6}{\varphi \cdot d60{,}0.001})^2(1-\varepsilon)\ 3{,}600(\frac{1-\varepsilon}{\varepsilon^3})\varepsilon^3 exp/(1-\varepsilon exp)}$
Ergun	$\frac{-b+\sqrt{\Delta}}{2\alpha} \cdot 3{,}600 \cdot ((1-\varepsilon)/\varepsilon^3) \cdot ({\varepsilon_{exp}}^3/1 - \varepsilon_{exp})$ $b = hk \cdot 36 \cdot (0.001 \cdot \mu)(1-\varepsilon))/(\Psi.d_{60}.0.001)^2$ $a = hb \cdot 6^2 \cdot \frac{1000 \cdot \rho w}{(0.001 \cdot \Psi \cdot d60)}$
Wen and Yu	$\frac{Re \cdot \mu \cdot 0.001(1-\varepsilon)\varepsilon 3_{exp}}{\frac{(d_{60}}{10^6 \cdot \rho w \cdot 3{,}600} \cdot \varepsilon^3(1-\varepsilon_{exp}))}$ $c = (\rho_{ac} - \rho_w) \cdot 1{,}000 \cdot g \cdot \varepsilon^3$ $\Delta = b^2 - 4ac$

Table 9.7. *Equations characterizing the fluidization and expansion speeds*

COMMENTS ON TABLE 9.7.– *Ar: Archimedean number; d_{60}: diameter of activated carbon particles at 60% passing (m); g: acceleration of gravity ($m \cdot s^{-2}$); Ga: Galileo number; hk: Kozeny's constant; hb: Ergun's constant; H: height of material at rest (m); Hexp: height of material in expansion (m); Lmf: bed height at minimum fluidization (m); Re: Reynolds number; Umf: minimum fluidization velocity ($m \cdot h^{-1}$); ρac: specific mass of activated carbon ($kg \cdot m^{-3}$); ρw: density of water ($kg \cdot m^{-3}$); Ø: form factor; μ: dynamic viscosity of water ($kg \cdot m^{-1} \cdot s^{-1}$); ε: bed porosity; ε_{exp}: porosity of the expanding bed; ε_{mf}: porosity of the bed at minimum fluidization.*

Table 9.8 gives the main size characteristics of the Microsorb 400R activated carbon, according to the laser diffraction methods described above. The information in Table 9.8 is given for a new AC after reactivation, taken directly from the storage "big bag".

Microsorb 400R activated carbon	
Effective size d_{10} (μm)	380
Average size d_{60} (μm)	520
Median size (μm)	560
Max size d_{90} (μm)	810

Table 9.8. *Size characteristics of micrograin activated carbon*

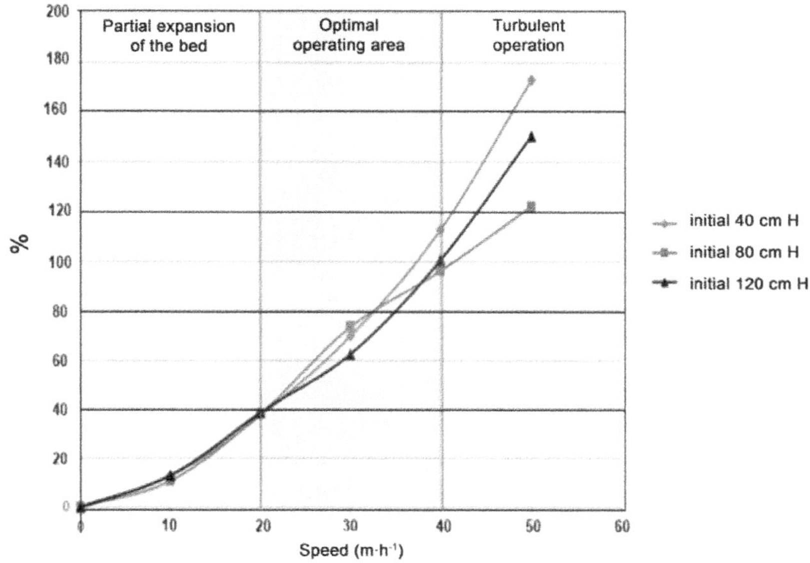

Figure 9.27. *Expansion of Microsorb 400R according to flow speed at 15°C (source: Veolia). For a color version of this figure, see www.iste.co.uk/gaid/watertreatment2.zip*

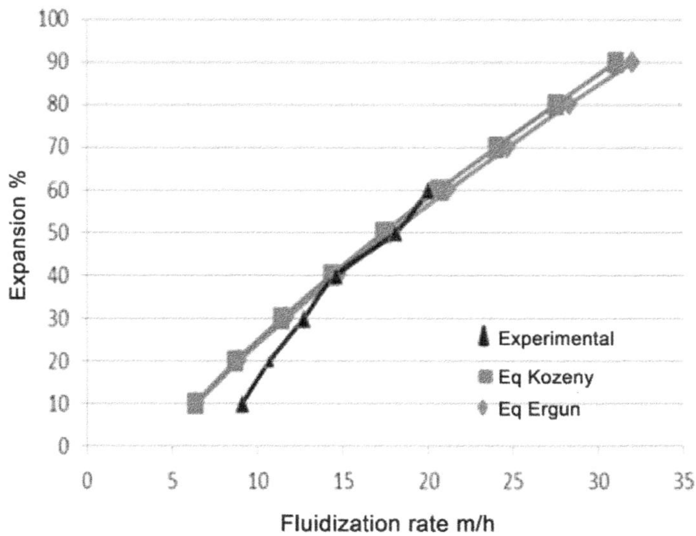

Figure 9.28. *Experimental results and theoretical equations. For a color version of this figure, see www.iste.co.uk/gaid/watertreatment2.zip*

An activated carbon reactor in fluidized mode (see Opacarb® FL sheet) is a reactor operating with an up flow, made up of a bed of micrograin activated carbon. The implementation of an up flow allows a controlled expansion of the activated carbon according to the applied speed. Indeed, this velocity is calculated to avoid excessive expansion, which could lead to a loss of media (turbulent flow) beyond 40 $m \cdot h^{-1}$.

The water to be treated is therefore fed from the bottom of the reactor using mechanical and hydraulic means, which ensure an optimal distribution of the flow through the media.

The flow rate is calculated in such a way as to obtain a compromise between an expansion zone of the AC bed and a sufficiently still area to avoid the presence of AC particles. The average expansion rates of the AC bed are between 10 and 100%. However, for the same ascension speed, this expansion rate varies according to the characteristics of the activated carbon. The parameters are defined according to the particle size of the activated carbon, the height of the desired structure and the concentration of activated carbon in the medium. The Opacarb®FL reactor is equipped with a lamellar pack located above the stilling zone (Figures 9.61 and 9.62) to trap any residual activated carbon particles that might attempt to be carried away with the outlet water.

9.6.3.3. *Calculation of the fluidization speed and expansion speed*

The minimum fluidization velocity and expansion rate of micrograin activated carbon are established experimentally at different temperatures and then compared with the Kozeny and Ergun equations. The results obtained at 10.5°C show a clear similarity between the different equations (Figure 9.28). The minimum fluidization velocities are between 3 and 6 $m \cdot h^{-1}$, depending on the temperature and thus on the viscosity of the water.

The expansion speed allows an expansion of approximately 10% at 10 $m \cdot h^{-1}$ and 10°C, while theoretical calculations show an expansion of approximately 20%. This can be explained by the variety of sizes of the activated carbon particles (d_{10}: 0.4 mm), which also depends on the particle sizes expressed in d_{60} and d_{90}, whereas theoretical calculations are most often based on a homogeneous size of the carbon particles.

For higher expansion velocities, the curves obtained are very similar and show that for the activated carbon tested beyond 15 $m \cdot h^{-1}$, the expansion no longer takes the differentiation between the particles into account because the velocity is high enough to drag all the particles. The Kozeny and Ergun equations lead to expansion rate values that are very close to those obtained experimentally, especially for expansions greater than 30%. The equations of Leva and Wen and Yu seem to lack a specific coefficient, such as that of Kozeny (hk) related to the permeability or that of Ergun (hb) related to the convective exchanges at the interface between the solid and liquid phases. We note that the equations of Leva and Wen and Yu become comparable to the experimental values with respective correction constants of 2.71 and 2.78, which are close to the Kozeny constant.

9.6.3.4. *Classification of the media in the expanding bed*

When the media particle size is based on a single diameter, the bed expands homogeneously, and all particles are expanded to the same height. When the media particle size is diversified, the 100% expansion allows the bed to be totally lifted. However, the lowest expansion leads to a stratification of the media according to its diameter (Table 9.9).

Heights (m)	**d_{10} (mm)**	**d_{60} (mm)**	**Micrograin concentrations ($g{\cdot}L^{-1}$)**
2	0.210	0.34	90–130
1	–	–	175–195
0.2	0.38	0.61	250

Table 9.9. *Concentration of micrograin carbon and d_{60} according to reactor heights*

For a 50% expansion, a classification of the media is observed with the fine particles located at the top of the expanded bed and the larger particles located at the bottom of the bed. The porosity varies with expansion, and the following relationship can be used to deduce the porosity of the expanding material:

$$\frac{Exp\ height}{Resting\ height} = \frac{(1 - \mathcal{E}o)}{(1 - \mathcal{E}exp)}$$

Thus, for a given expansion, the variation in expansion height induces a different porosity. Figure 9.29 shows the variation in expansion porosity for a d_{60} of 0.52 mm.

The implementation of activated carbon reactors in an expanded bed is possible, provided that the hydraulic parameters are controlled and that the activated carbon has suitable physical and chemical characteristics. Indeed, pulverized activated carbon (15–25 μm) implies a very low applied velocity (<1 $m{\cdot}h^{-1}$), which would lead to large structures, and it is necessary to consider the loss of the fine particles driven by the water. A GAC (1–1.4 mm) implies a very high applied velocity (>65 $m \cdot h^{-1}$) for a 40% expansion, and the attrition of the grains due to the turbulent regime would lead to fine particles at the outlet of the structure.

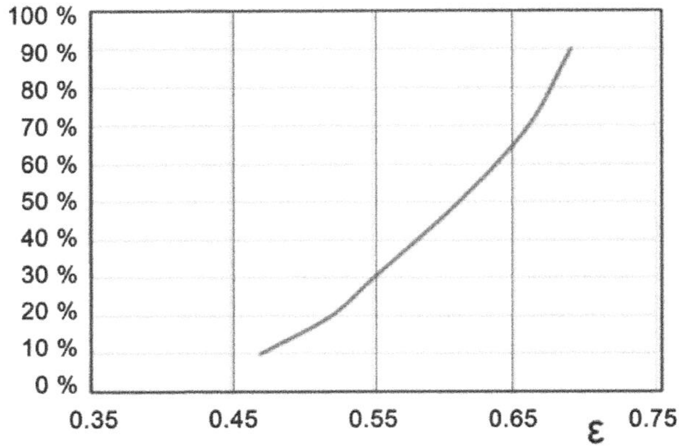

Figure 9.29. *Variation in the porosity during expansion (20°C)*

The minimum fluidization velocities are a function of the physicochemical characteristics of the media present in the reactor (particle diameter (dp), particle density (ρac), particle sphericity (φ), bed porosity (ε), expanding porosity (exp)) and the properties of the carrier fluid (density (ρw), dynamic viscosity (μ), T °C, etc.).

In expanded bed reactors, the expansion velocity guides the operation process both from a hydraulic and adsorption point of view. A 100% expansion requires operation at very high velocities (>50 $m \cdot h^{-1}$ at 20°C for a d_{10} of 0.4 mm), inducing a turbulent regime within the reactor, the consequences of which can be preferential circuits, carbon attrition, overly short contact times not allowing optimal kinetics to be obtained, etc.

9.6.4. *GAC-μgrain-PAC comparison*

Table 9.10 shows a comparison of PAC, micrograin activated carbon (μgrain), and GAC based on different applications.

	Granular activated carbon	Powdered activated carbon	Micrograin activated carbon
Physicochemical adsorption: **organic matter, TOC, DOC**	Limited to the first 6 months (beyond that, biological phenomena develop)	Good efficiency provided that the PAC is continuously added to the system (continuous dosage, or fresh PAC in the case of a recirculating system)	Good efficiency provided that fresh µgrain is continuously added to the process
Bioassimilation: **organic materials, BDOC, ammonia**	Permanent biomass on the material (grain colonization)	For an adsorption mode only (age < 10 d), no biomass	For an adsorption mode only (age < 10 d), no biomass
Suitable for chronic pollution: **natural organic matter**	Background treatment: no adjustment required	Recirculating PAC can be considered as a treatment of choice with daily additions of fresh PAC	µgrain with recirculation or in a fluidized bed can be considered as a treatment of choice with daily additions of fresh µgrain
Adapted to seasonal pollution: pesticides	Background treatment: no adjustment required An analytical follow-up makes it possible to anticipate and plan the reactivation and/or renewal of the material, and to maintain its performance over time	Need for a rigorous monitoring of the resource, in order to set up the PAC treatment In case of a permanent background, PAC with recirculation is a suitable process	In case of a permanent background, µgrain with recirculation is a suitable process
Protection against accidental/temporary pollution: pesticides, organic micropollutants	"Permanent" treatment: does not require installation	Need for a follow-up of the resource (alert station)	

	Granular activated carbon	Powdered activated carbon	Micrograin activated carbon
Residue	No residues: Possibility of fine carbon particles after washing Bacteria can be released, whether they are attached to the particles or not Full disposal of used carbon by the GAC supplier (for destruction) Limited regeneration	Obtained from black sludge	Regenerable material No sludge: taken care of by the supplier

Table 9.10. *Comparison of granular, powdered and micrograin activated carbon*

9.6.5. *Hybrid process with a PAC reactor associated with UF membranes*

Activated carbon can be used in combination with low-pressure membranes. Hybrid processes thus combine two complementary treatments with a PAC reactor, which is directly associated with membrane filtration. The first stage allows the activated carbon to remove organic substances. The second stage allows the activated carbon particles to be retained, as well as the usual parameters associated with UF membranes, such as turbidity and microorganisms. The Veolia processes associated with this hybrid configuration are Opaline® C and Opaline® B.

The PAC reactor operates on the same principles as described above. A fresh addition of PAC and a purge of the excess PAC make it possible to control the average age of the carbon in the reactor. The transmembrane flows correspond to the permeate flow rates divided by the membrane filtration surface. They are therefore defined as the volume of water passing per unit area of the membranes for a given time ($L{\cdot}m^{-2}{\cdot}h^{-1}$). The typical operation of a hybrid process consists of operating at a constant flow and variable pressure.

In the PAC reactor, the age of the PAC in the reactor is given by:

$$PAC\,age = \frac{M}{Q.PAC\ dosage}$$

with:

– *PAC age*: average retention time of PAC in the reactor (d);

– *M*: mass of PAC in the reactor (g);

– *PAC dosage*: dosage of PAC ($g \cdot m^{-3}$);

– *Q*: daily flow of treated water at the reactor outlet ($m^3 \cdot dj^{-1}$).

Flow related to carbon purges is not accounted for in this equation. However, it can be included if they are big purges. The flow of extracted activated carbon is given by:

$Q_{extraction}$ ($m^3.d^{-1}$) = reactor volume (m^3)/PAC age (d)

In terms of the membrane, the flow rate produced ($m^3 \cdot d^{-1}$) depends on the membrane filtration cycles:

$$Qp = \frac{Qf\ tf}{t} - \frac{Ql\ tl}{t}$$

with:

– *Qp*: product flow (daily produced water flow ($m^3 \cdot d^{-1}$);

– *Qf*: filtration flow ($m^3 \cdot d^{-1}$);

– *Ql*: backwash flow ($m^3 \cdot d^{-1}$);

– *tf*: time allocated to filtration (min);

– *t*: total time of the filtration cycle (min);

– *tl*: total backwash time (min).

Flow related to in-place and chemical cleaning is not accounted for in this equation. However, it can be included if there are significant volumes.

9.7. Veolia technologies: treatment process with PAC reactors

The integration of PAC into a treatment process depends on the quality of the raw water and the treatment objectives. Figure 9.30 summarizes some of the PAC arrangements that are installed in various plants built by Veolia.

These various arrangements are explained below.

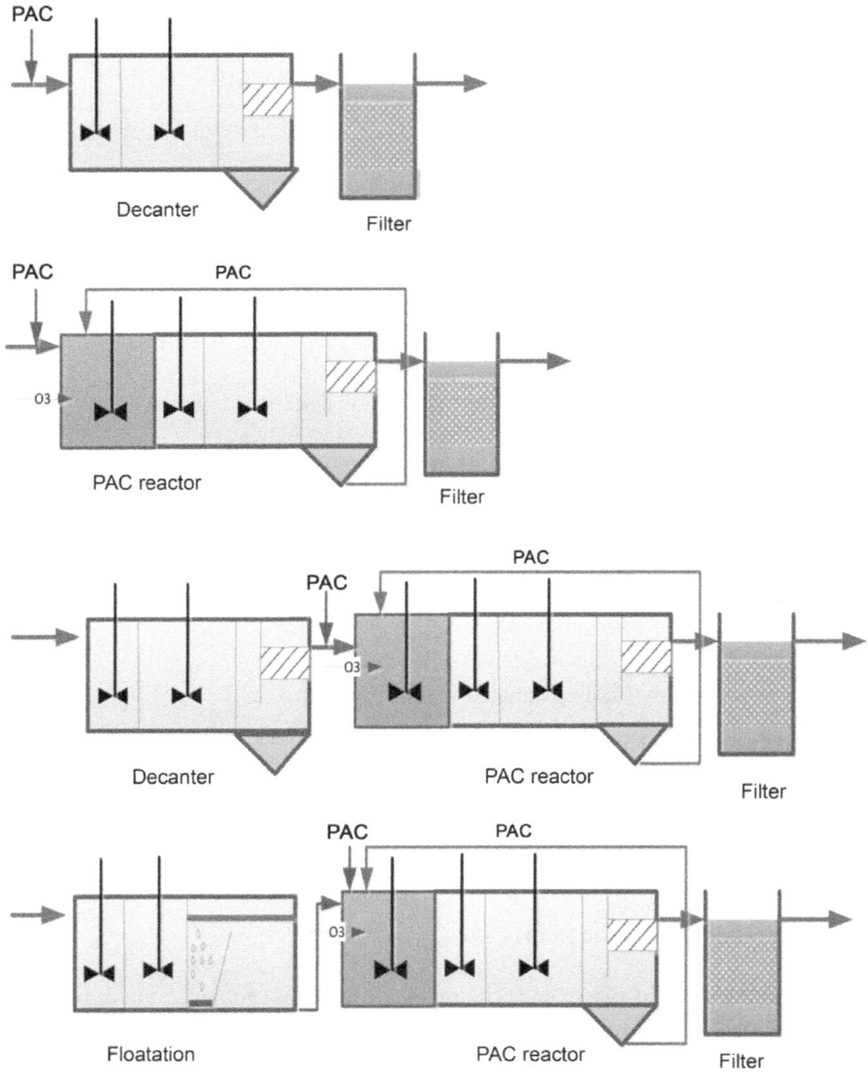

Figure 9.30. *Integration of powdered activated carbon in the treatment process. For a color version of this figure, see www.iste.co.uk/gaid/watertreatment2.zip*

9.7.1. *Implementation in a settler without PAC recirculation*

The use of PAC responds to timely treatment, following an outbreak period of pollution (storm, accidental spill), or to the presence of pesticides and/or organic substances at known periods of the year. It can also be injected into the decanter periodically as a preadsorption, upstream of a treatment plant with a GAC filter in the first or second stage, to increase the life of the GAC filter. The dosages, injection frequencies, and injection points of PAC are adapted to the characteristics of the raw water and the expected performance of the treated water during this stage.

The dosages can vary between 5 and 50 $g \cdot m^{-3}$. The PAC is collected at the bottom of the settler with the sludge.

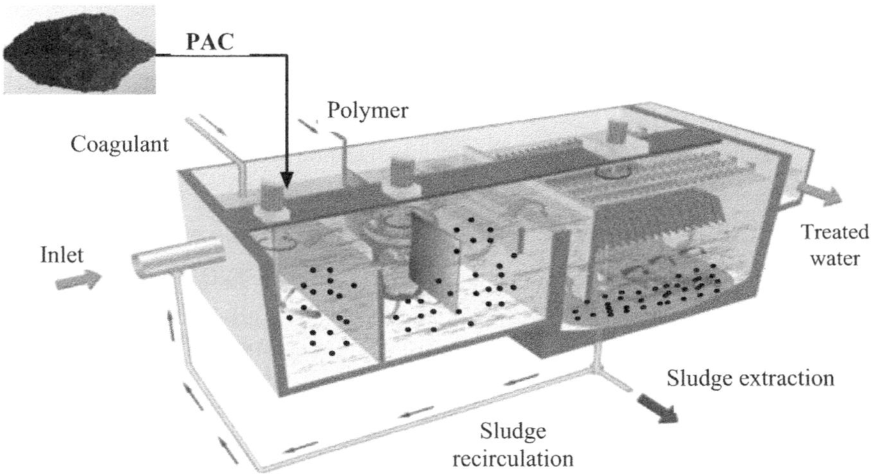

Figure 9.31. *PAC injection in a Multiflo®. For a color version of this figure, see www.iste.co.uk/gaid/watertreatment2.zip*

Table 9.11 gives the main size characteristics of the Microsorb 400R activated carbon according to the methods using laser diffraction, as described above. The information in Table 9.11 is given for a new activated carbon after reactivation, taken directly from the storage "big bag".

9.7.2. *Implementation with PAC recirculation*

The following are discussed here:

– Multiflo® Carb;

– Actiflo® Carb;

– Opacarb®MF.

9.7.2.1. *Multiflo® Carb*

Multiflo® Carb involves the use of a settler, in which the injected PAC contributes to the removal of micropollutants and organic matter.

The fields of application of the Opaline® solution with Multiflo® Carb are as follows:

– groundwater in the first stage of treatment, contaminated by pesticides and/or organic matter;

– surface water in the first stage of treatment on water with low turbidity and moderately loaded with organic matter or pesticides;

– surface water in the second stage of treatment on water with high turbidity and loaded with organic matter or pesticides.

Multiflo® Carb is characterized by the presence of a fluidized sludge bed (PAC and Fe hydroxides) in the flocculation tank. This accelerates the mechanisms involved in the flocculation stage during the passage of water and allows the production of denser flocs that can be settled more easily. The purpose of this flocculation stage is to increase the probability of the small flocs formed during the previous coagulation stage of encountering one another. This stage is thus intended to form large aggregates that can be settled quickly.

As such, the agitation of the mixture must be ensured in the most optimized way possible to avoid breakage of the flocs that have initially been formed. This is usually achieved by using mechanical agitators, which lead to speed gradients between 100 and 150 s^{-1}.

The flocculation kinetics, which control the probability of collisions between particles, are higher in Multiflo® Carb because of Turbomix™, which ensures an almost perfect homogeneity in this stage. The addition of a polymer in the flocculation tank promotes and accelerates the aggregation of flocs, leading to a rapid settling of the flocs.

The Multiflo® Carb process (Figure 9.32) is as follows:

– a first PAC contact tank, which favors the immediate PAC adsorption effect. It also receives the recirculated solution of part of the sludge, taken from the settler;

– a second tank for coagulation to destabilize the colloidal particles present in the raw water;

– a third tank for flocculation with a Turbomix™ is associated with intensive mixing, which ensures the formation of a homogeneous floc and the good distribution of the polymer. This tank receives an addition of anionic polymer, which agglomerates and masticates the flocs created in the second tank. These flocs are loaded with PAC. At the outlet of the flocculation tank, these formed and agglomerated flocs are directed toward the settler. The settler is based on the technique of countercurrent liquid–solid separation by means of lamellas arranged on the surface of the settler. Each settler comprises a set of lamellar modules formed by an assembly of PVC lamellas, an example of which is the hexagonal geometrical "honeycomb" configuration, without it being a general case.

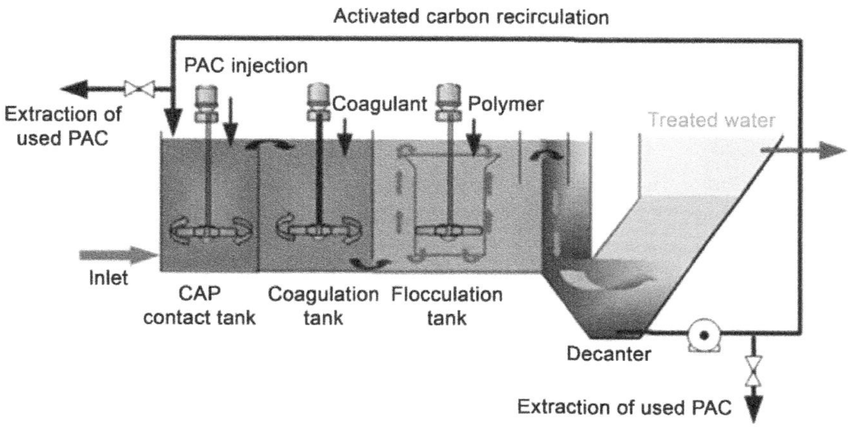

Figure 9.32. *How Multiflo® Carb works. For a color version of this figure, see www.iste.co.uk/gaid/watertreatment2.zip*

a) b) c)

Figure 9.33. *Multiflo®carb with contact tank (a), flocculation tank (b), settled water (c). For a color version of this figure, see www.iste.co.uk/gaid/watertreatment2.zip*

This stage is the object of various physical mechanisms, linked to one another by various modes of dynamic interactions, whose global process is the following: at the flocculation outlet, the particles are of various sizes, although good coalescence has already taken place. Nevertheless, given that PAC is more numerous than iron hydroxide flocs because it has been injected in a more continuous way than $FeCl_3$, the floc formed by PAC, iron hydroxides and polymer is homogeneous in terms of size. The floc flow (PAC-hydroxides) forms a sludge bed in the decanter. The flow in the sludge bed is uniformly distributed due to the coalescence that takes place during this stage and allows the particles to meet, forming larger flocs. It does not behave like a plug flow because the low velocity conditions of the mixture (scraper) reduce the intensity of the turbulent fluctuations and create homogeneity, which tends to decrease the porosity of the bed due to the agglomeration of the sludge.

This homogeneity results in the establishment of a balance between all the forces present (pressure forces, hydraulic forces), which gives this layer of sludge global cohesion.

The moderate fluctuations in turbulence continue to promote flocculation by bringing the flocs together at a minimal distance through Van der Waals forces.

An interesting consequence is that the sludge bed behaves like a filter, in which the new flocculated particles from the flocculation tank are retained on the surface of the flocs that are already present in the bed. In addition, the flow velocity through the sludge bed is calculated in such a way that it does not disturb the agglomerations of the flocs or the density of the flocs in the bed. It is observed that a momentary or long stop of the Multiflo®, followed by a restart at 16 $m \cdot h^{-1}$, does not affect the turbidity of the settled water (Figure 9.34).

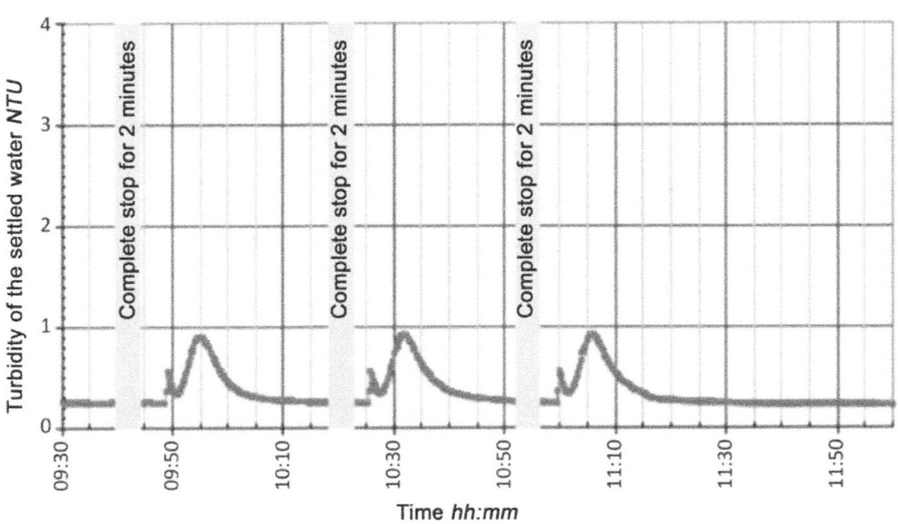

Figure 9.34. *Evolution of turbidity according to the stops (2 min) and immediate restarts of Multiflo® Carb. For a color version of this figure, see www.iste.co.uk/gaid/watertreatment2.zip*

The sludge deposited at the bottom of the Multiflo® Carb settler is collected and directed to the central thickening zone. A scraper promotes the harrowing effect, which leads to good homogenization of the sludge density within the bed. Due to the constant supply of PAC, as well as coagulated and flocculated colloids, purges are carried out regularly to ensure a constant height of the sludge bed. They are carried out at the bottom of the settler without generating dynamic disturbances in the bed.

To maintain a high mass of activated carbon in the contact tank and to obtain a high and stable performance of pollutant removal, a sludge recirculation system is implemented. This recirculation allows the sludge to be reinjected into the contact tank. The maximum recirculation rate is 5–10% of the raw water flow. The sludge is thickened at the bottom of the settler to a concentration between 20 and 30 $g \cdot L^{-1}$ (Figure 9.35). The depth of the structure allows a sludge bed to be maintained, in which the carbon concentration is always approximately 10–35 $g \cdot L^{-1}$.

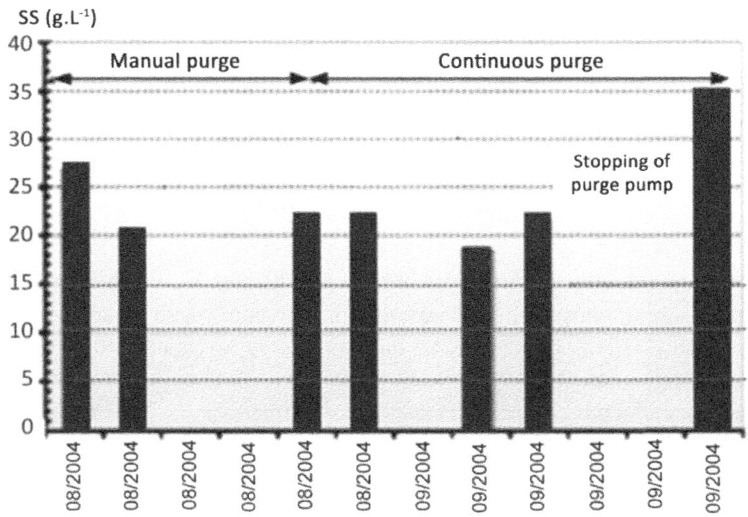

Figure 9.35. *Concentration of PAC sludge in Multiflo® Carb*

The sizing of the Multiflo®Carb takes several parameters into account:

– the hydraulic retention time (HRT), which concerns the total volume of the settler;

– the residence time of the sludge in the decanter, which only concerns the sludge thickening zone;

– the Hazen velocity;

– the mirror velocity.

It is the combined effect of these four parameters that guarantees the performance of Multiflo® Carb.

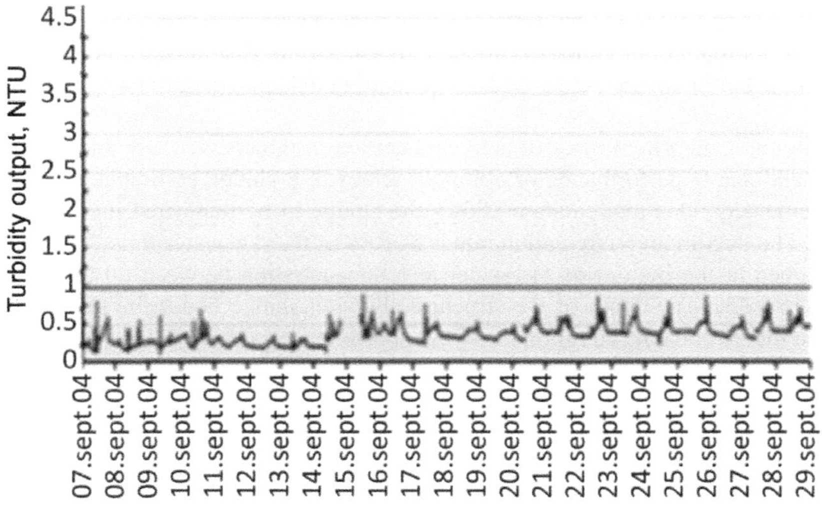

Figure 9.36. *Evolution of turbidity in settled water. For a color version of this figure, see www.iste.co.uk/gaid/watertreatment2.zip*

The mirror velocity on the settler is between 12 and 16.0 $m \cdot h^{-1}$ under nominal conditions, without causing PAC leakage, which would result in an increase in the turbidity of the settled water. This is, on average, approximately 0.5 NTU (Figure 9.36). PAC management is controlled by an initial PAC concentration, an applied treatment rate and a purge rate.

– Initial concentration of PAC ($g \cdot L^{-1}$) introduced into the system:

$$(PAC) = \frac{M}{V}$$

– Treatment rate ($g \cdot m^{-3}$): this corresponds to the quantity of new PAC injected according to the flow rate to be treated or the evolution of the concentration of organic matter:

$$PAC\ rate = \frac{(PAC)slurry\ .\ Q\ injection}{Q\ treated\ water}$$

– Purge flow rate ($m^3 \cdot h^{-1}$): this corresponds to the volume of PAC extracted for a given operating time. It depends on the concentration of the sludge bed, the quality of the water to be treated and the treatment rate:

$$Q\ purge = \frac{PAC\ rate\ .\ Q\ treated\ water}{[PAC]sludge\ bed}$$

Due to the high concentrations of PAC in the sludge bed (10–30 $g \cdot L^{-1}$) and therefore the immediate availability of a large adsorption capacity, activated carbon consumption can be reduced while maintaining treatment efficiency. In addition, the PAC stock allows for variations in micropollutant and organic matter concentrations, providing operators with a better understanding of seasonal peaks.

The quality of the treated water remains constant over time, as long as the PAC stock is properly managed. The flexibility of the operation provides security for the final quality of the treated water. By controlling the frequency and the volume of the purges, it is possible to increase the PAC concentration in the decanter to obtain a buffer effect against organic matter. The PAC renewal mode makes it possible to adapt the treatment rate to the quality of the water to be treated.

The PAC is chosen according to physical criteria (particle size, specific surface, mineral or plant origin, porous texture, activation type) and chemical criteria (adsorption capacity in relation to the organic substances to be removed, coefficients from the Freundlich and Langmuir equations).

Figure 9.37. *Multiflo® Carb photo, L'Haÿ-les-Roses plant (France). For a color version of this figure, see www.iste.co.uk/gaid/watertreatment2.zip*

9.7.2.1.1. Sizing parameters

Table 9.11 summarizes the Multiflo® Carb sizing parameters.

Parameters	Values
Reaction zones	
Contact time (PAC contact tank), min	10
Contact time (coagulation tank), min	1–2
Contact time (flocculation tank), min	10–30
Settling zone	
Mirror velocity, $m \cdot h^{-1}$ Hazen equivalent velocity (Greca, 1 m) Lamella type Orthogonal spacing, mm Inclined length (at 60°), m Settled-water intake Spacing between launders, m	12–15 0.75–1.07 Greca 0.35 1.0–1.4 Transverse launders + central channel 1.5
Recovery channel floating above the feed zone (water level raising) Presettling zone under the feed zone	
Sludge recirculation rate	2–10%, (continuously, depending on the target concentration in the CAP reactor and in the flocculator)
Circular scraper Maximum inlet speed under the plates Maximum sludge bed level Position of the probe (SS measurement) Average sludge bed concentration Recirculated sludge recovery	With harrow 3 × Mirror velocity 1 m under the lamellas High sludge bed level 15–30 $g \cdot L^{-1}$

Table 9.11. *Sizing parameters*

9.7.2.1.2. Operating parameters

Table 9.12 summarizes the operating parameters of the Multiflo® Carb.

Parameters	Values
PAC dosage, $g \cdot m^{-3}$	2–20
$FeCl_3$ dosage (41%), $g \cdot m^{-3}$	10–20
Polymer dosage, $g \cdot m^{-3}$	0.2–0.5
Average age of PAC, d	10–20
Concentration of settled sludge, $g \cdot L^{-1}$	15–30

Table 9.12. *Operating parameters*

9.7.2.2. *Actiflo® Carb*

In addition to its ability to absorb variations in raw water quality, the Actiflo® process is known for its compactness, as it is a ballasted floc process that allows settling speeds 10 times higher than those of traditional lamellar settlers. The process derived from Actiflo® to remove organic matter, pesticides, drug residues, endocrine disruptors and toxins is called Actiflo® Carb.

9.7.2.2.1. General description

According to the general principle of Actiflo® Carb operation, the colloidal matter of the water to be treated is destabilized by the injection of a coagulant reagent and then agglomerated by means of a polymer around a granular support of sand. The floc thus formed is considerably heavier than a standard floc resulting from a traditional physicochemical treatment. The floc is "ballasted", which, in the case of separation carried out in a lamellar settler (Figure 9.38), allows operation at very high mirror velocities of between 50 and 80 $m \cdot h^{-1}$ for the water to be treated.

The operating principle of the Actiflo® Carb is identical to that of the Actiflo® Carb described in Chapter 4. Upon arrival at the Actiflo® Carb unit, the raw water goes through in order. The first tank, known as the contact tank (or injection tank) with the PAC, is the one in which the adsorption reaction between the activated carbon, the organic matter and the micropollutants takes place (Figure 9.38).

The concentration of PAC is between 1.2 and 2.0 g·L^{-1} and results in maximum use of the PAC's efficiency. The PAC is permanently "refreshed" by new inputs (from 2 to 15 g·m^{-3}, depending on the period). The adsorption kinetics are very fast, and the organic matter and micropollutants are quickly adsorbed. This maintenance of a permanent and significant PAC bed in the system makes it possible to keep a quantity of PAC that is always ready to adsorb the possible peaks in organic matter or pesticides. The kinetics continued during the water journey in the different tanks, constituting the Actiflo® Carb. Thus, in the second tank, which is used as a coagulation tank, the destabilization reactions of the colloids take place with the injected coagulant, as well as the reagents necessary for pH adjustment.

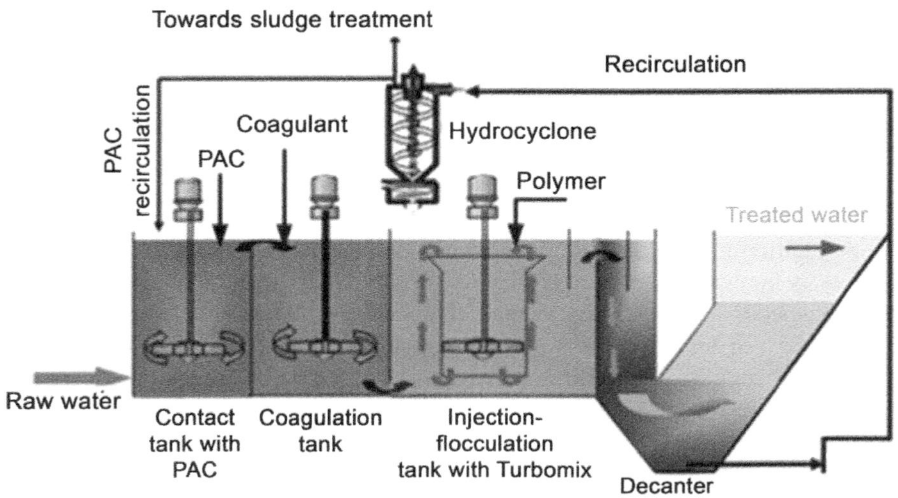

Figure 9.38. *Operating principle of the Actiflo® Carb (source: Veolia). For a color version of this figure, see www.iste.co.uk/gaid/watertreatment2.zip*

The coagulated water containing the PAC reaches a third tank, known as the injection and flocculation tank, in which the polymer and the recirculated micro-sand are injected. Each tank is equipped with a vertical axis agitator with adjustable speed. The micro-sand injected into the tank initiates the formation of the PAC-sludge-micro-sand complex for quick settling. This mixture reaches the lamellar settling section for liquid–solid separation. The PAC, hydroxide sludge and micro-sand deposited at the bottom of the settler are mixed by a bottom scraper and then collected and pumped to hydrocyclones (Figure 9.39). As a result of the low turbidity, the hydroxide sludge only represents a very small part of the PAC/micro-sand mixture.

Figure 9.39. *Hydrocyclones for the separation of sand and PAC. For a color version of this figure, see www.iste.co.uk/gaid/watertreatment2.zip*

The hydrocyclones ensure the separation of the micro-sand from the PAC/sludge mixture. The hydrocyclone's underflow, essentially made up of micro-sand, is reinjected into the injection tank. The polymer is introduced into this tank. The hydrocyclone's overflow, essentially made up of PAC (and some hydroxides), is reinjected into the PAC contact tank. The new PAC is also injected at this level, which initiates the adsorption kinetics.

The clarified water is collected in the upper part of the decanter by recovery channels, which evacuate the water to the next stage (gravity or membrane filtration). These channels are equipped with spillways, which ensure a good distribution of the flows in the structure (Figure 9.40). The sludge settles at the bottom of the structure and is collected by a bottom scraper and then pumped to the hydrocyclones.

The high density of the sand-PAC-sludge complex leads to high Hazen velocities and makes it possible to operate at mirror speeds between 35 and 45 $m \cdot h^{-1}$.

Fields of application of the Actiflo® Carb solution are as follows:

– in the first stage of groundwatertreatment that is contaminated with pesticides and/or organic matter;

– in the first stage of treatment of surface water with low turbidity and loaded with organic matter or pesticides;

– in the second stage of treatment downstream of a first clarification: traditional settling, Multiflo® decanter or flotation (Spidflow®) and Actiflo® ballasted settler.

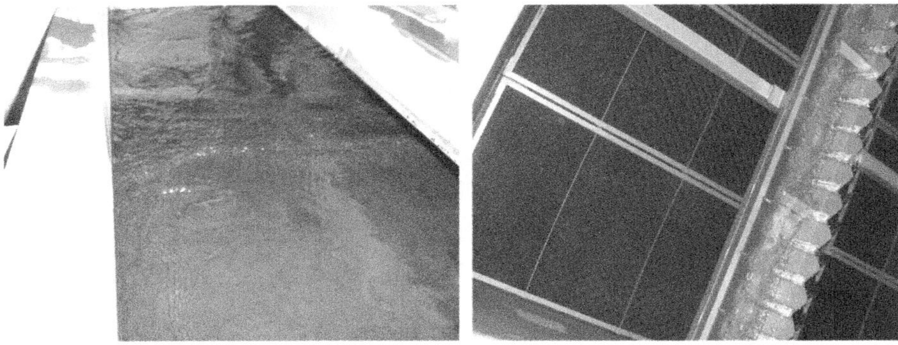

a) b)

Figure 9.40. *(a) PAC contact tank and (b) Actiflo® Carb settled water. For a color version of this figure, see www.iste.co.uk/gaid/watertreatment2.zip*

a) b)

Figures 9.41. *Actiflo® Carb in the (a) Lucien Grand and Moselle plant and (b) Madon plant. For a color version of this figure, see www.iste.co.uk/gaid/watertreatment2.zip*

Depending on the quality of the water to be treated and the performance to be achieved, the Actiflo® Carb process is available in an Actiflo®TwinCarb version. This unique configuration consists of a double treatment stage in series, allowing the removal of NOM to be amplified and reducing land use. This double treatment stage consists of a clarification stage (Actiflo® or Actiflo®Turbo), followed by a refining stage (Actiflo® Carb). Actiflo®TwinCarb (Figure 9.42) is particularly well suited to the treatment of water with a high pollution load, allowing the TOC content of the raw water to be reduced from approximately 15 $mg \cdot L^{-1}$ to less than 2 $mg \cdot L^{-1}$ in the treated water.

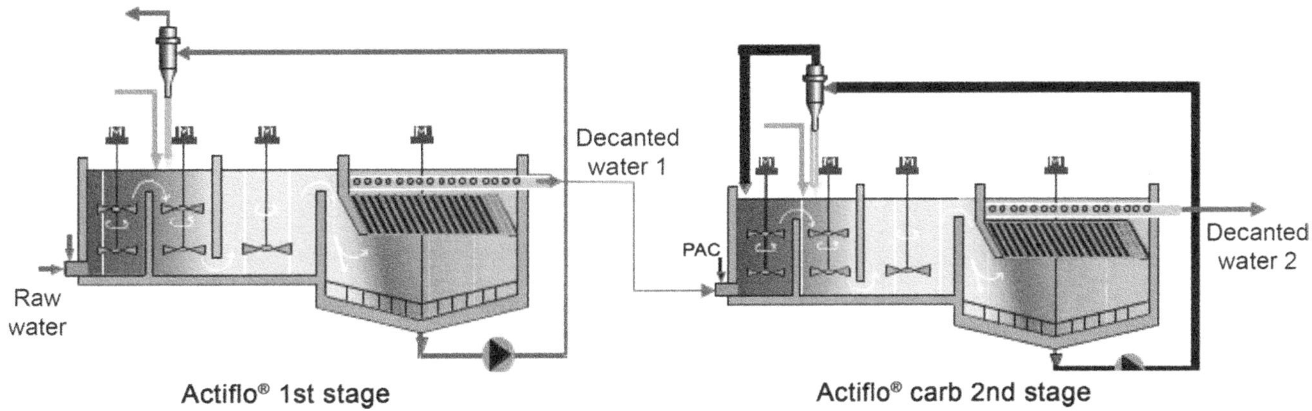

Figure 9.42. *Actiflo® and Actiflo® Carb in series for the removal of turbidity, color, algae, organic matter and micropollutants. For a color version of this figure, see www.iste.co.uk/gaid/watertreatment2.zip*

9.7.2.2.2. Sizing parameters

PAC contact tank	
Design	Contact time: 10 min
PAC dosage	15–25 $g \cdot m^{-3}$ to obtain 45–55% organic carbon removal below PAC concentration in the contact tank: 1.2–1.5 $g \cdot L^{-1}$
Coagulation tank	
Design	Contact time: 2 min
Coagulant	Al^{3+} or Fe^{3} coagulant dosage at 10–15 $g \cdot m^{-3}$
Injection–flocculation tank	
Design	Contact time: 6 min
Polymer	Anionic polymer dosage: 0.5 $g \cdot m^{-3}$ Polymer concentration: 2 $g \cdot m^{-3}$
Decanter	
Design	Greca type lamellas (L = 1 m) Mirror velocity: 40 $m \cdot h^{-1}$ Hazen velocity: 3 $m \cdot h^{-1}$ Residence time: 16 min Recirculation rate: 6–9%

Table 9.13. *Sizing parameters*

9.7.2.2.3. Performance

Due to the high volume of flocs formed in the various reaction tanks (coagulant injection, polymer) and micro-sand, the PAC is trapped within these flocs, and the turbidity of the settled water is usually less than 1 NTU. There is no leakage of fine PAC (Figure 9.43 and Figure 9.44) into the produced water. UV 254 nm absorbance measurements are always less than 3 m^{-1}, which correspond to dissolved organic carbon (DOC) concentrations less than 2 $mg \cdot L^{-1}$.

The advantages of Actiflo® Carb are as follows:

– advanced treatment with PAC;

– maximum removal of NOM and micropollutants;

– refining of treated water;

– compatibility with other upstream clarification processes such as Actiflo®, Actiflo®Turbo, Multiflo®, other decanters, conventional flotation device, Spidflow®;

– high settling velocity: 40 m·h^{-1};

– small footprint;

– easy to start up: start-up in just a few minutes;

– easy rehabilitation of existing installations.

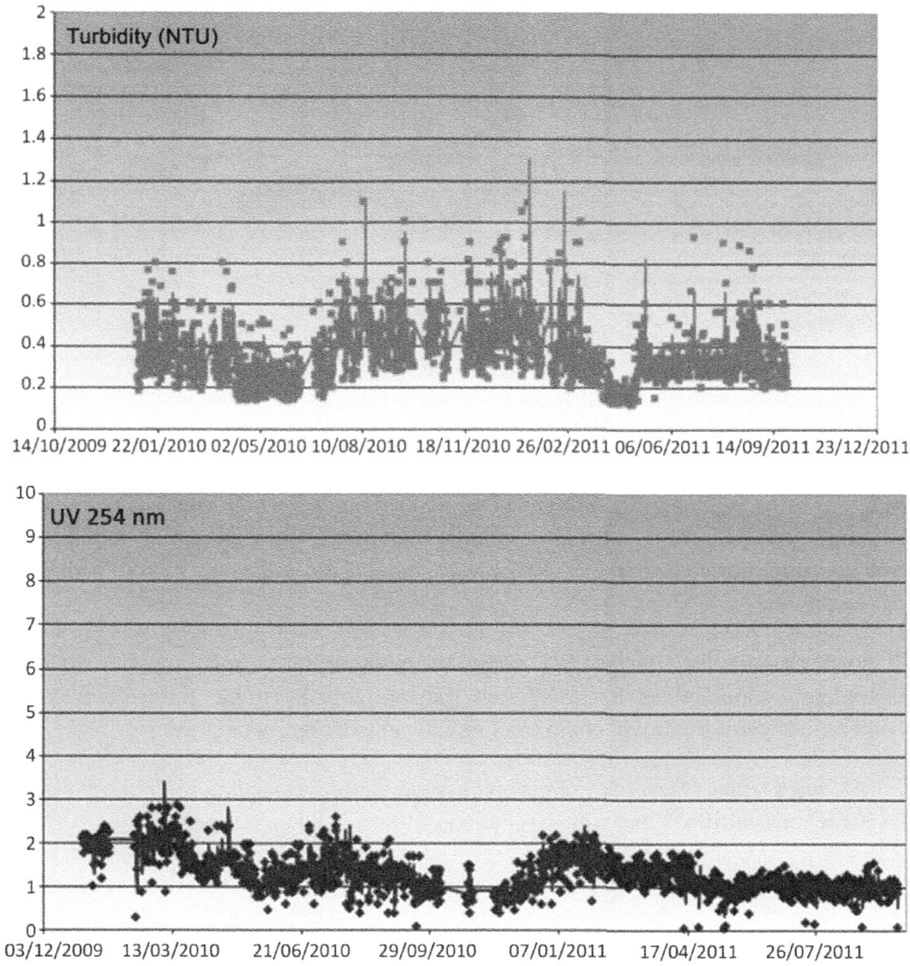

Figure 9.43. *(A) Turbidity and (B) UV 254 nm at the Actiflo® Carb outlet. For a color version of this figure, see www.iste.co.uk/gaid/watertreatment2.zip*

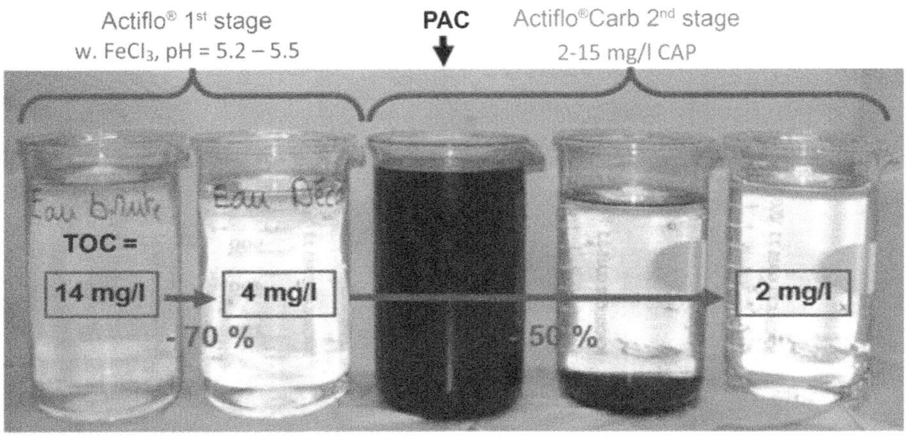

Figure 9.44. *TOC and water quality with the Actiflo® Twin Carb mode. For a color version of this figure, see www.iste.co.uk/gaid/watertreatment2.zip*

9.7.2.2.4. Solutions packaged in Actiflo® Carb and Multiflo® Carb

For many years, OTV Veolia has been developing a range of prefabricated modular units produced in series. These standard units, which can complement an existing process, also provide turnkey solutions for industrial companies. Standardized products make it possible to reduce the costs associated with their development.

OTV Veolia has designed a range of modular units and compact stations. Completely mounted on a chassis and transportable by truck, these devices can constitute a water treatment chain or complete an existing installation.

As such, the Opasep™ range is a coagulation–flocculation–settling system (Actiflo® or Multiflo®) that can also include filtration, when necessary.

Actiflo® Carb can be delivered in a packaged solution for flow rates up to more than 1000 $m^3 \cdot h^{-1}$.The sizing and operating conditions are identical to the concrete solutions.The advantage of this solution is that it can be quickly implemented.

Figure 9.45. *Photos of plants with Actiflo® Carb: (a) Lucien Grand, (b) Montlucon, (c) Cholet (France) and (d) Zhejiang Fuyang (China). For a color version of this figure, see www.iste.co.uk/gaid/watertreatment2.zip*

Figure 9.46. *Packaged solutions with Multiflo® and Actiflo®. For a color version of this figure, see www.iste.co.uk/gaid/watertreatment2.zip*

Figure 9.47. *Packaged solution (Actiflo® Carb and filter) for the Moselle and Maddon plant (France). For a color version of this figure, see www.iste.co.uk/gaid/watertreatment2.zip*

9.7.2.3. *Opacarb®MF*

The Opacarb®MF process is used for the treatment of water containing a concentration of suspended solids lower than 20 mg·L^{-1} and an above-standard concentration of organic substances and micropollutants.

This process works in a series with two stages:

– A first stage, including a contact tank containing PAC (5–100 μm) with a concentration between 150 and 1000 mg·L^{-1}. This tank is agitated with a mechanical mixer or an immersed pump. This contact tank constitutes the PAC reactor. The agitation created inside the PAC reactor is conducive to the suspension of the PAC, in such a way that the PAC concentration in the reactor is homogeneous. The contact time between the PAC and the water to be treated is sufficient (10–18 min) for the optimal use of the PAC adsorption capacities.

– The second stage includes a direct filtration of the water coming from the PAC reactor. The mechanical filtration unit is made up of filtering media, with a cutoff value lower than 7 μm absolute. This stage allows solid–liquid separation between the activated carbon and the water.

A recirculation of the activated carbon separated by mechanical filtration is implemented toward the contact tank for all or part of the flow (Figure 9.48). This increases the average residence time of a PAC particle in the reactor and allows it to maintain a high concentration of PAC, as well as optimize the use of its adsorption capacity.

Due to its low cutoff, mechanical filtration makes it possible to retain almost all the PAC present in the mixture at the outlet of the PAC reactor. It is done according to the external–internal mode, without PAC particle intrusion in the meshes of the filter, as it is not a membrane filtration.

One of the mechanical filtration systems comprises filter cassettes, each of which consists of a wire wound around a frame of an interior cavity, for the flow of said filtrate. This type of filter cassette has the advantage of retaining almost all of the PAC contained in the water to be filtered, clogging slowly and reversibly and being relatively easy to clean.

During the operation time, the PAC particles accumulate on and in the microfibers, forming a "filter cake". This creates a pressure differential (Dp). The allowable pressure differential (Dpa) is set at 0.4–0.6 bar, and when this Dpa is reached, the filter is automatically washed. The washing sequence is either triggered by a pressure differential or a time interval, which is approximately 45–60 min, depending on the PAC concentration in the reactor. The filter unit is cleaned with filtered water or service water. It is not necessary to use chemical products for deep cleaning, as clogging occurs on the surface of the filtering material. Depending on the concentration of organic substances, the used PAC is removed from the contact tank or the recirculation channels at intervals. It is replaced with fresh PAC. It is thus possible to keep the average age of the PAC in the reactor substantially constant while making maximum use of its adsorption capacities.

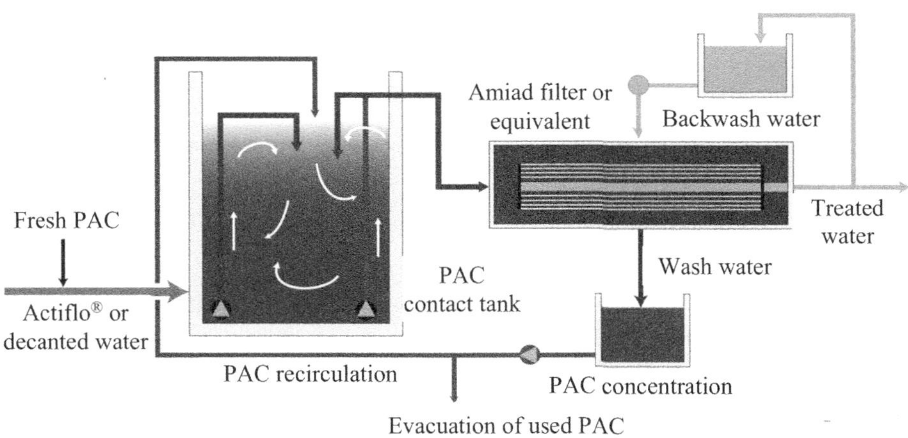

Figure 9.48. *Operating diagram of the Opacarb®MF process (source: Veolia). For a color version of this figure, see www.iste.co.uk/gaid/watertreatment2.zip*

As in all PAC or µgrain reactors, an ozonation stage of the activated carbon in the contact tank is possible. This solution increases the surface functions on the surface of the activated carbon and achieves better removal of organic substances and micropollutants. The Opacarb®MF process does not require the injection of coagulant and/or flocculant into the activated carbon reactor or into the water to be treated.

PAC dosages are between 10 and 20 $g \cdot m^{-3}$. The flow applied on the mechanical filtration is approximately 800–1,200 $L \cdot m^{-2} \cdot h^{-1}$. The global water loss is 6%.

The performance of this system with respect to physicochemical parameters is summarized in Table 9.14.

	Turbidity NTU	**TOC $mg \cdot L^{-1}$**	**Mn $\mu g \cdot L^{-1}$**	**Fe $\mu g \cdot L^{-1}$**	**Ca mg $CaCO_3 \cdot L^{-1}$**	**Alkalinity mg $CaCO_3 \cdot L^{-1}$**
Raw water	9–14	3.6–4.8	60–180	1,000–3,000	330–350	260–280
Treated water	<0.3	1.2–1.9	<10	<30	182–200	100–123

Table 9.14. *Performance of the chemical parameters of the La Bergerie plant (France)*

The fields of application of the Opacarb®MF process are as follows:

– in the first stage of treatment, groundwater is contaminated by pesticides and/or organic matter;

– in the first stage of treatment, surface water with low turbidity was loaded with organic matter or pesticides;

– in the second treatment stage, downstream of a coagulation–flocculation--settling stage (Multiflo® or Actiflo®) or the flotation stage (Spiflow®), when the surface water is loaded with turbidity, algae, color, organic matter and pesticides. The first stage of the treatment process allows the treatment of turbidity, algae, color and some organic matter. The Opacarb® MF, operating mainly with PAC, makes it possible to treat a large part of the organic matter and most of the adsorbable pesticides.

Figure 9.49. *La Bergerie plant (Donville, France). For a color version of this figure, see www.iste.co.uk/gaid/watertreatment2.zip*

Figure 9.50. *Installed mechanical filtration equipment. For a color version of this figure, see www.iste.co.uk/gaid/watertreatment2.zip*

9.7.3. *Opaline® C process: PAC-membrane hybrid*

Veolia's Opaline® C process uses a combination of adsorption on PACassociated with an ultrafiltration membrane.

It consists of a PAC injection stage in the water to be treated, followed by a reaction stage between the PAC and the water in a specific reactor and, finally, a filtration stage of the mixture coming from the reactor in a membrane filtration unit, which carries out the liquid–solid separation. The filtered water is sent to the disinfection stage. The concentrated activated carbon is recirculated to the PAC reactor (Figure 9.51).

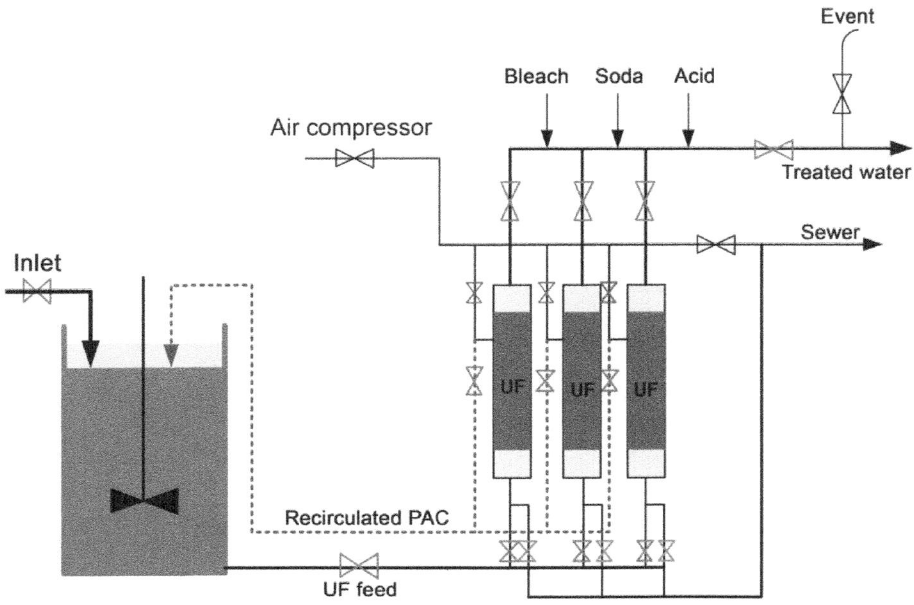

Figure 9.51. *Operating principle of the Opaline® C process. For a color version of this figure, see www.iste.co.uk/gaid/watertreatment2.zip*

The PAC contact tank makes it possible to remove pesticides, micropollutants and organic matter, prevent accidental pollution and improve the organoleptic qualities of the water. It is equipped with an agitator to mix the raw water and the activated carbon, according to a minimum contact time of 10 min.

The close combination with the ultrafiltration membranes allows them to act as the ultimate physical barrier, guaranteeing safety against microorganisms, activated carbon particles and turbidity. The filtration direction of the membranes is internal–external. They work in a frontal or tangential mode, depending on the PAC concentration. In the absence of PAC or in the presence of a low concentration of PAC, the frontal mode is used. The PAC particles are retained on the internal surface of the membrane module fibers. They are regularly evacuated by backwashing with water.

For a high concentration of PAC, the tangential mode is preferred. PAC particles do not accumulate on the inner surface of the fibers because they are continuously removed by sweeping water that flows tangentially to the membrane. A medium flow speed imposes a shear gradient within the fibers and thus limits the accumulation of material. This makes it possible to fight effectively against deposits, which induce an increase in pressure loss. This operating mode is adopted when the concentration of organic substances and/or micropollutants is high and requires high dosages of PAC.

A recirculation loop is created between the concentrate outlet and the raw water inlet. This loop creates a constant flow along the membrane fibers and prevents their early clogging. The concentrate containing the PAC is then directed toward the contact tank. This recirculation increases the PAC concentration in the reactor and increases the adsorption kinetics. We find the previously described PAC reactor mechanism with the same sizing and operating parameters. The recirculation loop makes it possible to control the PAC concentration in the contact tank, and a PAC material balance quantifies the PAC concentration in the reactor.

The membrane skids are designed to treat an instantaneous flow of raw water, with the addition of a recirculation flow of concentrate (which can represent up to 20–25% of the raw water flow fed to the membranes).

Figure 9.52. *Ultrafiltration membrane workshop in the Clermont–Ferrand plant (France). For a color version of this figure, see www.iste.co.uk/gaid/watertreatment2.zip*

Figure 9.53. *Ultrafiltration membrane workshop in the Alcay plant (France). For a color version of this figure, see www.iste.co.uk/gaid/watertreatment2.zip*

Ultrafiltration delivers a quality of treated water that guarantees a treated water turbidity ≤0.2 NTU, as well as a "mechanical" disinfection of the distributed water (retention of bacteria, parasites and viruses).

9.7.3.1. *Design parameters*

The PAC reactor has a contact time of 10–20 min, depending on the concentrations of organic substances and micropollutants to be removed. The PAC dosages can vary between 2 and 50 $g \cdot m^{-3}$. The recirculation flow rate can vary between 5 and 25% of the raw water flow rate. The transmembrane flow is 40–60 $m^{-2} \cdot h^{-1}$.

9.7.3.2. *Operating parameters*

The operating parameters for the PAC concentrations in the reactor are given by:

$$Qin.[PAC]dosage = Q\ wash\ water\ .[PAC]wash\ water$$

$$[PAC]tangentiel = [PAC]wash\ water$$

$$[PAC]reactor = \frac{Qin.[PAC]dosage\ + Qtangential\ .[PAC]tangential}{Q\ UF\ feed}$$

$$[PAC]tangential = \frac{Qin.[PAC]dosage}{Q\ wash\ water}$$

Flow rates are expressed in $m^3 \cdot h^{-1}$, and PAC is expressed in $g \cdot m^{-3}$.

	Dry apparent density, 20 °C	**Specific surface $m^2 \cdot g^{-1}$**	**Iodine index $mg \cdot g^{-1}$**	**Origins**	**Particle sizes µm**
			Jacobi		
Aquasorb MP23	0.48	1,100	1,100	Lignite	15–35
Aquasorb AFP21	0.50	1,100	1,000	–	15–35
Aquasorb XP17	0.50	800	700	–	15–35
Aquasorb 16	0.45	1,000	950	–	15–35
Aquasorb MP25 UF	0.48	1,100	1,100	Lignite	6–15

Cabot Norit					
SA super	0.35	1,150	950	Bituminous coal	8 (d50)
W 35	0.425	875	850	Plant	15 (d50)
SA UF	0.16	1,112	950	Bituminous coal	6 (d50)
20B	–	1,240	800	Bituminous coal	28 (d50)
Chemviron Calgon					
Pulsorb RD 90	0.25	>1,000	1,100	Bituminous coal	10 (d50)
Pulsorb W 260-90	0.36	>1,000	1,000	Bituminous coal	
Pulsorb WP235	0.38	900	900	Bituminous coal	25 (d50)

Table 9.15. *Characteristics of some powdered activated carbons used in PAC reactors*

9.8. Micrograin activated carbon reactors

To improve the performance of activated carbon reactors and to optimize the operating costs, new reactors have appeared on the market, in which a fluidization of the adsorbent media is performed. These reactors use a large mass of mobile carbon, which allows smoothing the peaks of contamination, potentially reducing the contact times and adapting the activated carbon dosages.

9.8.1. *Integration of µgrain activated carbon in the treatment process*

The integration of micrograin (also called µgrain) activated carbon into a treatment process depends on the raw water quality and treatment objectives. Figure 9.54 summarizes some of the configurations of µgrain activated carbon.

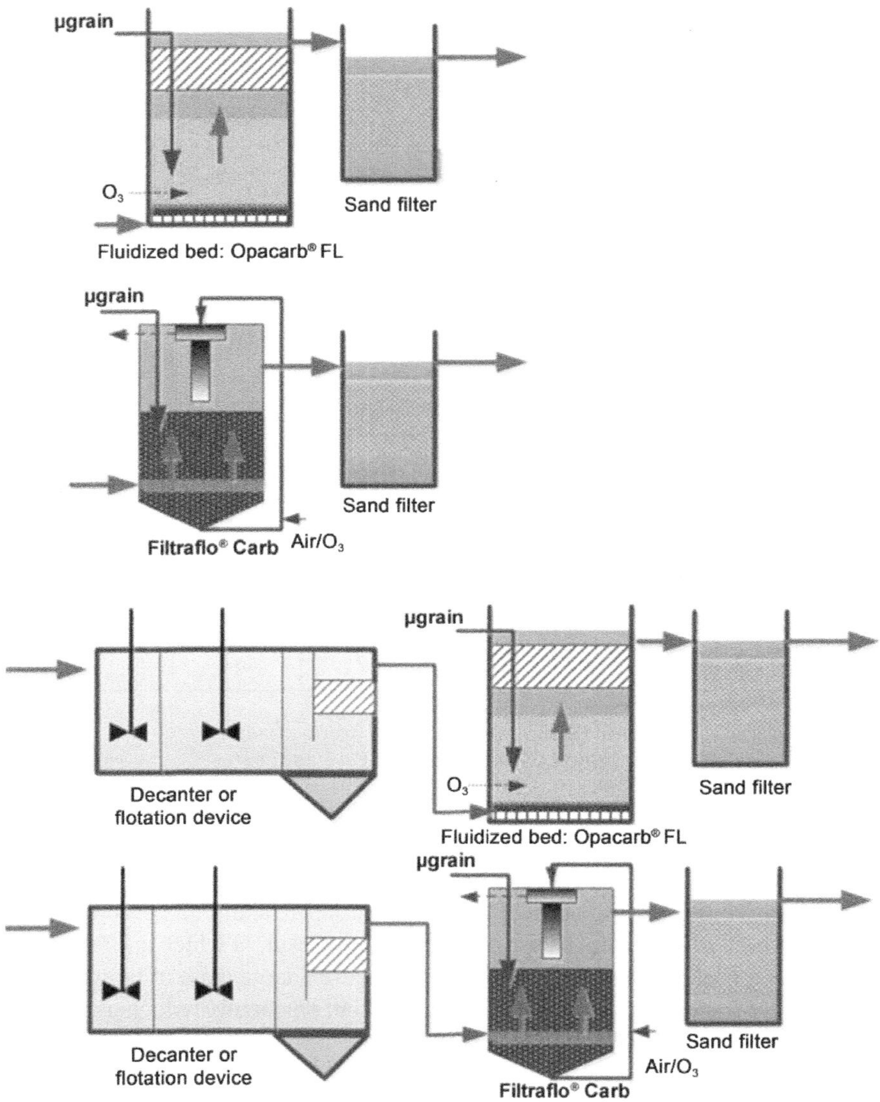

Figure 9.54. *Integration of µgrain activated carbon in the treatment process. For a color version of this figure, see www.iste.co.uk/gaid/watertreatment2.zip*

9.8.2. *The Filtraflo® Carb*

The Filtraflo® Carb is a micrograin activated carbon reactor operating in up flow filtration, which allows simultaneous adsorption and retention of organic substances and micropollutants, as well as suspended solids, which are present in the water to be treated.

The operating principle consists of passing the water to be treated through a reactor containing the µgrain carbon filter bed in an upward flow at a speed that does not cause the bed to fluidize but allows the µgrain activated carbon to migrate as it is filtered toward the lower part of the bed structure. The water to be treated is distributed within the bed by an equal-distribution ramp, which allows the distribution of water in the reactor to be uniform (Figure 9.55).

The µgrain activated carbon at the bottom of the filter bed is continuously removed and recirculated to the top of the bed by means of an air lift. As a result, the energy consumption is very low. This operation allows the µgrain activated carbon to be continuously cleaned of flocs and other impurities that may have agglomerated around the material. The material is ready to be used again for the adsorption of water contaminants.

According to continuous or intermittent frequencies, the µgrain activated carbon extracted at the bottom of the bed is evacuated and compensated by an equivalent amount of new µgrain at the top of the bed (Figure 9.55).

The advantage of a process such as Filtraflo® Carb is that it can quickly find this balance, as it permanently combines the injection of new activated carbon with a considerable mass of activated carbon that is already present in the reactor. The process can be compared to two reactors arranged in series, the first of which contains new activated carbon and the second of which contains a large quantity of new but partly used activated carbon.

The Filtraflo® Carb combines two treatments in a single unit: physical filtration of suspended solids and adsorption of organic matter. This characteristic of the process allows the renewal of the activated carbon without having to stop the operation of the Filtraflo® Carb and makes it possible to obtain a permanent organic substance and micropollutant removal efficiency. The recirculation of the filtering material also allows for self-washing of the material.

This process has the triple benefit of being able to be implemented on smaller installations, of not leading to the formation of sludge and of not requiring the addition of chemical products such as coagulants or flocculants. It is not only energy efficient but also easy to operate.

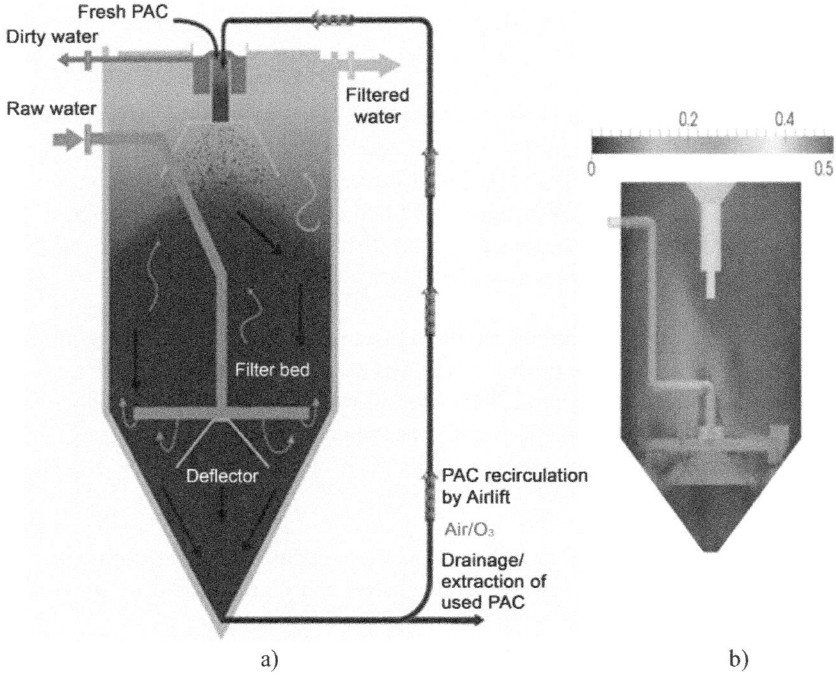

Figure 9.55. *(a) Operating principle of the Filtraflo® Carb and (b) fluent view of the speed distribution. For a color version of this figure, see www.iste.co.uk/gaid/watertreatment2.zip*

Micrograin activated carbon has an average size of 300–800 µm and a specific surface of approximately 900–1000 $m^2 \cdot g^{-1}$. This type of µgrain activated carbon is easily injected into a filter bed, is self-draining and is reactivated in the same way as granular activated carbon (GAC). The contact time of the water in the Filtraflo® Carb is between 5 and 20 min, depending on the concentration of the organic substances to be removed. The filtration rates are between 5 and 15 $m \cdot h^{-1}$. The height of the µgrain activated carbon varies between 1 and 3 m.

Once extracted, the µgrain activated carbon undergoes simple draining and is stored in drums for transport to a thermal regeneration unit. Water treatment plants using Filtraflo® Carb do not require a specific sludge treatment channel. This reduces operating and maintenance costs.

The key to the Filtraflo® Carb's continuous washing process is in the upper part of the process. The "dirty" carbon is brought from the bottom of the reactor to the top *via* the air lift. From there, it falls into a chimney equipped with a screw system. Some of the filtered water flows back up this chimney and backwashes the carbon.

This dirty water is extracted in the upper part. Water losses are less than 3%. The recirculation flow rate varies between 5 and 10%.

The renewal rate of the μgrain activated carbon is between 2 and 30 $g \cdot m^{-3}$ of water, depending on the concentration of organic substances and micropollutants that are present in the raw water, as well as the quality objectives of the treated water.

The initial concentration of μgrain activated carbon ($g \cdot L^{-1}$) introduced into the system is:

$$[\mu grain] = \frac{M}{V}$$

where:

– M: mass of activated carbon in the reactor (kg);

– V: volume of the reactor (m^3).

The treatment rate ($g \cdot m^{-3}$) corresponds to the quantity of new μgrain AC injected as a function of the flow rate to be treated or the evolution of the organic matter concentration:

$$\mu\text{grain rate} = \frac{[\mu\text{grain}]\ injected\,.\,Q\ injection}{Q\ \text{treated water}}$$

The purge rate ($m^3 \cdot h^{-1}$) corresponds to the volume of μgrain AC extracted (m^3) for a given operating time (tf, h). It depends on the concentration of the sludge bed, the quality of the water to be treated and the treatment rate:

$$Q\ purge = \frac{Extracted\ volume}{tf\ of\ Filtraflo®Carb}$$

The Filtraflo® Carb is characterized by:

– the constant addition of new activated carbon in the filter bed;

– an extraction of the μgrain activated carbon proportional to the addition of new μgrain;

– the continuous adsorption of organic substances and micropollutants;

– the continuous filtration of water through the mobile activated carbon bed;

– permanent movement of the μgrain extracted at the bottom of the filter toward the top of the filter, allowing the homogenization of the filtering mass;

– a self-cleaning of the filtering bed by means of a backwash system in turbulent condition, allowing a removal of the suspended matter trapped by the filtering bed through turbulence created in the high part of the bed, thanks to the recirculation;

– the injection of O_3 at the air lift, which supports the creation of new surface functions and increases the performance of the μgrain activated carbon.

In summary, the benefits of Filtraflo® Carb are as follows:

– low water loss of 1–3% (similar to a gravity filter);

– no washing equipment or storage tank for filtered water;

– no washing control or filter regulation equipment;

– no short circuiting possible through the structure;

– no coagulant or polymer;

– very good filtered water quality;

– low maintenance because there is no electromechanical equipment (agitator, scraper, etc.);

– no sludge;

– no N-1 filter operation;

– low energy consumption.

9.8.2.1. *Filtraflo® Carb performance*

The use of Filtraflo® Carb has several advantages: due to the high concentration of PAC in the filter bed, and thus the immediate availability of a high adsorption capacity, activated carbon consumption can be reduced while maintaining treatment efficiency.

In addition, this μgrain activated carbon stock makes it possible to accept variations in pesticide and organic matter concentrations, offering operators a better understanding of peak seasonal variations.

The quality of the treated water remains constant over time, as long as the management of the μgrain stock is properly done. The size of the micrograin implies that it can be injected as a dry product, without the need for processing systems such as a slurry or hydroejector. By controlling the frequency and method of μgrain renewal, the treatment rate can be easily adapted to the quality of the water to be treated.

9.8.2.2. *DOC removal*

DOC removal is illustrated in Figure 9.56, in which 50% DOC removal is achieved with a PAC dosage of 20 g·m^{-3} and a passage velocity of 15 m·h^{-1}.

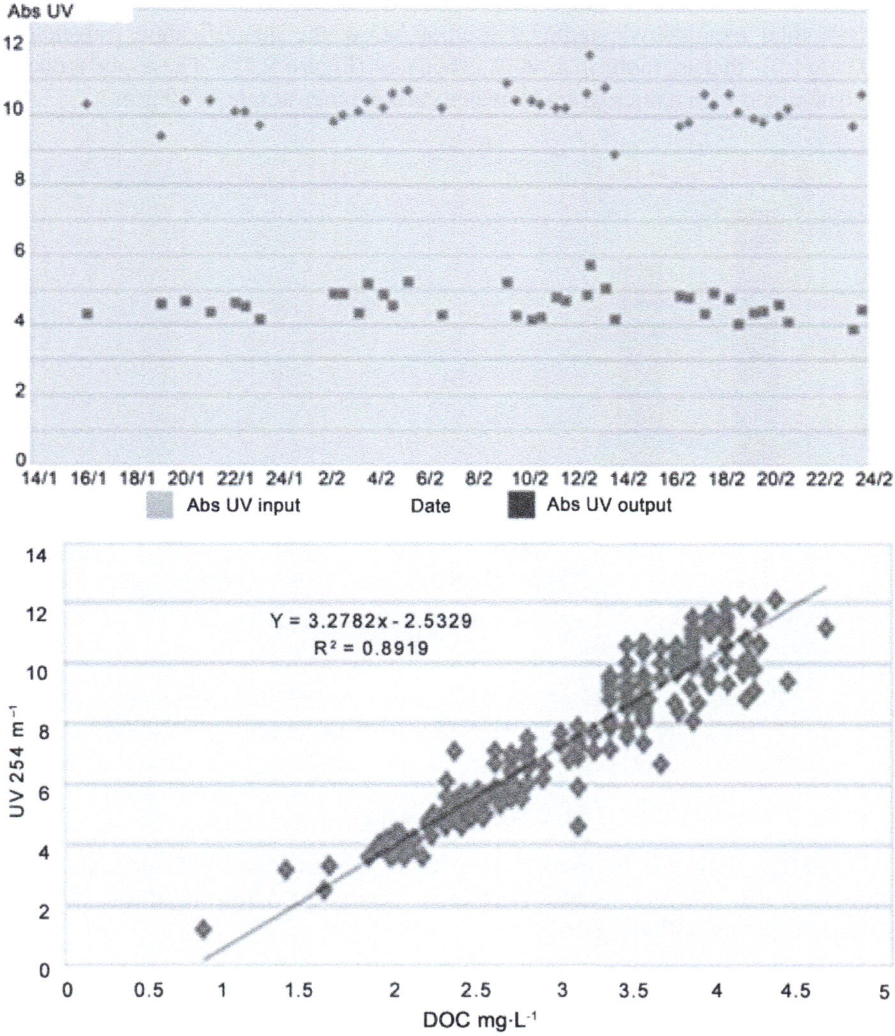

Figure 9.56. *UV absorbance 254 nm removal on Filtraflo® Carb and UV254 versus DOC correlation. For a color version of this figure, see www.iste.co.uk/gaid/watertreatment2.zip*

We note an organic substance removal of more than 50%. The monitoring of the turbidity at the outlet shows a refinement of approximately 0.5–1 NTU compared to the inlet value under the operating conditions of the Filtraflo® Carb.

Monitoring of the reactor's performance with respect to metaldehydes shows almost total metaldehyde removal (output below the quantification threshold of 0.02 $\mu g \cdot L^{-1}$), that is, more than 90% efficiency (Figure 9.57). These performances were obtained with a micrograin activated carbon renewal rate of 20 $g \cdot m^{-3}$.

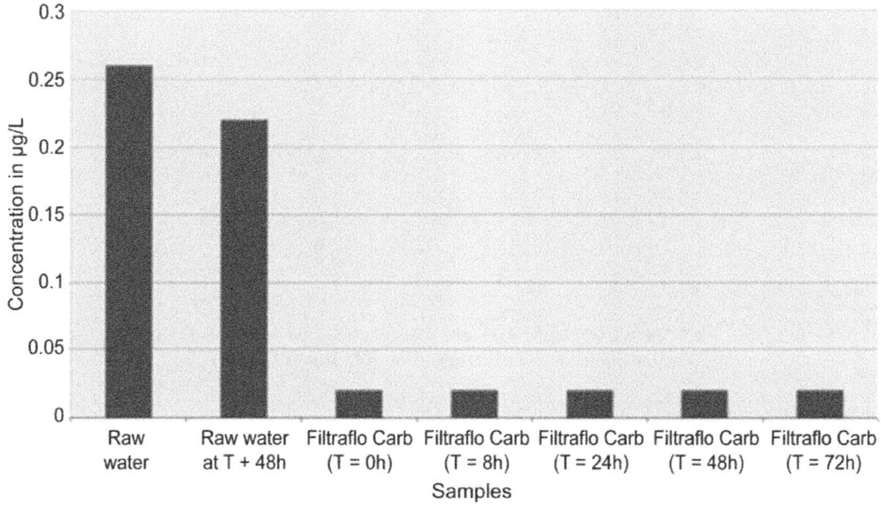

Figure 9.57. *Metaldehyde removal on Filtraflo® Carb. For a color version of this figure, see www.iste.co.uk/gaid/watertreatment2.zip*

9.8.2.2.1. The Filtraflo® Carb fields of application

Filtraflo® Carb can be used to treat water contaminated with organic matter, pesticides and micropollutants (endocrine disruptors, drug residues, industrial substances) in the following areas:

– groundwater inthe first stage of treatment;

– surface water in the first stage of treatment on water with low turbidity;

– surface water in the second stage of treatment, after removal of particulate pollution;

– (turbidity, algae) by flotation or settling;

– clarified wastewater in tertiary treatment.

Figure 9.58. *Filtraflo® Carb from the Craon plant (France). For a color version of this figure, see www.iste.co.uk/gaid/watertreatment2.zip*

a) b)

Figure 9.59. *(a) Gahard plant and (b) Barregant plant model. For a color version of this figure, see www.iste.co.uk/gaid/watertreatment2.zip*

The process can be advantageously installed on small- and medium-sized plants. It also has the advantage of being compact and very low energy-consuming. This process does not require the installation of washing equipment or a storage tank for the filtered water. There is no short circuit through the material bed, and maintenance is reduced to a minimum due to the absence of electromechanical equipment (agitator, scraper, etc.). One of the major advantages of using microcarb (μgrain) activated carbon is the absence of activated carbon sludge, the disposal of which is always a cause for concern, especially in the case of agricultural spreading.

9.8.3. *Opacarb®FL*

The Opacarb®FL is an upflow reactor made of a fluidized bed of micrograin activated carbon. The implementation of an ascending flow allows a controlled expansion of the activated carbon, according to the applied speed. Indeed, this speed is calculated to avoid excessive expansion, which could lead to a loss of media.

The water to be treated is therefore fed from the bottom of the reactor using mechanical and hydraulic means that ensure optimum distribution of the flow through the media (Figure 9.60). It circulates through three zones:

– a **partial expansion zone**: the flow rate is less than 20 $m \cdot h^{-1}$, the variation in the height of the activated carbon is low, and fluidization is limited;

– an **effective expansion zone**: the flow rate is 20–$m \cdot h^{-1}$. The expansion height increases rapidly, and fluidization is effective;

– a **quiet zone**, which is related to the expansion rate of the media and the total height of the structure. In this zone, the turbulence of the medium is null and avoids leakage of activated carbon particles.

The minimum fluidization velocity is between 3 and 10 $m \cdot h^{-1}$, depending on the temperature of the water to be treated. It depends on the physicochemical characteristics of the media present in the reactor (particle diameter (dp), particle density (ρp), particle sphericity (φ)), bed porosity (ε), expanding porosity (exp), etc., and the properties of the carrier fluid (density (ρw), dynamic viscosity (μ), T °C, etc.). However, in expanded bed reactors, it is the expansion velocity that guides the operating process from both a hydraulic and adsorptive perspective. A 100% expansion requires operating at very high velocities (>50 $m \cdot h^{-1}$ at 20°C for a d_{10} of 0.4 mm), inducing a turbulent regime within the reactor, the consequences of which may be preferential circuits, carbon attrition, contact times that are too short to obtain optimal kinetics, etc.

The flow rate is calculated in such a way as to, on the one hand, obtain a compromise between an expansion zone of the AC bed and, on the other hand, to ensure an adequate quiet zone to avoid the escape of AC particles. The average expansion rates of the AC bed are between 10 and 100%. However, for the same ascension speed, this expansion rate varies according to the characteristics of the activated carbon. The parameters are defined according to the particle size of the activated carbon, the height of the desired structure and the concentration of activated carbon in the medium. A turbulence zone exists when the passage speed is higher than 40 $m \cdot h^{-1}$, the expansion height is high (>100%), and a part of the grains is dragged with the flow. This is called turbulent flow.

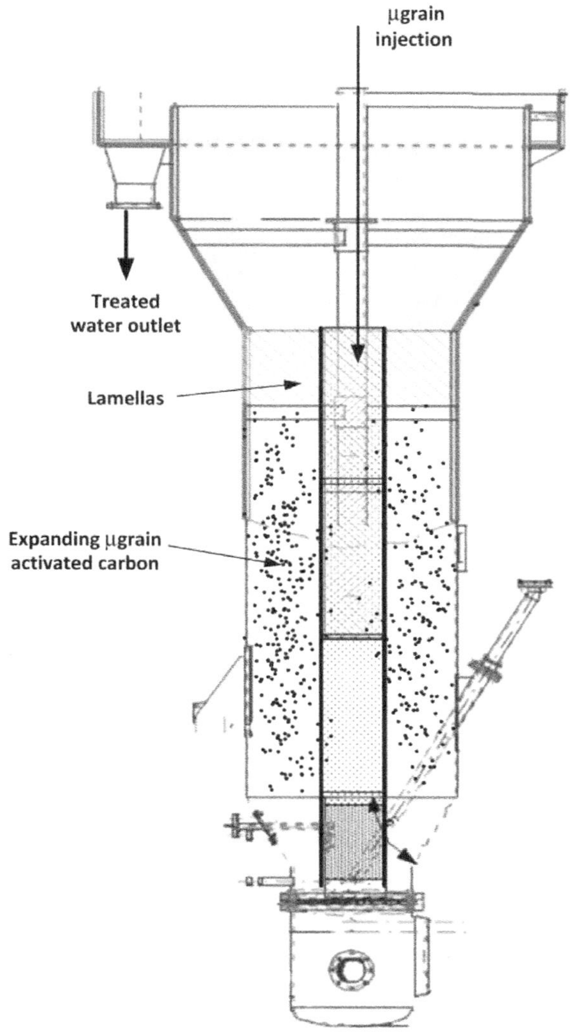

Figure 9.60. *Operating principle diagram of the Opacarb®FL process. For a color version of this figure, see www.iste.co.uk/gaid/watertreatment2.zip*

In its upper part, the Opacarb®FL reactor has means for a physical separation of the particles that could reach the upper part of the reactor. These means are either a lamellar block (Figure 9.61) or a deflector (Figure 9.62). In this case, the small-sized activated carbon particles that are carried to the top of the reactor come up against the lamellar block or the deflector and, due to the slowing down of the speed of the ascending flux, fall back down to the bottom of the reactor. The treated water becomes almost free of activated carbon particles.

Opacarb®FL injects new activated carbon and extracts used activated carbon. The used activated carbon is withdrawn at the bottom of the reactor and directed to a hydrocyclone, which allows recirculation of the particles within the reactor. The activated carbon used is sent to thermal regeneration.

Opacarb®FL can include a recirculation of the treated water, which makes it possible to decrease the concentration of polluting substances in the raw water, as well as reduce the dosages of activated carbon to be sent to the reactor. In addition, an analysis unit can be installed to measure the adsorption capacity of micrograin samples from the reactor, such as the iodine index.

The implementation of activated carbon reactors in expanded beds is possible, provided that the hydraulic parameters are controlled and that the activated carbon has suitable physicochemical characteristics. Indeed, a pulverized activated carbon (15–25 μm) would imply a very low applied velocity ($<1\ m \cdot h^{-1}$), which would lead to large structures with the loss of fine particles carried by the water. A GAC (1–1.4 mm) would imply a very high applied velocity ($>65\ m \cdot h^{-1}$) for a 40% expansion, and attrition of the grains due to the turbulent regime would lead to fine particles at the outlet of the structure.

Opacarb®FL works without the addition of a coagulant or polymer.

The controlled 40% expansion in the Opacarb®FL promotes a stratification of the activated carbon along the reactor. This stratification classifies the activated carbon particle sizes from the bottom to the top from the largest to the smallest particles. Theoretically, the adsorption kinetics take place more slowly at the bottom of the reactor, where the larger particles accumulate. The adsorption kinetics are faster in the middle and top of the reactor, where the particles are the finest. However, the mass of carbon is so high that the kinetics are also fast throughout the reactor, and the contact time does not become a determining parameter as in powdered activated carbon reactors.

In addition, it is possible to advantageously combine this process with an injection of ozone in the lower part of the AC bed. The purpose of this ozone injection will be to improve the surface functions of the activated carbon while simultaneously improving the treatment performance without generating by-products.

9.8.3.1. *Operating parameters*

The sizing of the reactor is based on a variable contact time between 5 and 20 min. The concentration of activated carbon in the reactor can be between 100 and 300 g·L^{-1}. The particle size of the activated carbon is between 100 and 800 µm. The upflow velocity varies between 10 and 20 m·h^{-1} but can be as high as 40 m·h^{-1}.

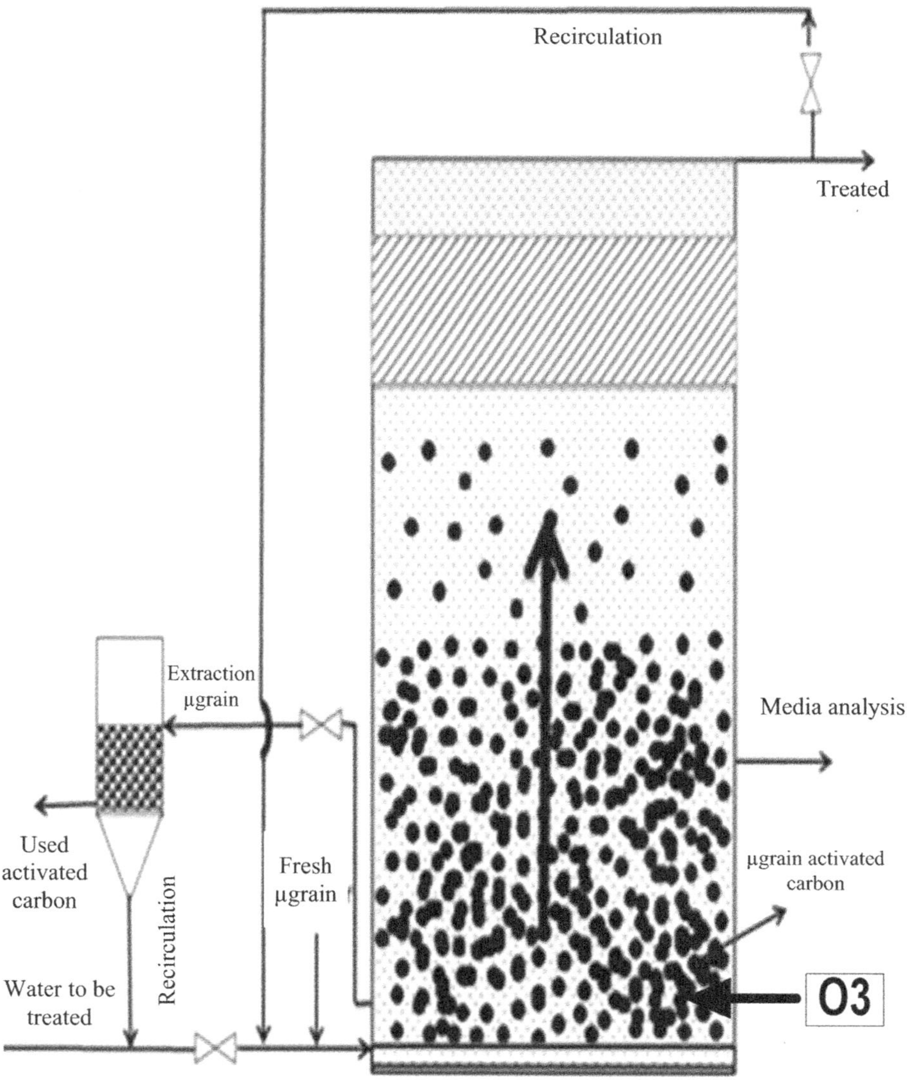

Figure 9.61. *Opacarb®FL with optional deflector lamellas. For a color version of this figure, see www.iste.co.uk/gaid/watertreatment2.zip*

The velocity inside the reactor must be chosen while considering the best compromise between the parameters of treated water turbidity, organic matter removal and carbon bed expansion. Figure 9.63 shows the evolution of the expansion percentage according to the passage speed.

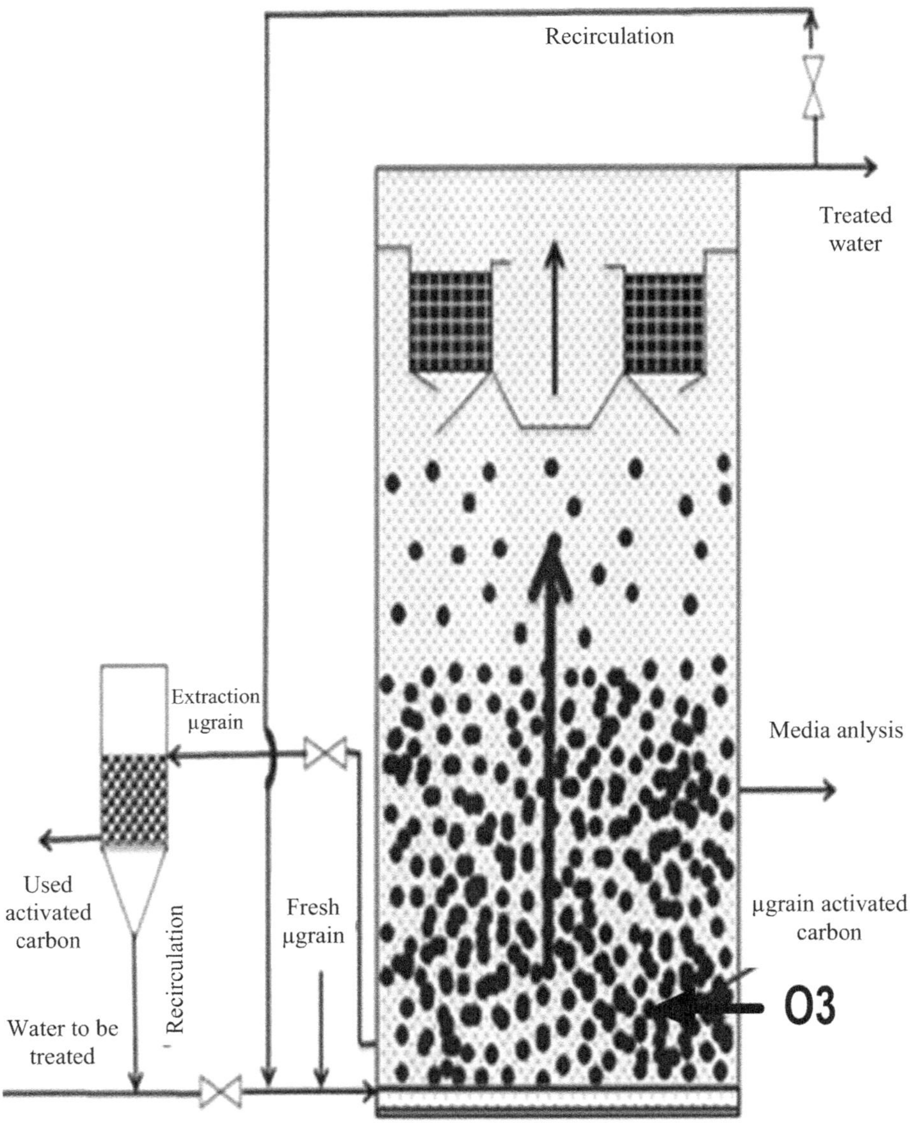

Figure 9.62. *Opacarb®F with in situ ozonation option. For a color version of this figure, see www.iste.co.uk/gaid/watertreatment2.zip*

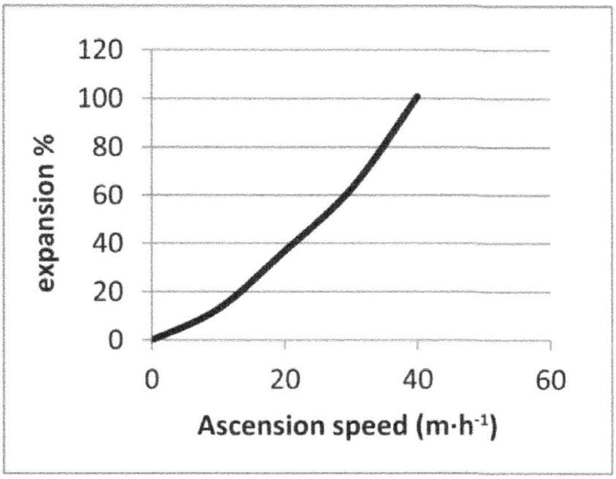

Figure 9.63. *Expansion percentage according to the ascension speed*

9.8.3.2. *Fields of application*

The Opacarb®FL fields of application are as follows:

– as a first stage in the treatment of groundwater (with very low turbidity) that is contaminated with pesticides and/or organic matter and organic micropollutants;

– integrated as a refining stage for surface water that is contaminated with pesticides, organic matter and/or organic micropollutants;

– integrated as a refining stage in tertiary treatment of wastewater that is loaded with pesticides, organic matter and organic micropollutants.

9.8.3.3. *Performance*

The results obtained show that there is no increase in turbidity due to possible fine activated carbon particles due to the separation and capture means implemented. The DOC removal efficiency varies between 50 and 70%, depending on the activated carbon dosage and the age of the activated carbon in the reactor. The removal of pesticides and emerging micropollutants is also very high and depends on the dosage of activated carbon.

The advantages of the Opacarb®FL process are as follows:

– the Opacarb®FL process brings a real technological contribution because:

- it works without the addition of coagulant or polymer to the water,
- the upflow velocities can vary between 10 and 40 $m \cdot h^{-1}$,

  - the activated carbon used can be regenerated, so there is no sludge to treat;
- its implementation is simple, and this technology makes it possible:
  - to obtain decontaminated water,
  - to better manage the activated carbon by reducing the dosage,
  - to generate substantial savings thanks to this operating mode,
  - to reduce water losses.

9.8.4. *Opacarb®MG*

The Opacarb®MG process is based on the adsorption of organic matter and organic micropollutants in a reactor containing micrograin activated carbon. The reactor is an open structure in which an agitator allows a thorough mixing of the activated carbon and thus ensures the principle of integral mixing. In this way, all activated carbon particles participate in the adsorption of the organic substances simultaneously.

The raw water therefore passes through the contact tank (Figure 9.64), only carrying particles smaller than 200 μm with it. The contact tank is equipped with sieves that have a cutoff point between 100 and 200 μm. These sieves are porous tubes (Figure 9.65) made of high-density polyethylene (HDPE), which allow the water to pass through the reactor while retaining the activated carbon. The filtration is done from the exterior to the interior of the tube, and the means of evacuation of said contact tank is connected to the inside of the tube. Such tubes have the advantage of being commercially available and can be easily replaced.

This sieve allows the turbidity to pass through the water, which is then treated downstream. These implemented devices avoid its propagation or its dissemination in treated water. These tubes are unclogged by an injection of air in an internal–external direction, which allows the evacuation of the activated carbon cake that may have been created. The contact time is adapted according to the substances to be removed and can vary between 2 and 5 min.

The used activated carbon is extracted in the mixing zone or can also be extracted at the bottom of the tank when the agitation is stopped. This is the case, for example, during plant shutdowns for productions that are not running 24 h a day. New carbon is then injected into the contact tank in agitated mode.

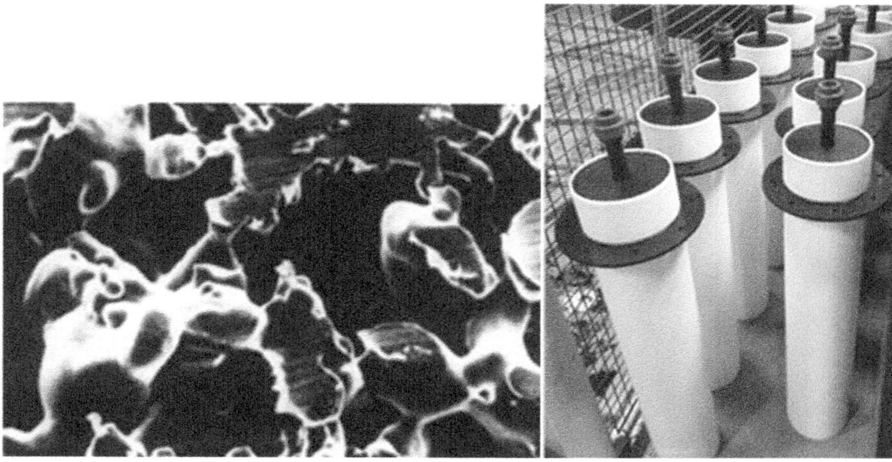

a) b)

Figure 9.64. *Microscopic view (a) and porous sieve (b). For a color version of this figure, see www.iste.co.uk/gaid/watertreatment2.zip*

Figure 9.65. *Porous sieves installed in the contact tank. For a color version of this figure, see www.iste.co.uk/gaid/watertreatment2.zip*

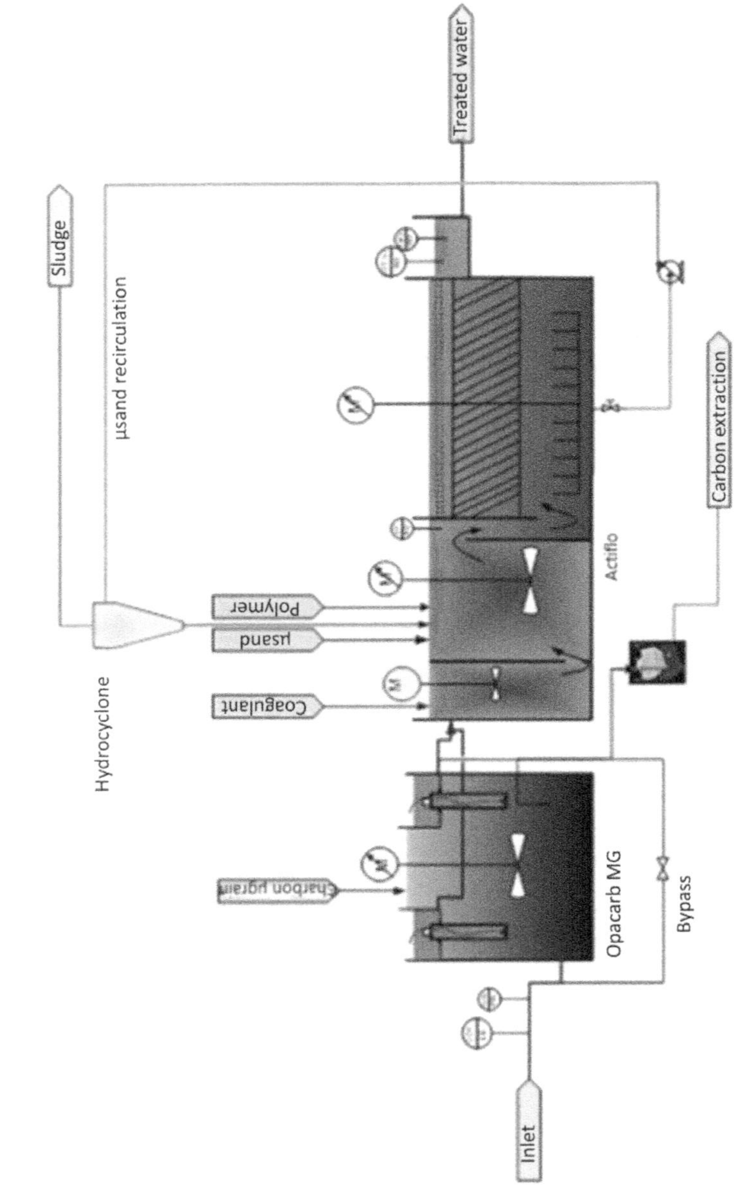

Figure 9.66. *Application example with Opacarb®MG and Actiflo® downstream. For a color version of this figure, see www.iste.co.uk/gaid/watertreatment2.zip*

Opacarb®MG does not feature activated carbon recyclingor coagulant or polymer injection. The activated carbon is thermally regenerated without any treatment other than simple draining.

The concentration of activated carbon in Opacarb®MG varies between 50 and 300 $g \cdot L^{-1}$, depending on the organic matter and micropollutant load of the water to be treated.

The contact tank may have a hopper in its lower part, including the means to purge. Thus, the agitation is designed in such a way that the contents of the upper part can be agitated without agitating the contents in the hopper. In this case, the used carbon is extracted by this means.

9.8.4.1. *Fields of application*

Opacarb®MG can be arranged as follows:

– directly on the raw water by means of prefiltration with a cutoff between 1 and 5 mm, when necessary;

– downstream of a coagulation–flocculation–settling or flotation stage intended to remove algae, high turbidity or part of the organic substances according to the usual operating conditions (coagulant dosage, acid conditions, if necessary, etc.);

– for waters with low or medium turbidity, Opacarb®MG can be installed directly on the raw water and then followed by a coagulation–flocculation–settling stage or sand filtration, dual media filtration or a membrane filtrationsuch as ultrafiltration.

The advantage of an installation with Opacarb®MG, followed by ultrafiltration, involves a treatment process without sludge or chemical reagents. An installation with Opacarb®MG followed by a coagulation–flocculation–settling stage reduces the consumption of reagents and the production of sludge because the adsorption efficiency installed in the first stage avoids high dosages, as well as the adsorption conditions of chemical reagents.

9.8.4.2. *Performances*

Results from a pilot plant (5 $m^3 \cdot h^{-1}$) with a turbidity input ranging from 5 to 20 NTU and a DOC concentration ranging from 2.5 to 3.7 $mg \cdot L^{-1}$, confirming the effectiveness of activated carbon in removing natural organic matter (Figure 9.67). TOC is removed by more than 50%. The concentration of activated carbon in Opacarb®MG was 100 $g.L^{-1}$ with a contact time of 10 min.

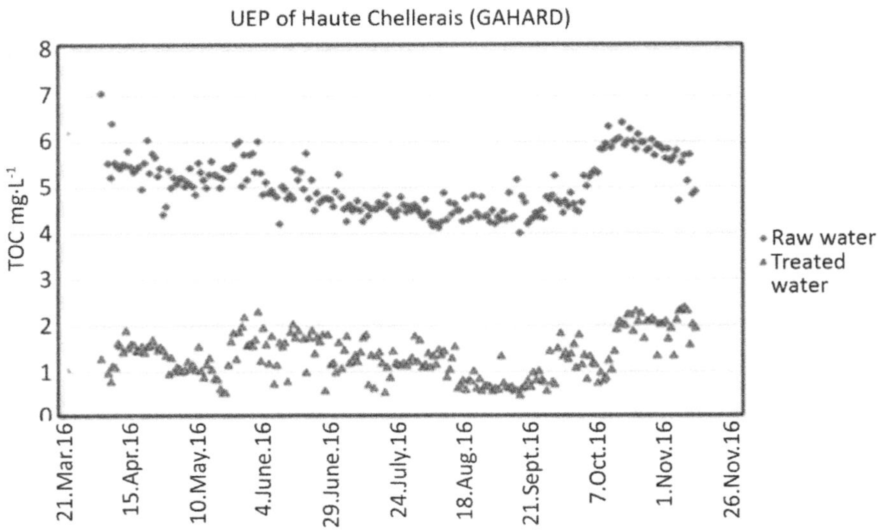

Figure 9.67. *TOC removal on Opacarb®MG (source: Veolia). For a color version of this figure, see www.iste.co.uk/gaid/watertreatment2.zip*

	Apparent dry density, 20°C	Specific surface $m^2 \cdot g^{-1}$	Iodine index $mg \cdot g^{-1}$	Origin	Particle size mm
Jacobi					
Aquasorb® 2000	0.48	1,000	1,000	Lignite	0.45–0.51 (d_{50})
Aquasorb® 6300	0.50	1,000	1,000	Lignite	0.42–0.48 (d_{50})
Chemviron					
Microsorb 400 R	0.425	900	850	Coal	0.35–0.80

Table 9.16. *Characteristics of the micrograin activated carbons used*

9.9. Fixed bed reactors – GAC filters

The integration of a GAC filter into a treatment system depends on the quality of the raw water and the treatment objectives. Figure 9.68 summarizes some of the GAC filter configurations installed at various plants built by Veolia.

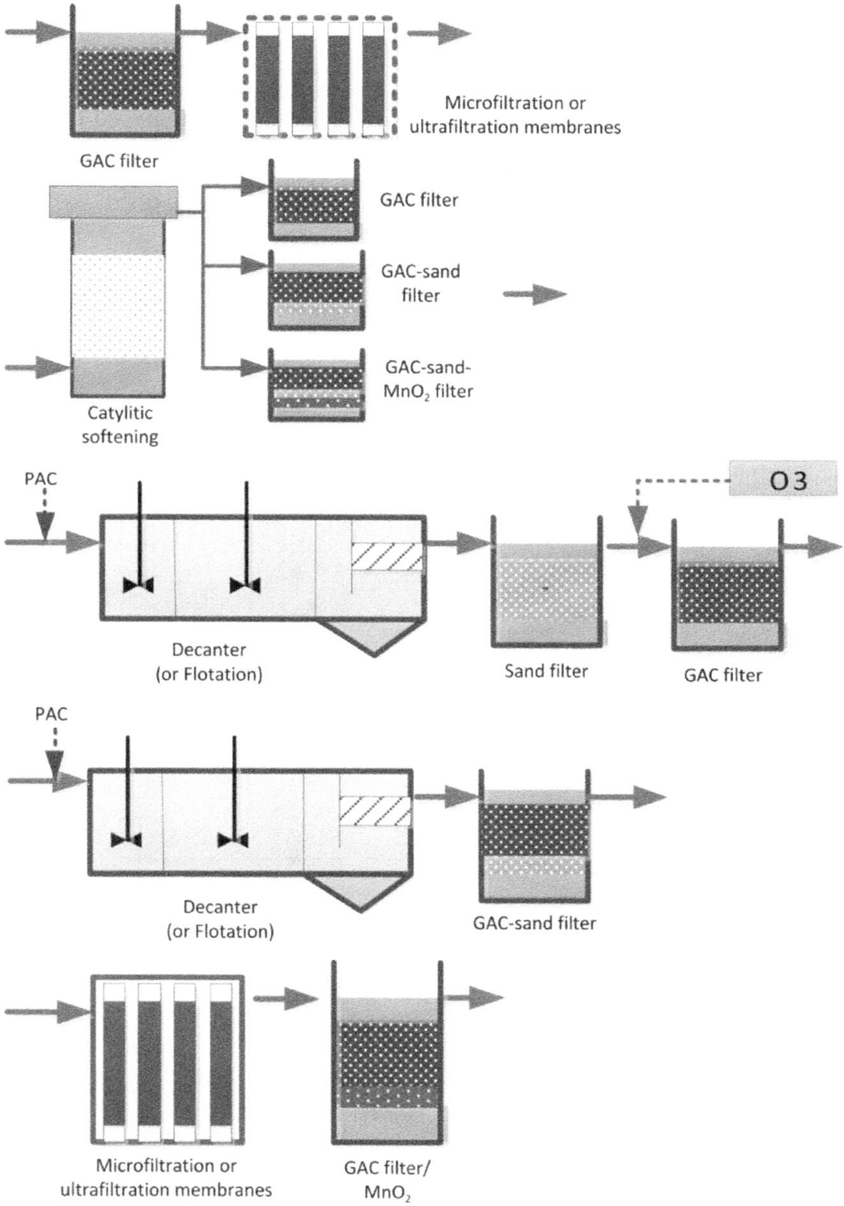

Figure 9.68. *Integration of the GAC filter in a treatment system. For a color version of this figure, see www.iste.co.uk/gaid/watertreatment2.zip*

9.9.1. *The GAC filter (Veolia)*

The GAC filter, designed as an open or closed filter, is sized according to the contact time, material height and passage speed. The water to be treated is admitted to the media layer through a flooded channel to avoid scouring of the activated carbon. The admission is done through the pipe leading into the longitudinal channel, which is also used for the draining of wash water. This solution avoids scouring of the filtering mass.

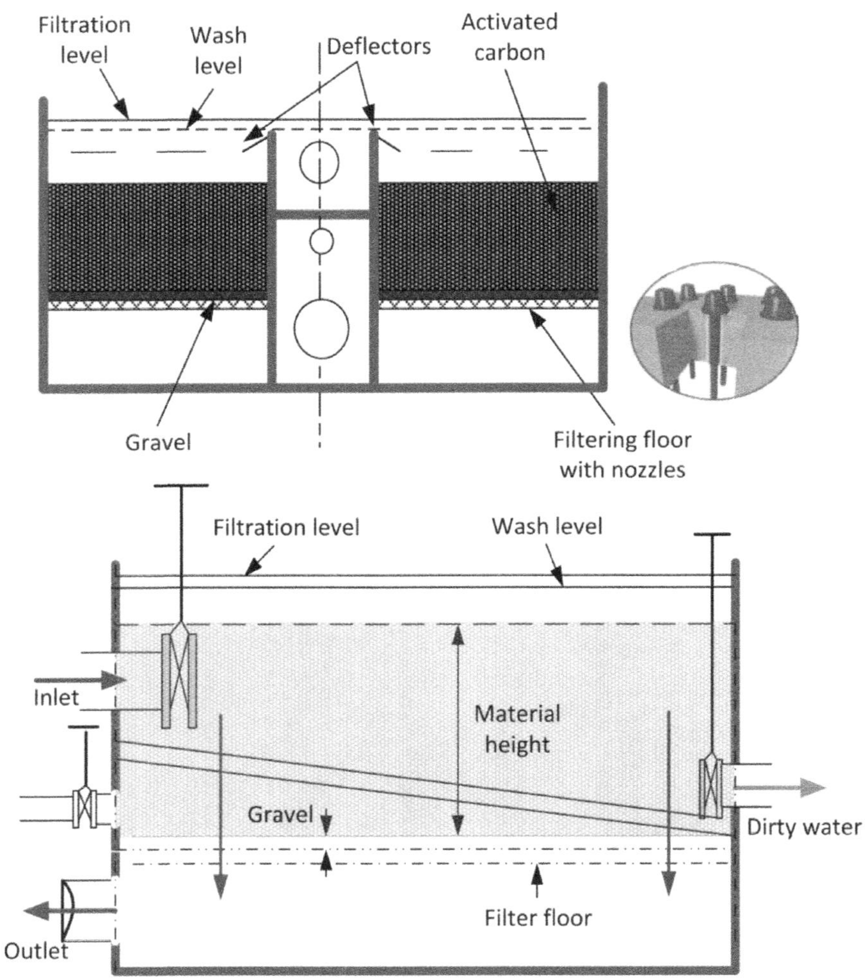

Figure 9.69. *Description of a GAC filter. For a color version of this figure, see www.iste.co.uk/gaid/watertreatment2.zip*

The role of the draining floor is to ensure the perfect homogeneity of the distribution of the fluid flows within the filtering mass, both during the filtration itself (water only) and during the backwash (water and air).

A 4–6 mm layer of gravel must be placed at a height of 100 mm under the GAC layer. This arrangement avoids the attrition of activated carbon close to the nozzles and the lifting of the floor if they are clogged. The heads of the nozzles are pierced with vertical slots, preventing any passage of filtering media under the floor, as well as ensuring a rigorous distribution of leaks, both in normal operation and during backwashing. These heads are embedded in the support gravel layer.

Figure 9.70. *GAC filter floor under construction. For a color version of this figure, see www.iste.co.uk/gaid/watertreatment2.zip*

The filtered water is discharged through a pipe from the bottom of the filter to the tank of treated water.

The filtration mechanism consists of the adsorption of micropollutants and the retention of SS. In specific cases of biological GAC filters, a biomass develops on the material. In case of a strong development of bacteria around the activated carbon grains, it reduces the size of the interstitial spaces and the passage section provided to the water. The "clogging" increases the pressure loss and results in a progressive decrease in the flow produced. The pressure loss is created by the difference between the water level on the filter and the water level in the filtered water tank. The regulator maintains a constant water level on the filter. The movements of the pressure loss compensator are controlled by the water level on the filter.

Each filter is equipped with a pressure loss compensating modulating valve, and its opening/closing is controlled by a hydrostatic level sensor placed in the filter above the filtering material.

Depending on the area of use, the following criteria help to optimize the choice of GAC:

– water quality, especially turbidity, TOC and micropollutants;

– temperature and contact time;

– the client's context (particular demand, operating conditions).

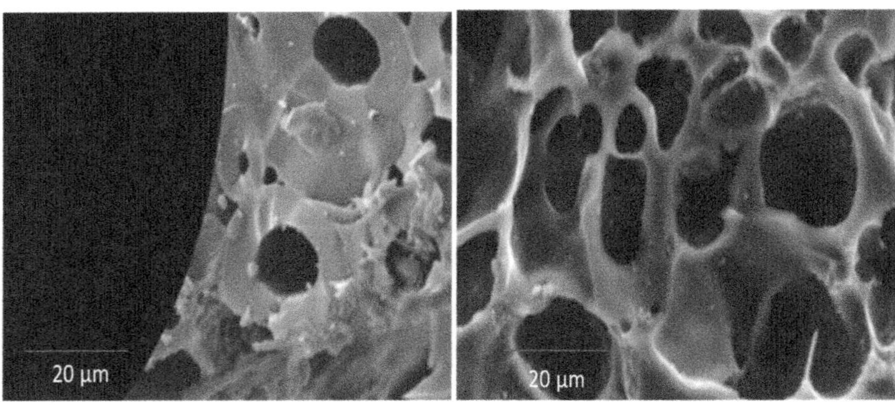

a) b)

Figure 9.71. *Structure of (a) picabiol (Jacobi) activated carbon and (b) Filtrasorb FTL (Chemviron) activated carbon. For a color version of this figure, see www.iste.co.uk/gaid/watertreatment2.zip*

The removal and implementation of activated carbon when renewing are carried out by means of a mobile hydroejector that is installed on the filter when the carbon is removed and on the delivery truck at the time of the installation (the hydroejector is supplied by the GAC supplier on renewal) or by a fixed carbon transport network installed at the foot of the filters.

Figure 9.72. *GAC renewal by hydroejector. For a color version of this figure, see www.iste.co.uk/gaid/watertreatment2.zip*

Figure 9.73. *GAC filters in Iver (UK) (a), Moulins-lès-Metz (France) (b), and Epernay (France-Pays de champagne) (c). For a color version of this figure, see www.iste.co.uk/gaid/watertreatment2.zip*

9.9.1.1. *Design parameters: filtration rate and media height*

The contact time is the sizing parameter, and it should be between 10 and 20 min, depending on the concentration of the organic substances to be removed and the treatment objectives. The height of the material and the filtration rate are directly associated with it. Nevertheless, the recommendations regarding these two parameters are as follows:

– maximum media height = 3 m;

– minimum media height = 0.8 m;

– maximum filtration rate = 10–15 $m \cdot h^{-1}$ for direct filtration and 20 $m \cdot h^{-1}$ for filtration after clarification.

The limiting dimensions are length ≤15 m and a filtration area less than 110 m^2 if $v < 16\ m \cdot h^{-1}$.

9.9.1.2. *Operating parameters: backwash procedure*

Backwash procedures differ depending on the position of the GAC filtration in the treatment file. It can be located in the first or second stage.

For effective filter cleaning, a 25–30% expansion is recommended.

For each family and each particle size of activated carbon, the expansion curves show the expansion of the carbon bed as a function of the water temperature, as well as the velocity and the conditions of the backwash.

For the same backwash velocity, the expansion percentage decreases as the water temperature increases. The protocol must then be adapted to the temperature of the water.

The large difference in density between activated carbon and sand or activated carbon, sand and MnO_2 guarantees a reclassification during washing.

In the case of GAC in the first stage, three phases (air, air/water and water alone) are required because it is necessary to remove the SS resulting from settling (as is the case for the treatment of surface water).

The respective frequencies of a backwash with water alone and backwash with air + water depend on the degree of clogging with SS and/or biological development. Biological development on media requires a backwash including air and water. This can be frequent during the hot season.

	Firststage single-media	**First stage single-media**	**First stage dual media**	**First stage three-media**	**Second stage single-media**	**Second stage dual media**
	Groundwater (turbidity < 0.5 NTU)	**After a settler (turbidity 0.8–2 NTU)**	**After a Settler (turbidity 0.8–2 NTU)**	**After a settler (turbidity 0.8–2 NTU)**	**After a sand or dual media filter or an ultrafiltration membrane**	**After a sand or dual media anthracite/sand filter**
Objectives	Micropollutant taste and odor dechlorination	Turbidity (≤ 0.5 NTU) NOM micropollutant	Turbidity (≤ 0.2 NTU) NOM micropollutant	Turbidity NOM micropollutant Fe and Mn metals	NOM micropollutant	NOM micropollutant Fe and Mn metals
Materials	GAC	GAC	GAC-sand	GAC-sand-MnO_2	GAC	GAC-MnO_2
Air phase	35 $Nm^3 \cdot m^{-2} \cdot h^{-1}$ 1–2 min	50 $Nm^3 \cdot m^{-2} \cdot h^{-1}$ 1–2 min	50 $Nm^3 \cdot m^{-2} \cdot h^{-1}$ 1–2 min	50 $Nm^3 \cdot m^{-2} \cdot h^{-1}$ 1–2 min	35 $Nm^3 \cdot m^{-2} \cdot h^{-1}$ 1–2 min[1]	35 $Nm^3 \cdot m^{-2} \cdot h^{-1}$ 1–2 min[1]
Air–water phase		Air: 50 $Nm^3 \cdot m^{-2} \cdot h^{-1}$ Water: 6-8 $m \cdot h^{-1}$ 8–10 min**	Air: 50 $Nm^3 \cdot m^{-2} \cdot h^{-1}$ Water: 6-8 $m \cdot h^{-1}$ 8–10 min**	Air: 50 $Nm^3 \cdot m^{-2} \cdot h^{-1}$ Water: 6-8 $m \cdot h^{-1}$ 8–10 min**	*	*
Water alone phase	Water: 25 $m \cdot h^{-1}$ 7–10 min	Water: 35 $m \cdot h^{-1}$ 7–12 min	Water: 35 $m \cdot h^{-1}$ 7–12 min	Water: 35 $m \cdot h^{-1}$ 7–12 min	Water: 25 $m \cdot h^{-1}$ 7–10 min	Water: 25 $m \cdot h^{-1}$ 7–10 min
Frequency	1–2 once/month	24–36 h	24–36 h	24–36 h	1–2 once/month	1–2 once/month

Table 9·17. *Washing conditions of GAC filters according to their place in the treatment system and according to the configuration of the material filters*

COMMENTS ON TABLE 9.17.– **The air and air/water phases are carried out only once a month or less, depending on the water temperature and the presence of assimilable organic matter. **The air/water flow must not lead to an overflow of the filter bases.*

The backwashes are triggered according to the following priority and parameters:

– manually by the operator;

– by indication of clogging on the filters;

– by clock;

– by indication of turbidity.

However, backwashing a filter is not possible if:

– the water level in the wash water tank is too low;

– the water level in the dirty water tank is too high.

The nominal-speed rise in the water level causes a complete degassing of the filter bed before discharge. To avoid losing the water above the material, it is recommended to lower the water level to the upper level of the filter bed (5–10 cm) just before a wash. If the control system is equipped with a pressure sensor in the filter, the lowering of the water level is controlled by this sensor.

After a wash, it is recommended that the first filtered water be discharged with a volume of water equivalent to 1.5–2 volumes of material. It is recommended that water be recirculated through the filters during extended plant shutdowns. This will prevent the formation of nitrites during these shutdowns (longer than 2 h).

9.9.1.3. *Monitoring the evolution of activated carbon*

Activated carbon is a material that evolves over time, depending on the conditions in which it is used (quality of the water to be treated, accidental or seasonal pollution, etc.). Its characteristics are important to follow to determine its adsorption capacity and the evolution of its lifespan. The monitoring of the evolution of the AC characteristics is based on an analysis of one GAC sample per filter per year of operation:

– the carbon tetrachloride index (CCl_4);

– iodine index;

– volatile matter content;

– ash content;

– mineral elements;

– particle size;

– apparent density.

Each parameter has particular significance, and their evolution over time makes it possible to estimate the saturation rate of the carbon, as well as its potential life under the operating conditions of the plant.

Depending on the condition of the GAC (iodine index, mineral load, volatile matter, particle size, etc.), the treatment operations (regeneration, reactivation, etc.) must be defined at the optimal time.

9.9.2. *GAC filters in series*

GAC filters that are arranged in series (Figure 9.74) can continuously produce water free of pollutants and micropollutants. When the first filter is saturated and the activated carbon needs to be replaced, the second filter takes over because it still has a high adsorption capacity, as most of the contaminants have been removed in the first filter. In this case, the feed water is fed directly to the second filter. Once the activated carbon has been renewed in the first filter, the water from the second filter is returned to the first filter. Thus, each of the two filters takes turns being a first-stage filter or a second-stage refining filter.

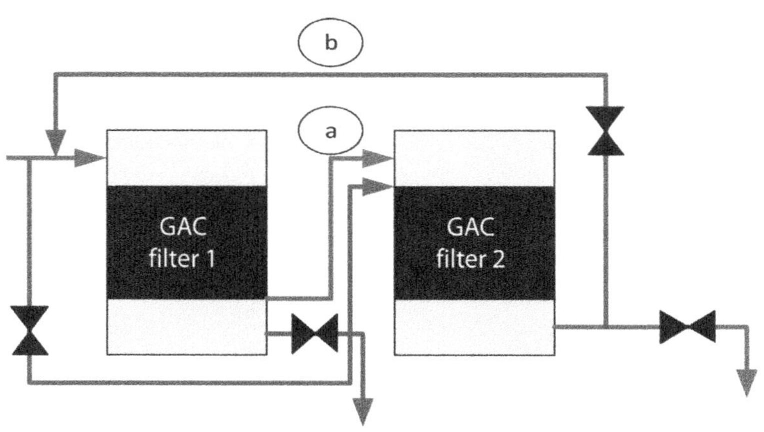

Figure 9.74. *GAC filters in series. For a color version of this figure, see www.iste.co.uk/gaid/watertreatment2.zip*

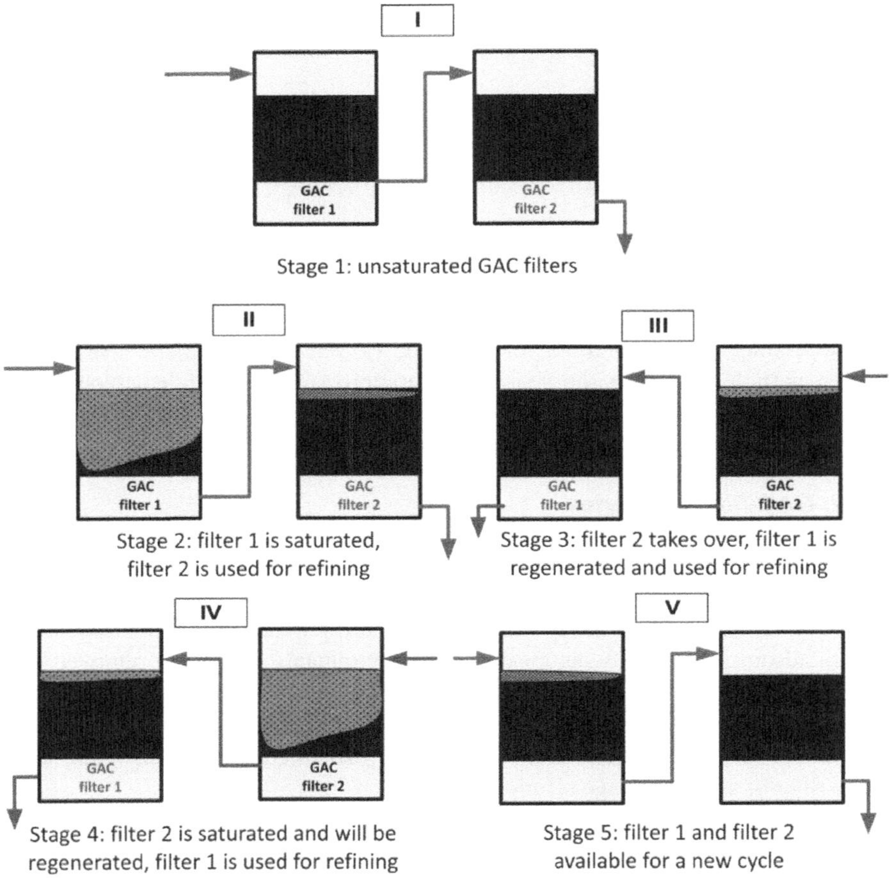

Figure 9.75. *Operating diagram of the GAC filters in series. For a color version of this figure, see www.iste.co.uk/gaid/watertreatment2.zip*

9.9.3. *Implementation modes for GAC filters*

The different modes of implementation of GAC filters are:

– single-media GAC filter;

– dual media GAC-sand filter;

– dual media GAC-MnO_2 filter;

– three-media GAC-sand-MnO_2 filter.

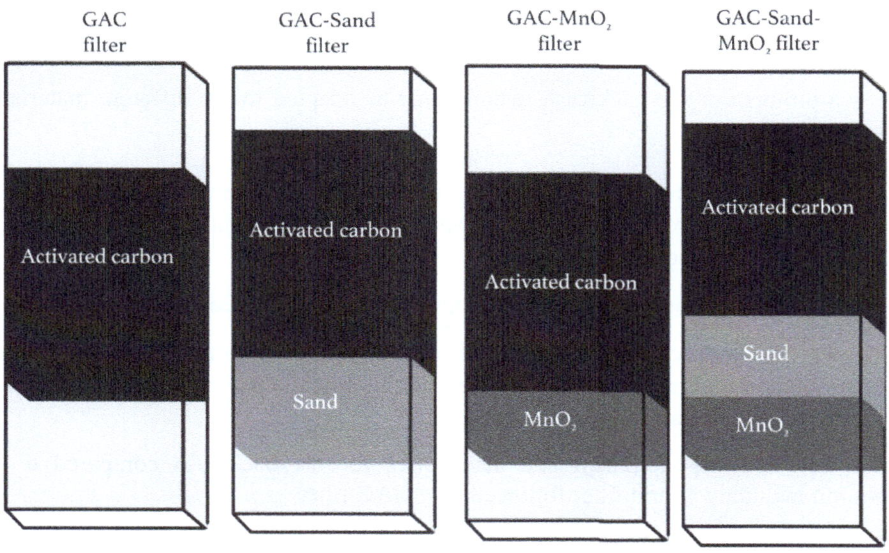

Figure 9.76. *Different implementation modes of GAC filters. For a color version of this figure, see www.iste.co.uk/gaid/watertreatment2.zip*

9.9.3.1. *The dual media GAC-sand filter*

The dual media GAC-sand filter can be installed as a first filtration stage behind a settler or a flotation device. This stage is mainly aimed at reducing the residual content of organic matter and micropollutants and possibly performing nitrification and refining the final turbidity.

This dual media combines the qualities of adsorption and retention.

In a dual media filter, the activated carbon has a contact time of 10–15 min. Its role is to remove pesticides, micropollutants and organic matter to prevent accidental pollution and to improve the organoleptic quality of the water.

The sand (TE of 0.65–0.85 mm) refines the turbidity. Located under the GAC layer, it also has a functional role by retaining any fine carbon particles that may be present during the first stages of filtration, after the GAC is replaced or after backwashing. The sand is not replaced, andonly the GAC is renewed.

This type of filter is characterized by:

– filtering layers of activated carbon and sand, adapted to the desired treatment;

– a high-water level above the filtering mass;

– a feeder through a channel, which is also used to drain the dirty water after a backwash;

– a protocol of the backwash, which must be adapted to the different materials used.

In all cases, filter design considers the loss of activated carbon during the backwash. There are numerous advantages of the implementation of dual media GAC-sand filters, such as:

– there is only one filtration stage to operate and no intermediate pumping;

– the maintenance of the GAC adsorption function (removal of pesticides, organic matter and any other adsorbable micropollutants) and the final barrier, due to the sand layer located under the carbon;

– there is less equipment and accessories for the backwash compared to a solution including a sand filter followed by a GAC filter.

Figure 9.77. *Dual media filter (Picotalen - la Montagne noire, France). For a color version of this figure, see www.iste.co.uk/gaid/watertreatment2.zip*

9.9.3.2. *The dual media GAC-MnO_2 filter*

The dual media GAC-MnO_2 filter is installed in the second stage behind a sand filtration system, upstream or downstream of a microfiltration or ultrafiltration membrane or, in some specific cases, in the first stage behind a settler or flotation. It is also placed in the first or second stage behind a softening.

This stage is mainly aimed at reducing the residual organic matter content and removing micropollutants, as well as manganese if it is still present in the water. It combines the adsorption, retention and catalytic oxidation qualities of Mn. The presence of other metals, such as Fe or Al, isalso retained on manganese dioxide.

Activated carbon, with a contact time of 10–15 min, removes pesticides, micropollutants and organic matter, preventing accidental pollution and improving the organoleptic quality of the water.

Manganese dioxide (MnO_2), with a contact time of 3–6 min, retains various metals (by catalytic oxidation), including Fe and Mn. The underlayer of the filters is made of Mangaflo® (Manganese dioxide). This material is a natural manganese dioxide, standardized in its chemical composition as well as in its granulometry. Its high density allows it to maintain a uniform and stable layer at the bottom of the filter during the washing phases. During the treatment, this material is coated with a layer of Mn_3O_4 resulting from the oxidation-catalysis reaction of manganese.

The initial state of the MnO_2 is recovered after a wash, where the Mn_3O_4 coating is removed by friction during the backwash or by the action of the chlorine added to the wash water.

The main characteristics of this material are given in Table 9.18.

Apparent density	2.3–2.4
Effective size	0.85–1 mm
Uniformity coefficient	2.35–2.45

Table 9.18. *Characteristics of Mangaflo (MnO_2)*

Manganese dioxide is not replaced during its long lifetime (5–10 years). Only the GAC is renewed.

This type of filter is characterized by:

– filtering layers of activated carbon and MnO_2, adapted to the desired treatment;

– a high-water level above the filtering mass;

– a feeder through a channel, which is also used to drain the washing water;

– a wash adapted to the different materials used.

In all cases, filter design considers the loss of material during the backwashes.

9.9.3.3. *The three-media GAC-sand-MnO_2 filter*

The three-media GAC-sand-MnO_2 filter is installed in the first or second stage after settling, flotation or softening.

In a three-media GAC-sand-MnO_2 filter, the different functions of the materials are as follows:

– activated carbon, with a contact time of 10–15 min, removes pesticides, micropollutants and organic matter, preventing accidental pollution and improving the organoleptic quality of the water;

– sand (TE: 0.85 mm) refines turbidity. Located below the GAC, it also has a functional role by retaining any fine carbon particles that may be present during the initial filtration period after the GAC has been replaced;

– manganese dioxide (MnO_2), with a contact time of 3–6 min, retains various metals (by catalytic oxidation), including Fe and Mn. It is placed below the sand;

– the filter sand and the manganese dioxide are not replaced, and only the GAC is renewed.

This type of filter is characterized by:

– filtering layers of activated carbon, sand and MnO_2, adapted to the desired treatment;

– a high-water level above the filtering mass;

– a protocol of backwash adapted to the different materials used. The large difference in density between these materials guaranteesgood reclassification during the backwashes.

9.10. Pressurized GAC filters (Opacarb™ filters)

For many years, OTV Veolia has been developing a range of prefabricated, mass-produced modular units. These standard units, which can complement an existing process, also provide turnkey solutions for industrial companies. Standardized products make it possible to reduce the costs associated with their development. OTV Veolia has designed a range of modular units and compact stations. Fully mounted on a chassis and transportable by truck, these units can constitute a real drinking water treatment chain or complete an existing installation. With a maximum production of 100 $m^3 \cdot h^{-1}$, Opalium has six different ranges whose specificities are correlated with the quality of the water. As such, the Opacarb™ range is a GAC filtration system for the removal of pesticides and micropollutants, with flow rates ranging from 10 to 100 $m^3 \cdot h^{-1}$ per unit. Opacarb™ is ideal for drinking water treatment and taste and odor reduction. The Opacarb™ activated carbon filter is environmentally friendly, skid-mounted, compact and can easily be integrated into a drinking water treatment system.

a)

b)

Figure 9.78. *Opacarb™ GAC filters at the Champigny-les-Langres plant (a) and in South Africa (b). For a color version of this figure, see www.iste.co.uk/gaid/watertreatment2.zip*

	Apparent dry density, 20°C	Specific surface $m^2 \cdot g^{-1}$	Iodine index	Origin	Particle size mm
	Jacobi				
Aquasorb®2000	0.52	950	1,000	Coal	1.4 (d50)
Aquasorb® Picabiol 2	0.32	1,400	900	Wood	1.4 (d50)
Aquasorb®6300	0.52	1,000	970	Coal	1.4 (d50)
Aquasorb®CS	0.54	1,050	1,000	Coconut	1.4 (d50)
	Cabot Norit				
N1240	0.48	1,110	950	Bituminous coal	0.8–2.0
PKST 1-3	0.25–0.28	–	–	Bituminous coal	1.0–3.0
HD 4000	0.40	700	–	Lignite	0.8–1.8
Row supra	0.39	1,150	950	Peat	0.8–1.6
	Chemviron				
Aquacarb 207C	0.46	1,100	1,000	Coconut	0.9 (TE)
Filtrasorb 300	0.56	1,000	950	Bituminous coal	0.8–1.0
Filtrasorb 400	0.425	1,050	1,000	Bituminous coal	0.6–0.7 (TE)
Filtrasorb TL 830	0.43	1,050	1,000	Bituminous coal	0.9–1.1

Table 9.19. *Characteristics of some granular activated carbons used in drinking water installations*

9.11. References

Al-Dibouni, M.R. and Garside, J. (1979). Particle mixing and classification in liquid fluidized beds: Transactions. *Institute of Chemical Engineers*, 57, 94–103.

Barbosa, M.O., Moreira, N.F., Pereira, M.F., Silva, A.M. (2016). Occurrence and removal of organic micropollutants: An overview of the watch list of EU Decision 2015/495. *Water Research*, 94, 257–279.

Barrett, E.P., Joyner, L.G., Halenda, P.P. (1951). The determination of pore volume and area distributions in porous substances I. Computations from nitrogen isotherms. *Journal of the American Chemical Society*, 73(1), 373–380.

Baudu, M., Le Cloirec, P., Martin, G. (1989). Modélisation des isothermes d'adsorption sur charbon actif de composés organiques en solution aqueuse. *Chemical Engineering*, 41, 81–89.

Bjelopavlic, M., Newcombe, G., Hayes, R. (1999). Adsorption of NOM onto activated carbon: Effect of surface charge, ionic strength, and pore volume distribution. *Journal of Colloid and Interface Science*, 210(2), 271–280.

Boehm, H.P. (1966). Chemical identification of surface groups. *Advances in Catalysis*, 16, 179–274.

Bord, G., Sauvignet, P., Rousselot, O., Gonçalvès, A., Rocher, V. (2015). Traitement des polluants émergents dans les rejets de STEP, étude du réacteur CAP couplé avec la clarification lestée (Actiflo®). In *Congès Astee 2015*, Montauban.

Brasquet, C., Oussy, J., Subrenat, E., Le Cloirec, P. (1996). Adsorption and selectivity of activated carbon fibers application to organics. *Environmental Technology*, 17, 1245–1252.

Brunauer, S., Emmett, P.H., Teller, E. (1938). Adsorption of gases in multimolecular layers. *Journal of the American Chemical Society*, 60, 309–319.

Carman, P.C. (1937). Fluid flow through granular beds. *Trans. Inst. Chem. Eng.*, 15, 150–166.

Clements, M. and Haarhoff, J. (2004). Practical experiences with granular activated carbon (GAC) at the Rietvlei water treatment plant. *Water SA*, January 30(1), 89–96.

Coltters, R. and Rivas, A.L. (2004). Minimum fluidization velocity correlations in particulate systems. *Powder Technology*, 147, 34–48.

Crittenden, J.C., Luft, P., Hand, D.W., Oravitz, J.L., Loper, S.W. (1985). Prediction of multicomponent adsorption equilibrium using ideal adsorbed solution theory. *Environmental Science & Technology*, 29(11), 1037–1043.

Crittenden, J.C., Reddy, P.S., Hand, D.W., Arora, H. (1989). Prediction of GAC performance using rapid small-scale column tests. Paper, American Water Works Association.

Dastgheib, S.A., Karanfil, T., Cheng, W. (2004). Tailoring activated carbons for enhanced removal of natural organic matter from natural waters. *Carbon*, 42(3), 547–557.

De Laat, J. (1988). Contribution à l'étude du mode d'élimination de molécules organiques modèles sur charbon actif en grain – Interaction entre les processus d'adsorption et de biodégradation. PhD Thesis, Université de Poitiers.

Delhumeau, V.P. (2000). Élimination de pesticides sur charbon actif en grain (CAG) : adsorption compétitive et évaluation des tests de sélection des CAG. PhD Thesis, Université de Poitiers.

Dharmarajah, A.H. and Cleasby, J.L. (1986). Predicting the expansion of filter media. *J. Am. Water Works Assoc.*, 1(11), 66–76.

Di Felice, R. (1995). Hydrodynamics of liquid fluidization. *Chemical Engineering Science*, 50(8), 1213–1245.

Dubinin, M.M. (1955). Surface and porosity of adsorbents. *Uspekki Khim*, 24, 3.

Duris, M., Garic-Grulovic, R., Arsenijevic, Z., Jacimovski, D., Grbavcic, Z. (2013). Segregation in water fluidized beds of sand particles. *Powder Technology*, 235, 173–179.

Ebie, K., Li, F., Azuma, Y., Yuasa, A., Hagishita, T. (2001). Pore distribution effect of activated carbon in adsorbing organic micropollutants from natural water. *Water Res.*, 35(1), 167–179.

Ergun, S. (1952). Fluid flow through packed columns. *Chemical Engineering Progress*, 48, 89–94.

Faur, C., Cognaud, A., Dreyfus, G., Le Cloirec, P. (2008). Modeling the breakthrough of activated carbon filters by pesticides in surface waters with static and recurrent neural networks. *Chemical Engineering Journal*, 145(1), 7–15.

Gaid, K. and Sauvignet, P. (2012). Élimination des micropolluants en eaux usées avec réacteur charbon actif en poudre (Actiflo® Carb) et suivi de la qualité du traitement, prés. In *Journées Informations Eaux*, Poitiers.

Gaid, K. and Sauvignet, P. (2013). High rate PAC reactor combined with ozone oxidation for enhanced micropollutants removal from urban waste water. In *International Water Association*, Zurich.

Gaid, K. and Sauvignet, P. (2015). Substances organiques, perturbateurs endocriniens, micropolluants : le Filtraflo® carb une solution efficace. *L'eau, l'industrie, les nuisances*, 381, 79–83.

Gaid, K. and Sauvignet, P. (2018). Opacarb® FL, une nouvelle génération de réacteur à charbon actif. *Eaux, industrie et nuisances*, 416, 71–76.

Gaid, K., Sauvignet, P., Tazi-Pain, A., Houssais, B. (2007a). Les réacteurs CAP : un plus pour l'élimination des matières organiques et pesticides. *Eaux, industries & nuisances*, 307, 49–56.

Gaid, K., Sauvignet, P., Chesneau, M., Bro Capron, R. (2007b). Veolia Water introduces the Actiflo® Carb process (PAC reactor combined with Actiflo®) delivering NOM (Natural Organic Matter), pesticides & endocrine disruptors removal for potable water treatment plants. In *Organic Matter Removal in Drinking Water Conference*, London.

Gaid, K., Sauvignet, P., Koh, W.K. (2012). Investigation and comparison of the performance of the magnetic ion exchange process and the powder activated carbon with ballasted clarification process for NOM removal from high TOC water source. In *Improvement of Technologies for Drinking Water*, 6 April, Singapore.

Gaid, K., Sauvignet, P., Bourdon, F., Rigot, M., Ingrand, V. (2015). Les réacteurs CAP "Multiflo-Actiflo" : la solution pour l'élimination des perturbateurs endocriniens et substances dangereuses. *L'eau, l'industrie, les nuisances*, 353, 68–75.

Gaid, K., Sauvignet, P., Marlin, P. (2017). Use of PAC (powder activated carbon) combined with Actiflo & UF membrane for enhanced NOM removal from a high DOC (dissolved organic carbon) water source. In *Organic Matter Removal in Drinking Water Conference*, Birmingham.

Gaid, K., Sauvignet, Ph., Jouaffre, P., Sabourdi, C. (2018). Réacteurs fluidisés à charbon actif. Théorie et applications. *L'eau, l'industrie, les nuisances*, 422, 55–62.

Gicquel, L. (1996). Élimination par adsorption sur charbon actif de l'atrazine contenue dans les eaux à potabiliser : étude sur la compétition avec les matières organiques et minérales dissoutes et modélisation. PhD Thesis, Université de Rennes.

Gupta, S.K., Agarwal, V.K., Singh, S.N., Seshadri, V., Mills, D., Singh, J., Prakash, C. (2009). Prediction of minimum fluidization velocity for fine tailings materials. *Powder Technology*, 196, 263–271.

Gur-Reznik, S., Katz, I., Dosoretz, C.G. (2008). Removal of dissolved organic matter by granular-activated carbon adsorption as a pretreatment to reverse osmosis of membrane bioreactor effluents. *Water Res.*, 42(6/7), 1595–1605.

Hazourli, S., Ziati, M., Hazourli, A., Cherifi, M. (2007). Valorisation d'un résidu naturel ligno-cellulosique en charbon actif – exemple des noyaux de dattes. *Revue des énergies renouvelables ICRESD-07 Tlemcen*, 187–192.

Karanfil, T. and Kilduff, J.E. (1999). Role of granular activated carbon surface chemistry on the adsorption of organic compounds 1. Priority pollutants. *Environmental Science and Technology*, 33(18), 3217–3224.

Leva, M., Shirai, T., Wen, C.Y. (1956). Prediction of onset of fluidization in beds of granular solids. *Génie chimique*, 75(2), 33–43.

Li, Q., Snoeyink, V.L., Mariñas, B.J., Campos, C. (2003b). Pore blockage effect of NOM on atrazineadsorption kinetics of PAC: The roles of PAC pore size distribution and NOM molecular weight. *Water Res.*, 37(20), 4863–4872.

Lippens, B.B. and Mulder, J. (1993). Prediction of the minimum fluidization velocity. *Powder Technology*, 75(1), 67–78.

Matilainen, A., Vieno, N., Tuhkanen, T. (2006). Efficiency of the activated carbon filtration in the natural organic matter removal. *Environment International*, 32(3), 324–331.

Matsui, Y., Fukuda, Y., Inoue, T., Matsushita, T. (2003). Effect of natural organic matter on powdered activated carbon adsorption of traces contaminants. *Water Res.*, 37, 4413–4424.

McCreary, J.J. and Snoeyink V.L. (1980). Characterization and activated carbon adsorption of several humic substances. *Water Res.*, 14(2), 151–160.

Morenocastilla, C. (2004). Adsorption of organic molecules from aqueous solutions on carbon materials. *Carbon*, 42(1), 83–94.

Newcombe, G., Morrison, J., Hepplewhite, C. (2002a). Simultaneous adsorption of MIB and NOM onto activated carbon I. Characterisation of the system and NOM adsorption. *Carbon*, 40(12), 2135–2146.

Newcombe, G., Morrison, J., Hepplewhite, C., Knappe, D.R.U. (2002b). Simultaneous adsorption of MIB and NOM onto activated carbon II. Competitive effects. *Carbon*, 40(12), 2147–2156.

Nikam, S.V. and Mandal, D. (2017). Hydrodynamic study of activated carbon for adsorption of volatile organic compounds in fluidization condition. *70th Annual Session of Indian Institute of Chemical Engineers CHEMCON*, Haldia.

Peck, G.C.A. and Watkins, S.B. (1956). The flow of fluids through beds of spherical particles. *The Industrial Chemist*, 32, 122–127.

Pelekani, C. and Snoeyink, V.L. (1999). Competitive adsorption in natural water role of activated carbon pore size. *Water Res.*, 33(5), 1209–1219.

Quinlivan, P.A., Li, L., Knappe, D.R.U. (2005). Effects of activated carbon characteristics on the simultaneous adsorption of aqueous organic micropollutants and natural organic matter. *Water Res.*, 39(8), 1663–1673.

Rowe, P.N. (1987). A convenient estimation of the Richardson-Zaki exponent. *Chem. Eng. Sci.*, 43, 2795–2796.

Sauvignet, P. and Mechouk, C. (2010). Élimination de la matière organique utilisant l'Actiflo® Carb sur une eau de nappe alluvialle de Loire. In *Journées informations eaux*, Poitiers, 12–14 October.

Slasli, A.M., Jorge, M., Stoeckli, F., Seaton, N.A. (2004). Modeling of water adsorption by activated carbons: Effects of microporous structure and oxygen content. *Carbon*, 42, 1947–1952.

Smith, E.H. (1994). Bench-scale tests and modeling of adsorption of natural organic matter by activated carbon. *Water Res.*, 28(8), 1693–1702.

Velten, S., Knappe, D.R.U., Traber, J., Kaiser, H.P., von Guten, U., Boller, M., Meylan, S. (2011). Characterization of natural organic matter adsorption in granular activated carbon adsorbers. *Water Res.*, 45(13), 3951–3959.

Weber, W.J. and Lian, G S. (1983a). A dual particle-diffusion model for porous adsorbents in fixed beds. *Environmental Progress*, 2(3), 167–175.

Wen, C.Y. and Yu, Y.H. (1966a). Mechanics of fluidization. *Chem. Engng Prog. Symp. Ser.*, 62, 100–111.

Wen, C.Y. and Yu, Y.H. (1966b). A generalized method for predicting the minimum fluidization velocity. *AIChE Journal*, 12, 610–612.

Wu, S.Y. and Baeyens, J. (1991). Effect of operating temperature on minimum fluidization velocity. *Powder Technology*, 67(2), 217–220.

Yaacoubi, A. and Ayele, J. (1999). Sorption de l'atrazine et du diuron sur charbon actif en poudre en présence de tensioactifs, ion calcium et bichromate – essai de modélisation. *Sciences de l'eau*, 12(2), 389–406.

Index

P, R

S, T, V

Summary of Volume 1

Chapter 5. Settling

Chapter 6. Flotation

Index

Summary of Volume 3

Chapter 12. Biological Removal of Ammonia

Chapter 13. Nitrate Removal

Chapter 14. Removal of Perchlorates

Chapter 15. Water Softening

Chapter 16. Metal Removal

Index

Summary of Volume 4

Chapter 17. Microfiltration and Ultrafiltration

Chapter 18. Nanofiltration and Reverse Osmosis

Chapter 19. Desalination by Reverse Osmosis

Summary of Volume 5

Chapter 22. Disinfection By-products

Chapter 23. Sludge Treatment

Printed and bound by CPI Group (UK) Ltd, Croydon, CR0 4YY

13/12/2023

08208008-0003